Isochronous Systems

Isochronous Systems

Francesco Calogero
Physics Department, University of Rome "La Sapienza"
Istituto Nazionale di Fisica Nucleare, Sezione di Roma

OXFORD
UNIVERSITY PRESS

Great Clarendon Street, Oxford OX2 6DP

Oxford University Press is a department of the University of Oxford.
It furthers the University's objective of excellence in research, scholarship,
and education by publishing worldwide in

Oxford New York

Auckland Cape Town Dar es Salaam Hong Kong Karachi
Kuala Lumpur Madrid Melbourne Mexico City Nairobi
New Delhi Shanghai Taipei Toronto

With offices in

Argentina Austria Brazil Chile Czech Republic France Greece
Guatemala Hungary Italy Japan Poland Portugal Singapore
South Korea Switzerland Thailand Turkey Ukraine Vietnam

Published in the United States
by Oxford University Press Inc., New York

First Published 2008

British Library Cataloguing in Publication Data
Data available

Library of Congress Cataloging in Publication Data
Data available

Typeset by Newgen Imaging Systems (P) Ltd., Chennai, India
Printed in Great Britain
on acid-free paper by
Biddles Ltd., www.biddles.co.uk

ISBN 978–0–19–953528–6

1 3 5 7 9 10 8 6 4 2

FOREWORD

A classical dynamical system is called *isochronous* if it features in its phase space an open, *fully dimensional*, region where *all* its solutions are *periodic* in *all* their degrees of freedom with the same, *fixed* period—independently from the initial data, provided they are inside the *isochrony* region. Recently a simple transformation (often referred to as "the trick") has been introduced, featuring a *real* parameter ω and reducing to the identity for $\omega = 0$. This transformation is applicable to a *quite large* class of dynamical systems and it yields ω-modified *autonomous* systems which are *isochronous*, with period $T = 2\pi / \omega$. Additional tricks have subsequently been found, suitable to modify a *Hamiltonian* dynamical systems so that the system thereby obtained is still *Hamiltonian* and *isochronous.* These developments justify the notion that *isochronous systems are not rare.*

Of course, in the *real* world the examples of purely *isochronous* behavior are rather rare—otherwise life would be pretty dull. Indeed, as it were rather appropriately, the anonymous referee of a paper dealing with this approach mentioned, as a kind of analogous explanation for the phenomenology in question, a broken vinyl record repeating over and over its tune. But it should as well be noted that, if the period of *isochrony* is *much larger* than all the natural timescales of the systems under scrutiny, then the modified systems featuring in the long run the *isochronous* behavior shall, nevertheless, continue to behave for quite some time essentially as the original, unmodified systems before the *isochronous* phenomenology takes over.

Moreover examples of *isochronous* phenomena abound in our environment, and their quantitative understanding belongs to several, more or less exact, sciences: physics, chemistry, biology, medicine, economics, and so on. The approach used in this book is rather that of mathematical physics, or perhaps of applied mathematics; but it is also meant, hopefully, to provide tools for experimenters and practitioners interested in eventual applications.

Hence, this monograph should be of interest to researchers working on dynamical systems, including integrable and nonintegrable models, with a finite or infinite number of degrees of freedom; but it should also appeal to experimenters and practitioners interested in *isochronous* phenomena. It might be used as basic textbook for an undergraduate or graduate course.

In this monograph—which covers work done over the last decade by its author and several collaborators—the technique to manufacture *isochronous* systems based on the trick is reviewed, and plenty of examples of such systems are provided, including many-body problems characterized by Newtonian equations of motion in spaces of one or more dimensions, Hamiltonian systems, and also nonlinear evolution equations (PDEs). We do not review in any detail this material in this Foreword, since this information can be gathered directly from the following Table of contents.

The rules employed below to identify references (mainly, albeit not exclusively, in the last section of each chapter) and for cross referencing equations are, we trust, sufficiently self-evident not to require any elaboration here. Likewise, the notation used is defined below whenever appropriate. And let us mention in this connection that in this monograph we also try and cater for the reader who does not go through books systematically from their beginning—although we do not recommend this practice. Hence, we made no effort to eliminate redundancies in notational definitions and also in the presentation of basic remarks—to the extent this did not entail an *excessive* waste of space. We rather practiced repeatedly the rule: *repetita iuvant.*

Finally: we make throughout an *extensive*—some might think *excessive*—use of *italics.* This typographical device is meant to focus the attention of the reader on what we consider *key* words (generally, or in the specific context they appear): we hope most readers will find this *emphasis* helpful (at least occasionally), and we beg the *tolerance* of the (hopefully) few who are disturbed by it and who should simply ignore this aspect of our presentation.

PREFACE

The material reported in this monograph reflects work performed with, and by many, collaborators, and it is a duty and a pleasure to thank them all: Mario Bruschi, Antonio Degasperis, Luca Di Cerbo, Riccardo Droghei, Marianna and Norbert Euler, Jean-Pierre Françoise, David Gómez-Ullate, Sandro Graffi, Santino Iona, Edwin Langmann, François Leyvraz, Mauro Mariani, Paolo Maria Santini, and Matteo Sommacal. Their contributions are detailed below: see in particular the sections of Notes at the end of each chapter. I also like to thank some colleagues and friends (and in particular for his thorough job Michele Bartuccelli) who read the manuscript of this monograph and helped me to eliminate misprints. It goes without saying that all shortcomings of this book are my own responsibility—including all the remaining misprints, since I typed it myself.

A complementary apology must be addressed to all the authors who investigated *isochronous* systems but whose work is not mentioned in this monograph, the purpose and scope of which is indeed mainly focused on (mainly recent) findings to which the author directly contributed.

Acknowledgements are also due to various institutions for supporting visits, by colleagues to Rome and by me abroad, which were instrumental to arrive at the findings reported in this book: the Research Project of National Interest (PRIN) "SINTESI" co-funded by the Italian Ministry for the University and for Research (MIUR), the exchange programs of my University with the Landau Institute in Chernogolovka near Moscow and with the Universidad Complutense in Madrid, the Centro Internacional de Ciencias in Cuernavaca (Mexico), and the program jointly funded by the EINSTEIN Consortium in Lecce and the Russian Foundation for Basic Research in Moscow (grant RFBR-06-01-92054-CEa).

This book was mainly drafted during August 2006, and then reviewed and updated one year later, in both cases while vacationing in our summer house in Capalbio with my wife Luisa, my daughter Anna and her family, and several friends. It is dedicated collectively to all of them—and in particular to our two grandchildren Cor and Stella Luna—for bearing with my bad habit of sitting in front of my computer to write this book rather than going to the beach with them.

Francesco Calogero
July 2007

CONTENTS

1

INTRODUCTION

In this monograph we tersely review recent results on *isochronous* systems. The presentation is largely based on recent papers, which are sometimes followed *verbatim*. Our main purpose and scope is to explain plainly the main ideas and to report tersely the main results, referring (with some guidance) to the literature for more detailed treatments. We also try to cater to the hasty or casual reader who will read some part of this monograph without having gone through everything that comes before it—to the limited extent this can be done without *excessive* repetitions.

Throughout this monograph we consider *classical* (i.e., nonquantal) systems (except for occasional lapses); a brief mention of the extension of the treatment to *quantum* systems is outlined in the last chapter ("Outlook"). The bulk of the treatment focusses on ordinary differential equations (ODEs), and systems of ODEs, with particular attention to those interpretable as *Newtonian* equations of motion for point particles; only the last but one chapter is devoted to partial differential equations (PDEs). The specific material treated is clearly indicated in the *Table of contents* (see above); hence we feel no need to review it here, and we trust the notation used below for cross-references to equations, for the bibliography (references to which are generally confined to the last sections of each chapter), and so on, to be sufficiently self-evident not to require any elaboration here.

An *isochronous* system is characterized by the property to possess a domain—*open*, hence having *full dimensionality* in phase space—such that *all* the motions evolving from a set of initial data in it remain in it throughout the time evolution and are *completely periodic* with the same *fixed* period. The natural measure of this *open* domain might, or it might not, be *infinite* when the measure of the entire phase space is itself *infinite*: for instance, if the entire phase space is the two-dimensional Euclidian plane, such a domain might be the exterior, or the interior, of a circle of finite radius.

It is justified to call such systems *superintegrable*, or perhaps *partially superintegrable*, inasmuch as the property of *isochrony* of all their motions holds only in a *subregion* of the *entire* phase space. This terminology is justified by the observation that *all* confined motions of a (Hamiltonian) *superintegrable* system—in which *all* the degrees of freedom are slave to the *same* time evolution due to the existence of the *maximal* possible number of (functionally independent) constants of motion—are *completely periodic*, although *not* necessarily *all* with the same *fixed* period—entailing that *isochrony* entails *superintegrability*, while the converse is not the case.

For instance, a well-known *isochronous* system is the one-dimensional N-body problem characterized by the (normal) Hamiltonian

$$H\left(\underline{z},\underline{p}\right)=\frac{1}{2}\sum_{n=1}^{N}\left[p_n^2+\left(\frac{\omega}{2}\right)^2 z_n^2\right]+\frac{1}{4}\sum_{m,n=1;m\neq n}^{N}\frac{g^2}{\left(z_n-z_m\right)^2} \tag{1.1a}$$

and correspondingly by the Newtonian equations of motion (with one-body and two-body velocity-independent forces)

$$\ddot{z}_n+\left(\frac{\omega}{2}\right)^2 z_n=\sum_{m=1,m\neq n}^{N}\frac{g^2}{\left(z_n-z_m\right)^3}. \tag{1.1b}$$

Here and hereafter ω is a *positive* constant, $\omega>0$, and the rest of the notation is, we trust, self-evident: in particular, superimposed dots denote differentiations with respect to the *real* independent variable t ("time"), and indices such as n generally range from 1 to N (note that, here and throughout, we set to *unity* the mass of the moving particles). Indeed, in the *real* domain, *all* the solutions of these equations of motion are *isochronous*: completely periodic,

$$z_n(t+T)=z_n(t), \tag{1.2}$$

with

$$T=\frac{2\,\pi}{\omega} \tag{1.3}$$

(provided $g\neq 0$; otherwise in (1.2) T should, of course, be replaced by $2\,T$).

This is not quite true in the *complex* domain, namely, if we consider the dependent variables (or "particle coordinates") $z_n\equiv z_n\left(t\right)$ to be *complex* rather than *real*, as we will generally do below (and the "coupling constant" g might then as well be *complex*, while we will always consider the constant ω to be *real*, indeed, without loss of generality, *positive*, $\omega>0$). Then *all nonsingular* motions, which of course, take place in the *complex* z-plane, are again *completely periodic*, but the period may be an *integer* multiple of T: indeed, also in this case, the overall particle configuration does repeat itself with period T, but the individual particles might exchange their roles through the motion, entailing that the period of the motion become an *integer* multiple of T (this cannot happen in the *real* case, when the motion takes place on the *real* line and the ordering of the particles cannot change throughout the motion due to the *singular* character of the repulsive two-body interaction, see the right-hand side of (1.1b)—assuming, of course, that g does *not* vanish). The *singular* motions (which, of course, can only occur in the *complex* case) correspond to a *lower dimensional* set of initial data, and are characterized by the occurrence, at a time $t_S<T$, of a *collision* among two, or more, particles (see the right-hand side of (1.1b)).

Hence, the many-body problem characterized by this Hamiltonian is *isochronous*, both in the *real* and in the *complex* contexts; and the open domain of initial data for which it possesses the *isochrony* property coincides in this case with the *entire* phase space, with the only exclusion of a *lower dimensional* set of initial data leading to particle collisions (this can, of course, only happen in the *complex* case). And it is well known (for over three decades) that this Hamiltonian N-body problem is *superintegrable*.

Hereafter, we call *entirely isochronous* the dynamical systems, such as this one, which are *isochronous* in their *entire* (natural) phase space—possibly up to a *lower dimensional* set of *singular* solutions, and possibly featuring a *finite* (generally small) number of different periods—all of them *integer* multiples of a basic one—in different, *fully dimensional*, regions of their phase space, separated from each other by the *lower dimensional* set of initial data yielding *singular* solutions.

Also *isochronous* (albeit only if it is considered in the *complex*) is the more general N-body problem characterized by the (normal) Hamiltonian

$$H\left(\underline{z},\underline{p}\right)=\frac{1}{2}\sum_{n=1}^{N}\left[p_n^2+\left(\frac{\omega}{2}\right)^2 z_n^2\right]+\frac{1}{4}\sum_{m,n=1;m\neq n}^{N}\frac{g_{nm}^2}{\left(z_n-z_m\right)^2} \qquad (1.4a)$$

and correspondingly by the Newtonian equations of motion

$$\ddot{z}_n+\left(\frac{\omega}{2}\right)^2 z_n=\sum_{m=1,m\neq n}^{N}\frac{g_{nm}^2}{\left(z_n-z_m\right)^3}. \qquad (1.4b)$$

This N-body problem differs from the previous one, (1.1), because the coupling constants g_{nm} are now allowed to be *arbitrarily different* (except for the obvious symmetry restriction $g_{nm}^2=g_{mn}^2$, see (1.4a)). As explained below, these *Newtonian* equations of motion yield again a *completely periodic* motion, see (1.2), provided the initial data fall in an appropriate (open, *fully dimensional*) domain of phase space, which, however, generally does *not* include only *real* data. For initial data falling instead *outside* the *isochrony* region the motion could again be *periodic* with a period which is again an *integer* multiple of T, see (1.3), but this period might now be *arbitrarily large*; or it could be *aperiodic*, indeed, possibly extremely complicated (perhaps, in some sense, "chaotic").

A second example of *isochronous* N-body problems is characterized by the (not normal) Hamiltonian

$$H\left(\underline{p},\underline{z}\right)=\sum_{n=1}^{N}\left[-\frac{i\,\omega\,z_n}{c}+\exp\left(c\,p_n\right)\prod_{m=1,m\neq n}^{N}\left(z_n-z_m\right)^{-a_{nm}}\right], \qquad (1.5a)$$

where c is an arbitrary (nonvanishing) constant and the $N\,(N-1)$ constants a_{nm} are *arbitrary*. The corresponding Hamiltonian equations of motion read

$$\dot{z}_n = \frac{\partial H}{\partial p_n} = c\,\exp\left(c\,p_n\right) \prod_{m=1,m\neq n}^{N} \left(z_n - z_m\right)^{-a_{nm}}, \tag{1.5b}$$

$$\dot{p}_n = -\frac{\partial H}{\partial z_n} = c^{-1}\left[i\,\omega + \sum_{m=1,m\neq n}^{N} \frac{a_{nm}\left(\dot{z}_n - \dot{z}_m\right)}{z_n - z_m}\right], \tag{1.5c}$$

and they yield the *Newtonian* equations of motion (with one-body and two-body *velocity-dependent* forces; note that the constant c has dropped out from these equations of motion)

$$\ddot{z}_n = i\,\omega\,\dot{z}_n + 2 \sum_{m=1,m\neq n}^{N} \frac{a_{nm}\,\dot{z}_n\,\dot{z}_m}{z_n - z_m}. \tag{1.5d}$$

In this case, the motion takes place necessarily in the *complex* z-plane, due to the *complex* character of these equations of motion (see the first term in the right-hand side of (1.5d)). But they can be reformulated as N *real* equations of motion satisfied by N *real* two-vectors $\vec{r}_n(t)$ describing N point particles moving in the (physical) horizontal plane, by setting

$$z_n = x_n + i\,y_n, \quad \vec{r}_n = (x_n,\, y_n,\, 0)\,, \quad \hat{k} = (0,0,1)\,, \quad a_{nm} = \alpha_{nm} + i\,\beta_{nm}. \tag{1.6}$$

Note that, for notational convenience (see below), we actually consider the two-vectors $\vec{r}_n(t)$ as three-vectors with identically vanishing third component.

The *Newtonian* equations of motion (1.5d) may then be written as follows:

$$\ddot{\vec{r}}_n = \omega\,\hat{k}\wedge\dot{\vec{r}}_n + 2 \sum_{m=1,m\neq n}^{N} \left\{ r_{nm}^{-2}\left(\alpha_{nm} + \beta_{nm}\,\hat{k}\wedge\right) \left[\dot{\vec{r}}_n\left(\dot{\vec{r}}_m\cdot\vec{r}_{nm}\right) + \dot{\vec{r}}_m\left(\dot{\vec{r}}_n\cdot\vec{r}_{nm}\right) - \vec{r}_{nm}\left(\dot{\vec{r}}_n\cdot\dot{\vec{r}}_m\right)\right]\right\}. \tag{1.7}$$

Here we use the short-hand notation $\vec{r}_{nm} = \vec{r}_n - \vec{r}_m$, entailing $r_{nm}^2 = r_n^2 + r_m^2 - 2\,\vec{r}_n\cdot\vec{r}_m$, and we trust the remaining notation to be self-evident: in particular, the symbols "$\cdot$" respectively "$\wedge$", sandwiched among two three-vectors denote the standard scalar respectively vector products in three-dimensional space, so that, for instance, $\hat{k}\wedge\vec{r}_n = (-y_n,\, x_n,\, 0)$, see (1.6).

Remarkably, these *Newtonian* equations of motion are both *translation-invariant* and *rotation-invariant*, and when *all* the coupling constants α_{nm}, β_{nm} vanish, they have a clear physical interpretation: they describe a "cyclotron", namely, the motion of N equal, electrically charged, point-like particles moving in the horizontal plane in the presence of a constant magnetic field orthogonal to that plane, in the approximation in which their mutual (electrostatic) interactions are neglected. And these *Newtonian* equations of motion are of course no less *Hamiltonian* than the (*complex*) equations of motion (1.5d): indeed a *real* Hamiltonian that generates directly the *real* equations of motion (1.7) is provided by the *real part* of the *complex* Hamiltonian that is obtained by inserting the assignment (1.6) in (1.5a), together with

$$p_n = p_{nx} - i\, p_{ny}, \quad \vec{p}_n = (p_{nx},\, p_{ny},\, 0)\,. \tag{1.8}$$

Note the minus sign in the first of these two formulas: where p_{nx}, respectively p_{ny} are of course the x-component, respectively the y-component of the three-vector $\vec{p}_n$ (whose third component vanishes identically, so that this vector always lies in the horizontal plane). This vector $\vec{p}_n$ plays of course the role of canonically conjugate momentum to the canonical vector variable $\vec{r}_n$.

Although the formulation in terms of the *real* two-vector variables $\vec{r}_n$ is perhaps "more physical," we will generally in the following stick with the "mathematically neater" formulation in terms of the *complex* variables z_n. But of course it should always be kept in mind that any system in the *complex* is completely equivalent to a system in the *real* with double the number of variables; a *real* system that can easily be obtained from the *complex* one by expressing every *complex* number via its *real* and *imaginary* parts, or via its *modulus* and *phase.* As shown by the above example, in some cases such a transition to *real* variables produces a model endowed with a rather evident physical significance.

The *isochronous* character of this N-body problem, characterized by the *Newtonian* equations of motion (1.5d), is implied—just as for the system described above, characterized by the *Hamiltonian* (1.4a) and by the *Newtonian* equations of motion (1.4b)—by the existence (demonstrated below, see **Example 4.1.2-2**) of an open (*fully dimensional*) region in its phase space where *all* motions are *completely periodic*, see (1.2) and (1.3). And, just as in the previous case, for initial data which fall instead *outside* the *isochrony* region the motion could be again *periodic* with a period which is an *integer* multiple of T, but this period might be *arbitrarily large*; or the motion could be *aperiodic*, indeed possibly extremely complicated (again perhaps, in some sense, "chaotic").

In the special case in which *all* the "coupling constants" a_{nm} are unity,

$$a_{nm} = 1; \quad \alpha_{nm} = 1, \quad \beta_{nm} = 0, \tag{1.9}$$

namely, when the Newtonian equations of motion (1.5d) read

$$\ddot{z}_n = i\,\omega\,\dot{z}_n + 2 \sum_{m=1, m\neq n}^{N} \frac{\dot{z}_n\,\dot{z}_m}{z_n - z_m}, \tag{1.10}$$

the region of *isochrony* coincides with the *entire* phase space, excluding only a *lower dimensional* set of initial data yielding motions that run into singularities associated with particle collisions. So in this special case, the system is *entirely isochronous*; and it is *superintegrable* indeed *solvable*, its initial-value problem being solved (as shown in Section 4.2.2) by the following simple rule: *the N coordinates $z_n(t)$ are the N solutions of the algebraic equation in the variable z*

$$\sum_{n=1}^{N} \frac{\dot{z}_n(0)}{z - z_n(0)} = \frac{i\,\omega}{\exp(i\,\omega\,t) - 1}. \tag{1.11}$$

Note that, by multiplying this formula by $\prod_{n=1}^{N} [z - z_n(0)]$, one reduces the search of its N solutions $z_n(t)$ to the identification of the N zeros of a *polynomial* of degree N in the variable z whose coefficients are *all periodic* in t with period T. Hence, these zeros $z_n(t)$ are *all* periodic functions of t with period T, or possibly with an *integer* multiple of the period T due to the possibility that some of its zeros get exchanged through the motion.

The last two examples of many-body problems we report in this introductory section are also *entirely isochronous*; indeed, they can be characterized as assemblies of *nonlinear harmonic oscillators*, inasmuch, as these two dynamical systems (which are actually special cases of more general systems: see Sections 2.2 and 4.5) have the remarkable property that their *generic* solutions—namely, *all* their solutions, except for a *lower dimensional* set of *singular* solutions in which one or more of the "moving particles" shoot off to infinity at a finite time—are *completely periodic* with the *same* period T, see (1.2) with (1.3). Their *Newtonian* equations of motion read

$$\ddot{\underline{z}}_{nm} - 3\,i\,\omega\,\dot{\underline{z}}_{nm} - 2\,\omega^2\,\underline{z}_{nm} = c\sum_{\nu=1}^{N}\sum_{\mu=1}^{M} \underline{z}_{n\mu}\left(\underline{z}_{\nu\mu}\cdot\underline{z}_{\nu m}\right), \tag{1.12}$$

$$\ddot{\underline{z}}_{nm} - 3\,i\,\omega\,\dot{\underline{z}}_{nm} - 2\,\omega^2\,\underline{z}_{nm} = c\sum_{\nu=1}^{N}\sum_{\mu=1}^{M} \underline{z}_{\nu\mu}\left(\underline{z}_{\nu\mu}\cdot\underline{z}_{nm}\right). \tag{1.13}$$

These are two (different) systems of NM Newtonian equations of motion satisfied by the NM *complex* S-vectors $\underline{z}_{nm}$ (with S an *arbitrary* positive integer); hence, here the index n runs from 1 to N and the index m runs from 1 to M, with N and M two *arbitrary* positive integers, while c is, of course, an *arbitrary* complex constant (which might actually be rescaled away). The dot sandwiched between two S-vectors denotes the standard (Euclidian) scalar product, entailing the *rotation-invariant* character, in S-dimensional space, of these equations of motion. Since these systems only feature *linear* (velocity-dependent) and *cubic* (velocity-independent) forces, these models are remarkably close to physics; and they become even more applicable if they are written in their *real* versions, which obtain in an obvious manner by setting

$$\underline{z}_{nm} = \underline{x}_{nm} + i\,\underline{y}_{nm}, \quad c = a + i\,b. \tag{1.14}$$

These few instances of *isochronous* N-body problems have been exhibited above in order to raise the interest of the reader in this kind of models. They are far from exhausting the vast universe of *isochronous* systems: indeed, the main purpose of this monograph is to advertise the notion that *isochronous systems are not rare*; as explained in the next chapter, and exemplified in the chapters that follow.

Let us end this introductory chapter by highlighting the very recent findings, according to which the most general (standard: autonomous, nonrelativistic, *Hamiltonian*) many-body problem—only restricted by the quite physical requirement to be *translation invariant*—can be Ω-*modified* to yield an (also *Hamiltonian*) many-body system which is *entirely isochronous* with period $T = 2\pi/\Omega$ yet behaves in a manner quite similar to the original system over timescales much smaller than T (and note that the choice of Ω, hence of T, is *arbitrary*). This phenomenology is remarkable, inasmuch as the standard many-body problem with *realistic* interparticle forces (for instance, in the context of molecular dynamics) is generally characterized by a *chaotic* behavior yielding, via statistical mechanics, the *irreversible* time evolution underlying the validity of the second principle of thermodynamics. It demonstrates the existence of (autonomous, *Hamiltonian*—if not normal) *entirely isochronous* many-body models behaving *chaotically* over *finite* (yet possibly *arbitrarily long*) times before the *isochronous* phenomenology takes over, and it also provides a full understanding of the mechanism yielding this outcome (which might therefore—*after* this understanding has been internalized—be considered *trivial* rather than *remarkable*).

1.N Notes to Chapter 1

For the notion of *integrable* and *superintegrable* systems, and for the *integrable* and *superintegrable* character of the N-body problems considered in this chapter see, for instance, [37] [36] [52].

The fact that *all* solutions of the dynamical system (1.1) are *completely periodic*, see (1.2), was conjectured when this Hamiltonian system was first solved in the quantal context and the equispaced character of its energy spectrum was ascertained [28]. I have been told—but unfortunately never had a chance to check personally the correctness of this information—that when this conjecture was brought to the attention of Jurgen Moser he was surprised and skeptical. In fact he later discovered [115] that a Lax pair [111] could be associated with the $\omega = 0$ case of this classical Hamiltonian system, entailing its complete integrability. The extension of this finding to include the $\omega \neq 0$ case soon followed [3] [4] [123] [120], yielding a proof of the *isochronous* character of this classical system. A simple way to prove this result was also introduced a little later by A. M. Perelomov [124]. This development—which might be considered a preliminary version of the trick, see Section 2.1—is discussed in some detail in Section 2.1.3.3 of [37], and see also [125].

The *isochronous* character of the equations of motion (1.4b) was first shown in [38], and subsequently investigated (also for initial data outside the *isochrony* region) in [86].

The *isochronous* character of the system (1.5) was first shown in [61], and subsequently investigated (also for initial data outside the *isochrony* region) in [66].

For the correspondence among systems evolving in the complex z-plane and systems characterized by real two-vectors $\vec{r}_n$ evolving in the (physical) horizontal plane see [34], [35] and Chapter 4 of the book [37].

The *solvable* character of the N-body problem (1.10) was first noted in [29]. The honorary title of "goldfish" was attributed to this system in [36]; the (rather presumptuous) motivation for doing so originated from the fact that this paper [36] was contributed to a meeting held to celebrate the 60th birthday of V. E. Zakharov, who had compared the search for *integrable* systems to the quest for the mythical goldfish (see page 622 of [143]; for a full quote of the relevant sentence see page 71, the paragraph after the *Remark 4.2.2-1*). For the place of this system, (1.10), in the context of the Ruijsenaars–Schneider class of integrable N-body problems see, for instance, [37].

The models (1.12) and (1.13) were introduced, and their *isochronous* character demonstrated, in [75] (a paper which relied heavily on previous results by V. I. Inozemtsev); this proof is provided below, see Section 4.5. The characterization [75] of these models as describing assemblies of *nonlinear harmonic oscillators* is clearly justified by the *nonlinear* character of the *Newtonian* equations of motion (1.12) and (1.13), and by the *isochrony* of (almost) *all* their solutions, which entails a maximally *harmonic* character of these systems, characterized as they are by a *single* frequency of oscillation.

References in which some of the material treated in this monograph has been previously reviewed include the following: [37] [43] [51] [53] [54] (warning: reference [43] is marred by many misprints).

Finally, concerning the notions mentioned in the last paragraph of this introductory chapter—which are based on recent findings obtained with François Leyvraz—the interested reader is referred to the treatment provided below (see Chapter 5, in particular Section 5.5) and to the bibliographic information provided in Section 5.N.

2

ISOCHRONOUS SYSTEMS ARE NOT RARE

In this chapter we describe a simple trick modifying a dynamical system so that the new system yielded by it is *isochronous*. This trick amounts to a change of dependent and independent variables. It features a *positive* parameter ω: the "ω-modified" model yielded by its application is *isochronous* inasmuch as it possesses an open, *fully dimensional*, region in its phase space where *all* its solutions are *completely periodic* —i.e., *periodic* in *all* degrees of freedom—with the same period $\tilde{T}$ which is a finite *integer* multiple of the basic period $T = 2\pi / \omega$, see (1.3) and below. The "ω-modified" system manufactured in this manner can generally be made *autonomous*—this being indeed the case we deem more interesting hence on which we focus hereafter. It obtains from (possibly *nonautonomous*) dynamical systems belonging to a *quite large* class, restricted mainly by the condition that it makes sense to extend by analytic continuation its time evolution to *complex* time, and moreover by a *scaling* requirement (see below). The fact that this class is *large* justifies the claim advertised by the title of this chapter.

2.1 The trick

Let us consider to begin with a quite general dynamical system, which we write as follows:

$$\zeta' = F\left(\zeta; \tau\right) . \tag{2.1}$$

Here $\zeta \equiv \zeta(\tau)$ is the dependent variable, which might be a scalar, a vector, a tensor, a matrix, you name it. The independent variable is τ, and we assume it is permissible to treat this variable as *complex*; this requires that the derivative with respect to this *complex* variable τ appearing in the left-hand side of this evolution equation (2.1) (denoted there, and below, by an appended prime) make sense, namely that this dynamical system (2.1) is *analytic*, entailing that the dependent variable ζ is an *analytic* function of the complex variable τ—but this does not require $\zeta(\tau)$ to be a *holomorphic* or a *meromorphic* function of τ; $\zeta\left(\tau\right)$ might feature all sorts of singularities, including branch points, in the *complex* τ-plane, indeed this will generally happen since we assume the evolution equation (2.1) to be *nonlinear*. Indeed the quantity F in the right-hand side of (2.1)—which has of course the same scalar, vector, matrix, . . . , character as ζ— might depend (arbitrarily but analytically) on ζ as well as on τ. (The possibility that this dynamical system might also feature other, "spacelike", independent variables—for instance, be a system of PDEs rather than ODEs—is taken into consideration below, see Chapter 7.)

In spite of the generality of this dynamical system, (2.1)—note that, without loss of generality due to the arbitrariness of the multicomponent character of ζ and F, it has been formulated as a *first-order* system—there generally holds a standard result ("Theorem of existence, uniqueness and analyticity") that characterizes the solution $\zeta\,(\tau)$ of its initial-value problem determined by the assignment

$$\zeta(0) = \zeta_0. \tag{2.2}$$

Here, for notational simplicity, we assign the initial datum ζ_0 at $\tau = 0$; and we assume of course that the right-hand side of (2.1) is *nonsingular* for $\tau = 0$ and $\zeta = \zeta_0$. The relevant, standard result guarantees then the existence of a circular disk in the *complex* τ-plane, centered at $\tau = 0$ (where the initial data are assigned) and having a *positive* radius ρ, $\rho > 0$, such that the solutions $\zeta\,(\tau)$ corresponding to these initial data are *holomorphic* in it, namely for $|\tau| < \rho$ (if $\zeta\,(\tau)$ is a *multicomponent* object, the property to be *holomorphic* is, of course, featured by *each* of its components). The value of ρ depends on the initial datum, and as well on the evolution equation (2.1), namely on the function $F\,(\zeta;\tau)$.

[A simple minded way to express/understand this fact is via the notion that *all* solutions of an analytic ODE are *holomorphic* functions of the independent variable in a sufficiently small neighborhood of the assigned initial data—provided (as hereafter assumed) these are assigned where the ODE is *not* singular].

Let us now introduce the following changes of dependent and independent variables:

$$z(t) = \exp\,(i\,\lambda\,\omega\,t)\;\zeta\,(\tau)\,, \tag{2.3a}$$

$$\tau \equiv \tau\,(t) = \frac{\exp\,(i\,\omega\,t) - 1}{i\,\omega}. \tag{2.3b}$$

This transformation is called "the trick". The essential part of it is the change of *independent* variable (2.3b): and let us re-emphasize that, here and hereafter, the new *independent* variable t is considered as the *real*, "physical time" variable. Note that (2.3) implies

$$\tau\,(0) = 0, \quad z\,(0) = \zeta\,(0)\,, \quad \dot{z}\,(0) - i\lambda\omega z\,(0) = \zeta'\,(0)\,, \tag{2.4}$$

and, most importantly, that $\tau(t)$ is a *periodic* function of t with period T, see (1.3). More specifically, as the time t increases from zero onwards, the *complex* variable $\tau \equiv \tau\,(t)$ travels counterclockwise round and round, making a full tour in the time interval T, on the circle C the diameter of which, of length $2\,/\,\omega$, lies on the imaginary axis in the *complex* τ-plane, with one extreme at the origin, $\tau = 0$, and the other at the point $\tau = 2\,i\,/\,\omega$ (it is instructive to take the time and draw for oneself the relevant graph). Hence, if $\zeta(\tau)$ is *holomorphic* in τ (namely, each of its components is a *holomorphic* function of the *complex* variable τ) in the (closed) disk D encircled by this circle C, then the trick relation (2.3) with (1.3) implies that $z(t)$ is "λ-*periodic*,"

$$z(t + T) = \exp\,(2\,\pi\,i\,\lambda)\;z\,(t)\,. \tag{2.5}$$

We moreover restrict hereafter the prefactor $\exp(i\,\lambda\,\omega\,t)$ that multiplies $\zeta(\tau)$ in the right-hand side of (2.3a)—whose purpose is, via an appropriate choice of the parameter λ, to allow the ω-modified system to be *autonomous* (see below)—by the requirement that λ be *real* and *rational*, say

$$\lambda = \frac{p}{q} \tag{2.6}$$

with p and q *coprime integers* and $q > 0$. This restriction entails, see (2.5) and (1.3), that if $\zeta(\tau)$ is *holomorphic* in τ in the (closed) disk D encircled by the circle C, then $z(t)$ is *completely periodic* (namely, *each* of its components is *periodic*) with the period

$$\tilde{T} = q\,T, \tag{2.7a}$$

$$z(t+\tilde{T}) = z\,(t)\,. \tag{2.7b}$$

The "ω-modified" dynamical system is the one that obtains from (2.1) via the trick (2.3). It clearly reads as follows:

$$\dot{z} = i\,\lambda\,\omega\,z + \exp\left[i\;(\lambda+1)\;\omega\,t\right]\;F\left(\exp\left(-i\,\lambda\,\omega\,t\right)\;z;\;\frac{\exp\left(i\,\omega\,t\right)-1}{i\,\omega}\right). \tag{2.8}$$

And it is plain, on the basis of the arguments we just gave, that this system is *isochronous*, a sufficient condition for the *complete periodicity* with period $\tilde{T}$, see (2.7), of its solutions being provided by the inequality

$$\frac{2}{\omega} < \rho, \tag{2.9}$$

which entails that the (closed) disk of radius ρ centered at the origin in the complex τ-plane, $|\tau| \leq \rho$, encloses the circle C, thereby guaranteeing the *holomorphy* of $\zeta(\tau)$ in the disk D encircled by the circle C. This inequality, (2.9), can clearly be satisfied by initial data, see (2.4), situated inside an *open* domain of such data, at least provided ω is not too small (actually, in all the examples reported below no restriction on the value of ω is required, namely such an open domain exists for any arbitrary value of $\omega > 0$; this is certainly the case whenever the unmodified system (2.1) is *autonomous*, see below).

The *isochrony* of this "ω-modified" dynamical system, (2.8), is a rather obvious consequence of the way it has been manufactured. But, trivial as the emergence of this property might be, remarkably the class of *isochronous* dynamical systems that can be manufactured in this manner is *vast* and it includes many interesting examples, as shown below. An important attribute of these *isochronous* models, which is generally indispensable to deem them *interesting*, is that they be *autonomous*. This can be achieved in many cases by assigning appropriately the value of the (*rational*) parameter λ. For instance this can be done if the original system (2.1) is itself *autonomous* to begin with,

$$\zeta' = F(\zeta), \tag{2.10}$$

and moreover the function $F(\zeta)$ satisfies the scaling property

$$F(c\zeta) = c^{\gamma} F(\zeta) \tag{2.11}$$

with γ a *rational* number ($\gamma \neq 1$). It is indeed then clear that the ω-modified system (2.8) becomes *autonomous*, taking the simple form

$$\dot{z} = i\,\lambda\,\omega\, z + F(z), \tag{2.12a}$$

provided one makes the assignment

$$\lambda = \frac{1}{\gamma - 1} \tag{2.12b}$$

(note that the assumed *rational* character of γ entails that λ is as well *rational*).

This is not the only possible mechanism to manufacture a dynamical system that possesses *both* (desirable) attributes, to be *isochronous* and to be *autonomous*: indeed one can also work backwards, namely identify to begin with an *autonomous* ω-modified system candidate to be obtainable via the trick (2.3) hence to be *isochronous,* and then identify an unmodified system such as (2.1) (possibly itself *not* autonomous) from which that ω-modified system can indeed be obtained via the trick (2.3) (or from a generalized version of this trick: for instance, different parameters λ could be introduced, in a generalized version of the relation (2.3a), for the different components of the multicomponent quantities z and ζ; but rather than elaborating on such possibilities at this stage, we postpone their exhibition to the cases treated below in which their introduction turns out to be appropriate).

In conclusion, we see that via the trick (2.3) a relationship is established among two systems, so that to the—generally valid—property of *holomorphy* of the solutions $\zeta(\tau)$, considered as functions of the *complex* independent variable τ, of the unmodified system (2.1) in the neighborhood of the value $\tau = 0$ where the initial data are assigned, there corresponds the property of *isochrony* of the ω-modified system, see (2.7).

Let us emphasize that—as clearly implied by this discussion—it is by no means necessary that the (unmodified) dynamical systems which constitute the starting point of our treatment be *integrable*, in order to conclude that the time evolution (in the, of course *real*, "physical time" variable t) of the "ω-modified" systems obtained from them via the trick (2.3) be *isochronous*. Indeed, let us reiterate that the only (very mild) essential requirement on the original systems is that it make sense to consider them for *complex* values of the independent variable τ. An additional requirement must be added if one wishes the ω-modified systems to be *autonomous*, this being a natural condition in order that they might be *interesting.* Of course this issue—whether or not the ω-modified systems obtained in this manner are *interesting*—calls into play a metamathematical value-judgment. Yet we are confident that every reader who will have the zeal to read through this monograph will eventually agree that the class of *isochronous* systems treated herein deserve this accolade.

In the special case in which the original, unmodified system is indeed *integrable*, the analytic dependence of its solutions on its independent variable τ is generally quite simple: indeed a class of *integrable* dynamical systems is characterized by the "Painlevé property" requiring *all* their solutions to be *meromorphic* functions of their independent variable τ in their *entire* (natural) phase space. It is then clear from the above treatment that every ω-modified system obtained via the trick (2.3) from such a system—featuring the Painlevé property—is *completely periodic* with period $\tilde{T}$, see (2.7), for *arbitrary* initial data in its *entire* (natural) phase space—except possibly for a *lower dimensional* set of initial data such that a pole of the solution of the unmodified system falls exactly on the circle C in the *complex* τ-plane, so that the corresponding solution of the ω-modified system, rather than being *periodic* in t with period $\tilde{T}$, becomes *singular* at a finite time $t_S < T$. And clearly an analogous property of *isochrony*—for *arbitrary* initial data in their *entire* phase space, except for a *lower dimensional* set of initial data—is featured by any ω-modified system obtained via the trick (2.3) from an unmodified dynamical system having the property that *all* its solutions, considered as functions of their (*complex*) independent variable τ, feature as singularities only a *finite* number of *rational* branch points in the *complex* τ-plane, possibly in addition to an arbitrary (possibly infinite) number of poles. Examples of such *integrable* dynamical systems—which do *not* possess the Painlevé property, but whose solutions only feature a *finite* number of *rational* branch points—are well known: two such systems are the N-body problems (1.1) and (1.10) (see the next section), and several others will be considered in the following chapters.

2.2 Examples

In this section we show how the trick (2.3) can be used to demonstrate the *isochronous* character of the examples reported in the introductory Chapter 1, and we also outline how the trick is moreover instrumental to understand the behavior of these systems *outside* their *isochrony* regions (this last discussion is of course not relevant for those systems which are *isochronous* in their *entire* phase space).

Firstly let us identify the unmodified system of ODEs from which the ω-modified equations of motion (1.4b) obtain. It just corresponds (up to the appropriate notational changes) to (1.4b) with $\omega = 0$:

$$\zeta_n'' = \sum_{m=1, m \neq n}^{N} \frac{g_{nm}^2}{\left(\zeta_n - \zeta_m\right)^3}. \tag{2.13}$$

As can be easily verified the version of the trick (2.3) relating this system of ODEs to (1.4b) is characterized by the assignment $\lambda = -1/2$. Note that this entails the following relations among the initial data for the *Newtonian* equations of motions (1.4b) and (2.13):

$$\zeta_n(0) = z_n(0), \quad \zeta_n'(0) = \dot{z}_n(0) + \frac{i\,\omega}{2}\, z_n(0). \tag{2.14}$$

It is then rather evident, see (2.13), that the parameter ρ (as defined in the preceding section) can be made *arbitrarily large*—so as to satisfy the condition (2.9), which then entails that the solutions of the ω-modified equations of motion (1.4b) are *completely periodic* with period $2T$—provided the initial data are assigned so that the quantity

$$v = \max_{n=1,\ldots,N} \left|\zeta_n'(0)\right| = \max_{n=1,\ldots,N} \left|\dot{z}_n(0) + \frac{i\,\omega}{2}\, z_n(0)\right| \tag{2.15}$$

be adequately *small* and the quantity

$$r = \min_{n,m=1,\ldots,N;\, n \neq m} \left|\zeta_n(0) - \zeta_m(0)\right| = \min_{n,m=1,\ldots,N;\, n \neq m} \left|z_n(0) - z_m(0)\right| \tag{2.16}$$

be adequately *large*. Clearly this can be achieved by a set of initial data $z_n(0)$, $\dot{z}_n(0)$ having *full dimensionality* in phase space (for a more detailed, quantitative treatment we refer to the literature, see Section 2.N). The *isochronous* character of the dynamical system (1.4) is thereby proven.

Let us also discuss, quite tersely, the behavior of the N-body problem (1.4) *outside* the *isochrony* region. The above treatment implies that it depends on the nature, and location, of the *singularities* of the corresponding solutions $\zeta_n(\tau)$ of the system of ODEs (2.13), considered as functions of the *complex* variable τ. These singularities can clearly occur—see (2.13)—*only* where two or more coordinates $\zeta_n(\tau)$ coincide. It is easy to ascertain that the *only* such singularities are *square-root branch points*. [This can be shown by a local analysis of the behavior of the solutions of the system of ODEs (2.13) in the neighborhood of such a singularity occurring, say, at $\tau = \tau_b$, namely, by verifying in such a neighborhood the consistency with the system of ODEs (2.13), and the generality, of the *ansatz*

$$\zeta_n(\tau) = b + \beta_n\left(\tau - \tau_b\right)^{1/2} + O\left(|\tau - \tau_b|\right), \quad n = 1, \ldots, M, \tag{2.17a}$$

$$\zeta_n(\tau) = b_n + O\left(|\tau - \tau_b|\right), \quad n = M+1, \ldots, N, \tag{2.17b}$$

where of course $b_n \neq b$, $b_n \neq b_m$]. Much more difficult—indeed, other than by painstaking numerical treatments, quite impossible in the generic case with *arbitrary* coupling constants g_{nm}^2—is to ascertain the *location* of these singularities in the *complex* τ-plane, and even more difficult would be to map out the detailed structure of the Riemann surface associated with the dependence of the functions $\zeta_n(\tau)$ on the *complex* variable τ (except in the *integrable* case when *all* the coupling constant coincide, see (1.1): then the number of square-root branch points is *finite*—in contrast to the generic case, when it is generally *infinite*). It is indeed clear—see the trick formulas (2.3)—that the behavior, as the time t evolves from the initial value $t = 0$ onwards, of the component $z_n(t)$ of the solution of our ω-modified system (1.4) coincides—up to the simple *periodic* prefactor $\exp(-i\,\omega\,t\,/\,2)$ (see (2.3b) with $\lambda = -1/2$)—with the behavior, as the *complex* variable τ travels round and round on the circle C in the *complex* τ-plane, of the corresponding component $\zeta_n(\tau)$ of the solution of the system of ODEs (2.13), which evolve then on its Riemann surface (having, in the *generic* case, an *infinite* number of sheets). Of course, the location of the branch points, and the detailed structure of these Riemann surfaces, depends on the initial data identifying the solution of the system of ODEs (2.13). If on the *main* sheet (that from which the travel starts) of the Riemann surfaces of all the functions $\zeta_n(\tau)$ there are *no* branch points in the (closed) disk D enclosed by the circle C (see above), then the travel over the Riemann surface does not occur (namely, no other sheet gets accessed), hence the corresponding solutions $z_n(t)$ are *all periodic* with period $2T$ (the factor 2 is due to the prefactor $\exp(-i\omega t/2)$, see (2.3) with $\lambda = -1/2$). This is the *isochronous* regime; and we know from the above discussion that there are initial data—indeed, occupying an open region having *full dimensionality* in phase space—entailing that this *does* happen. Another possibility is that, via this travel, only a *finite* number of sheets of the Riemann surface of $\zeta_n(\tau)$ get accessed before returning to the main sheet (note that the *square-root* character of the branch points entails that each of them only connects *two* sheets of the Riemann surface). To the initial data that yield such a structure for the component $\zeta_n(\tau)$ of the solution of the system of ODEs (2.13), there clearly correspond a component $z_n(t)$ of the solution of our ω-modified, "physical" system (1.4) that is again *periodic* with a period which is a *finite* (but possibly *quite large*) *integer* multiple of the basic period T, see (1.3). But there might also be initial data such that the travel over the corresponding (*infinitely* sheeted) Riemann surface entails *no* return to the *main* sheet, hence *no* periodicity; the time evolution of the corresponding solutions of our ω-modified, "physical" system (1.4) is then *aperiodic.* And these *aperiodic* motions shall generally be characterized by a *sensitive dependence* on their initial data, since even the tiniest change of such data shall generally cause some of the *infinitely many* branch points characterizing those features of the Riemann surface that determine the character of the relevant travel on it as τ goes round and round over C, to cross over, from being inside to outside the disk D (or viceversa), thereby causing a drastic change of these features, hence—from a certain critical time onwards—a drastic change in the time evolution of the corresponding solution $z_n(t)$ of our ω-modified, "physical" system (1.4). The corresponding *physical* explanation of this phenomenology is plain: clearly all motions of the N-body problem (1.4) are *confined* (indeed, if a particle coordinate $z_n(t)$ were to stray away towards infinity, the term in the right-hand side of its equation of motion (1.4b) representing the two-body forces would become extremely small, while the linear one-body term in the left-hand side would cause the particle to return towards the origin); but any *aperiodic* motion of an N-particle system *confined* in a region near the origin in the *complex* z-plane features an *infinite* number of *near misses*—namely, two particles sliding *very close* to each other, their motion being then dramatically influenced by their two-body interaction, which has then a major effect due to its singular character at zero separation (note that instead the probability of an actual *collision* among *point* particles moving in the plane is *vanishingly small*, so that *no* such collision occurs for *generic* initial data); hence any tiny change in the initial data shall eventually cause one *near miss* to change character, resulting essentially in an exchange of the trajectories of the two particles just after the *near miss*, which makes a substantial difference for the future evolution of the system provided the two particles are *different*—as they indeed generally are in the case under present discussion, with *different* coupling constants. [Note however that the *sensitive* dependence on the initial data caused by this mechanism refers to the different evolutions of two solutions, of the system under consideration, characterized by initial data differing by an *arbitrarily small* yet *finite* amount; excluding thereby the *infinitesimal* change of trajectory relevant to define Liapunov exponents].

Since this monograph focusses on *isochronous* motions, we limit our discussion of *aperiodic* (possibly, in some sense, "chaotic") motions to this terse mention, referring the interested reader who wishes to delve more deeply into this fascinating topic to the relevant recent literature (see Section 2.N).

Secondly we proceed to consider the Hamiltonian N-body problem (1.5). In this case as well the unmodified equations of motion correspond (again, up to the appropriate notational changes) to (1.5d) with $\omega = 0$:

$$\zeta_n'' = 2 \sum_{m=1, m \neq n}^{N} \frac{a_{nm}\, \zeta_n'\, \zeta_m'}{\zeta_n - \zeta_m}. \tag{2.18}$$

In this case the version of the trick (2.3) relating this system of ODEs to (1.5d) is even simpler, being characterized by the assignment $\lambda = 0$. These formulas entail therefore the following simpler relation among the initial data for the *Newtonian* equations of motions (1.4b) and (2.13):

$$\zeta_n(0) = z_n(0), \quad \zeta_n'(0) = \dot{z}_n(0). \tag{2.19}$$

As in the case treated just above, it is now rather obvious from (2.18) that the parameter ρ can again be made *arbitrarily large*—hence such as to satisfy the inequality (2.9), which then entails that the solutions of the ω-modified equations of motion (1.5d) are *completely periodic* with period T—provided the initial data are assigned so that the quantity

$$V = \max_{n=1,\dots,N} \left|\zeta'_n(0)\right| = \max_{n=1,\dots,N} \left|\dot{z}_n(0)\right| \qquad (2.20a)$$

be adequately *small* and the quantity

$$r = \min_{n,m=1,\dots,N;\, n\neq m} \left|\zeta_n(0) - \zeta_m(0)\right| = \min_{n,m=1,\dots,N;\, n\neq m} \left|z_n(0) - z_m(0)\right| \qquad (2.20b)$$

be adequately *large*. Indeed in this case there holds the simple inequality

$$\rho > \frac{A\,r}{V} \qquad (2.21a)$$

with

$$A > \frac{1}{4\,(1+2\,N)\,(1+2\,N\,\alpha)}, \qquad \alpha = \max_{n,m=1,\dots,N;\, n\neq m} |a_{nm}|\,. \qquad (2.21b)$$

Hence the goal to satisfy the inequality (2.9) can always be achieved by a set of initial data $z_n(0)\,,\ \dot{z}_n(0)$ having *full dimensionality* in phase space—entailing the *isochronous* character of the dynamical system (1.5). And in this case an analogous phenomenology to that outlined just above characterizes the motions of this system *outside* the *isochrony* region—albeit with significant differences: the nature of the branch points depends now on the values of the coupling constants a_{nm} (for a more detailed, quantitative treatment we refer to the literature, see Section 2.N).

Thirdly and lastly, let us consider a class of *isochronous* many-body problems which includes the two many-body problems, see (1.12) and (1.13), discussed at the end of the introductory Chapter 1.

The starting point is the most general class of *nonlinear* oscillators with *cubic* forces, characterized by the equations of motions (with the N^4 *arbitrary* coupling constants a_{nmjk})

$$\zeta''_n = \sum_{m,j,k=1}^{N} a_{nmjk}\,\zeta_m\,\zeta_j\,\zeta_k. \qquad (2.22)$$

One then uses the trick (2.3) with $\lambda = 1$, entailing the following Newtonian equations of motion characterizing the corresponding (*autonomous*) ω-modified N-body problem,

$$\ddot{z}_n - 3\,i\,\omega\,\dot{z}_n - 2\,\omega^2\,z_n = \sum_{m,j,k=1}^{N} a_{nmjk}\,z_m\,z_j\,z_k, \qquad (2.23)$$

as well as the following relations among the initial data for these two systems:

$$\zeta_n(0) = z_n(0), \quad \zeta'_n(0) = \dot{z}_n(0) - i\,\omega\, z_n(0). \tag{2.24}$$

It is then clear that the ω-modified N-body problem characterized by these Newtonian equations of motion (2.23) is *isochronous*. It is indeed plain from the structure of the system of ODEs (2.22) that its solutions $\zeta_n(\tau)$ are *holomorphic* in τ in a disk $|\tau| \le \rho$ whose radius ρ can be made *arbitrarily large* by assigning initial data such that the two quantities

$$W = \max_{n=1,\dots,N} \left|\zeta'_n(0)\right| = \max_{n=1,\dots,N} \left|\dot{z}_n(0) - i\,\omega\, z_n(0)\right|, \tag{2.25a}$$

$$R = \max_{n=1,\dots,N} \left|\zeta_n(0)\right| = \max_{n=1,\dots,N} \left|z_n(0)\right|, \tag{2.25b}$$

are both *adequately small* (the derivation of an explicit lower bound to ρ providing a rigorous proof of this statement is relegated to the end of this section). This implies that there is an open (*fully dimensional*) set of initial data $z_n(0)$, $\dot{z}_n(0)$ entailing the validity of the inequality (2.9), implying via the trick formula (2.3) (with $\lambda = 1$) the *complete periodicity* with period T, see (1.3), of the solutions $z_n(t)$ of (2.23) evolving from such initial data; thereby confirming the *isochronous* character of the N-body problem characterized by these Newtonian equations of motion (2.23).

Clearly this result implies that the two dynamical systems characterized by the *Newtonian* equations of motion (1.12) and (1.13) are as well *isochronous*, because these two systems of ODEs are merely special cases of the system of ODEs (2.23); of course for this to happen the number N characterizing the number of dependent variables for this system, (2.23), must coincide with the number SNM identifying the number of scalar dependent variables for each of the systems (1.12) and (1.13).

But this only guarantees the *complete periodicity* with period T of a *subset* of the solutions $z_n(t)$ of the Newtonian equations of motion (1.12) and (1.13): enough to conclude that these two dynamical systems are *isochronous*, but a weaker statement than that made above (at the end of the introductory Chapter 1), claiming that *all nonsingular* solutions of the two systems, (1.12) and (1.13), are *completely periodic* with period T (and note that the *generic* solutions of these models are indeed *nonsingular*: the *singular* solutions emerge from a special, *lower dimensional* set of initial data, see below).

Let us now proceed and prove this more general result: that *all nonsingular* solutions of the two systems, (1.12) and (1.13), are *completely periodic* with period T. The proof of this remarkable finding hinges upon a fundamental result due to V. I. Inozemtsev, implying the following

Lemma 2.2-1. Let $\zeta \equiv \zeta(\tau)$ be a *square matrix* of *arbitrary* order M satisfying the *nonlinear* second-order (matrix) ODE

$$\zeta'' = c\,\zeta^3, \tag{2.26}$$

where c is an arbitrary scalar constant (which might be rescaled away) and appended primes denote, of course, differentiations with respect to the independent variable τ. Then the *only* singularities of the (elements of the) matrix $\zeta(\tau)$ as regards their dependence on the independent variable τ are *simple poles*, namely *all* the solutions of this matrix ODE are *meromorphic* functions of the independent variable τ (in the *entire complex* τ-plane). ⊡

Let us now apply to this matrix ODE, (2.26), the trick (2.3) with $\lambda = 1$, namely let us set

$$z(t) = \exp(i\,\omega\,t)\;\zeta(\tau)\,, \tag{2.27}$$

with τ related to t by (2.3b). It is then plain that the matrix $z(t)$ satisfies the (matrix) ODE

$$\ddot{z} - 3\,i\,\omega\,\dot{z} - 2\,\omega^2\,z = c\,z^3, \tag{2.28}$$

and that, as a consequence of *Lemma 2.1-1* and the trick relation (2.27) with (2.3b) we may now assert that *all nonsingular* solutions of this matrix ODE are *completely periodic* with period T, see (1.3):

$$z(t+T) = z(t). \tag{2.29}$$

It is moreover clear that the *singular* solutions of this matrix evolution equation, (2.28), are exceptional (i.e. *nongeneric*): they correspond via (2.27) to the special solutions $\zeta(\tau)$ of (2.26) that feature a pole at a value $\tau = \tau_P$ corresponding via (2.3b) to a *real* value of t, requiring that τ_P satisfy the (*nongeneric*) condition

$$\operatorname{Im}\tau_P = \frac{\omega}{2}\left[(\operatorname{Re}\tau_P)^2 + (\operatorname{Im}\tau_P)^2\right]. \tag{2.30}$$

Our proof is then completed by observing that the *Newtonian* equations of motions (1.12) and (1.13) are merely two special cases of the *matrix* evolution equation (2.28), corresponding to appropriate *ansatzen* (compatible with (2.28)) for the matrix elements of the matrix z: for details see the literature (as surveyed in Section 2.N), where it is also indicated how other analogous *Newtonian* evolution equations can be obtained.

Finally let us provide the proof promised above: it is the prototype of several analogous proofs given in the recent literature on *isochronous* systems. The task is to find a lower bound to the value—depending of course on the initial data—of the radius ρ of the disk $|\tau| \le \rho$ in which the solution $\zeta_n(\tau)$ of (2.22) is *holomorphic*. Our first step—in order to then use standard results—is to reformulate this system of N *second-order* ODEs as the following *equivalent* system of $2\,N$ *first-order* ODEs:

$$\varphi_j' = f_j, \quad j = 1, \ldots, 2\,N, \tag{2.31a}$$

where we set

$$\varphi_n \equiv \varphi_n(\tau) = \zeta_n(\tau) - \zeta_n(0)\,, \quad \varphi_{N+n} \equiv \varphi_{N+n}(\tau) = \gamma\left[\zeta_n'(\tau) - \zeta_n'(0)\right], \quad n = 1, \ldots, N, \tag{2.31b}$$

$$f_n(\tau) = \zeta'_n(0) + \frac{\varphi_{N+n}(\tau)}{\gamma},$$

$$f_{N+n}(\tau) = \sum_{m,j,k=1}^{N} a_{nmjk} \left[\varphi_m(\tau) + \zeta_m(0)\right] \left[\varphi_j(\tau) + \zeta_j(0)\right] \left[\varphi_k(\tau) + \zeta_k(0)\right],$$

$$n = 1, \ldots, N. \tag{2.31c}$$

Here γ is an *a priori* arbitrary, but *positive*, constant, $\gamma > 0$: we reserve the privilege to assign its value at our convenience, see below. Note that these definitions entail that the *initial* data are now *vanishing* for *all* the $2N$ new dependent variables $\varphi_j(\tau)$, $\varphi_j(0) = 0$, and that the properties of *holomorphy* of the solutions $\zeta_n(\tau)$ of our original problem (2.22) *coincide* with the properties of *holomorphy* of the (first N of the) $2N$ new dependent variables $\varphi_j(\tau)$.

The standard result [105] provides then the following lower bound for the radius ρ of the circular disk $|\tau| \leq \rho$ within which the solutions $\varphi_j(\tau)$ of this system of $2N$ ODEs, hence *a fortiori* the solutions $\zeta_n(\tau)$ of the system of N ODEs (2.22), are *holomorphic*:

$$\rho > \frac{b}{(2N+1)M} \tag{2.32}$$

(this formula coincides with the last equation of Section 13.21 of [105] with the assignments $m = 2N$ and $a = 0$, the first of which is justified by the fact that the system (2.31a) features $2N$ ODEs, the second of which is justified by the *autonomous* character of our equations of motions, see (2.31)). The two *positive* quantities b and M appearing in the right-hand side of this inequality, (2.32), are defined as follows. The quantity b is required to guarantee that the right-hand sides of the ODEs (2.31a) with (2.31c) are *holomorphic* (when considered as functions of the $2N$ *complex* variables φ_j) provided there holds the $2N$ inequalities

$$|\varphi_j| \leq b; \tag{2.33}$$

this entails, in our case, no restriction at all on b, other than its positivity. The second quantity appearing in the right-hand side of the inequality (2.32), $M \equiv M(b)$, is the upper bound of the right-hand sides of the ODEs (2.31a) with (2.31c), when the quantities φ_j satisfy the restriction (2.33),

$$M = \max_{j=1,\ldots,2N;\ |\varphi_j| \leq b} [f_j]; \tag{2.34}$$

but of course the inequality (2.32) holds *a fortiori* if we *overestimate* M (as we shall now do).

It is indeed clear from (2.31c) and (2.25) that

$$M(b) \leq \max\left[W + \frac{b}{\gamma},\ N^3 \alpha (b + R)^3\right], \tag{2.35}$$

having set

$$\alpha = \max_{n,m,j,k=1,\dots,N} |a_{nmjk}|, \quad W = \max_{n=1,\dots,N} |\zeta_n'(0)|. \tag{2.36}$$

We now make the convenient assignment

$$\gamma = \frac{b}{N^3 \alpha (b+R)^3 - W}, \tag{2.37}$$

entailing

$$M(b) \leq N^3 \alpha (b+R)^3 \tag{2.38}$$

and

$$W < N^3 \alpha (b+R)^3, \tag{2.39}$$

the second inequality, (2.39), being required, see (2.37), to guarantee that γ be *positive*, $\gamma > 0$. Next we make the convenient assignment $b = R/2$, and from (2.32) and (2.38) we then get

$$\rho \geq \frac{4}{27 N^3 (2N+1) \alpha R^2}, \tag{2.40a}$$

while (2.39) reads

$$W < \left(\frac{3}{2} N\right)^3 \alpha R^3. \tag{2.40b}$$

The first of these two inequalities shows that ρ can indeed be made *arbitrarily large* by making R *sufficiently small*, while the second indicates that W is also then required to be *adequately small.* Q. E. D.

2.N Notes to Chapter 2

The trick was first introduced in [33] (in the present context; but, given its elementary nature, it is quite possible that something analogous was used previously in somewhat analogous contexts). It was then used quite extensively over the last few years, see most of the recent references quoted throughout this monograph.

For the standard theorem of existence, uniqueness and analyticity of (systems of) ODEs see any relevant textbook: for instance Section 13.21 of [105].

For a complete proof of the *isochronous* character of the N-body problem (1.4) see [38] (there is a misprint in eq. (3.10b) of this paper: the exponent 1/2 should be replaced by −1/2). For a more complete investigation of this Hamiltonian system—including an analytical and numerical study of its behavior for initial data *outside* the *isochrony* region—see [86].

The *isochronous* character of the Hamiltonian system (1.5d) (see (2.18) and the discussion following it) was first noted in [62]. A more satisfactory proof of the *isochronous* character of this system is provided in [66]—which includes also an analytical and numerical study of the behavior of this system for initial data *outside* the *isochrony* region and a derivation (in Appendix A) of the formula (2.21).

The many-body systems (1.12) and (1.13) (as well as several other analogous ones) were identified—and their properties demonstrated—in [75], where the notion of *nonlinear harmonic oscillators* was also introduced. The basic paper by V. I. Inozemtsev reporting the solution of the matrix evolution equation (2.26) and thereby entailing the validity of *Lemma 2.2-1* is [106]. For convenient *ansatzen* of the matrix z allowing to transform the *isochronous matrix* evolution equation (2.28) into systems of *Newtonian* evolution equations having a more "physical" look (including the two systems (1.12) and (1.13)) see, in addition to [75], the following references: [15], [17], [108], Section 5.6.5 of [37], [18], [19], [57].

In Section 2.2 we tersely outlined some notions relevant to understand the behavior of *isochronous* systems *outside* their *isochrony* region, and above (in this section) we have indicated the references [86] and [66] where such investigations have been pursued. Let us end this section by mentioning that these ideas—which extend beyond the scope of this monograph—are currently being investigated inasmuch as they provide a somewhat novel paradigm to understand the transition (as the initial data, or the parameters of the models under consideration, are varied) from regular ("periodical", "integrable") to irregular (perhaps, in some sense, "chaotic") behaviors of classical dynamical systems—and possibly to understand certain features of "deterministic chaos"—in terms of travel over Riemann surfaces: see [72] [101] [95] [73].

3

A SINGLE ODE OF ARBITRARY ORDER

In this chapter we exhibit several ω-modified ODEs involving a *single* dependent variable, that because of the way they have been obtained (via the trick described in Section 2.1) are generally *isochronous*: they possess *periodic* solutions living in an *open* (hence *fully dimensional*) region of their phase space—or equivalently, in the context of the *initial-value* problem, evolving from initial data situated in such *open* (hence *fully dimensional*) region of their phase space. In several (special) cases these ODEs are *entirely isochronous*, namely (see Chapter 1) *all* their *nonsingular* solutions are *periodic*, their *isochrony* region encompassing their *entire* (natural) phase space, with the only, possible, exception of some *lower dimensional* subregion(s) where the solutions are *singular:* then these *lower dimensional* subregion(s) where the solutions are *singular* generally separate(s) *fully dimensional* regions—in which the *entire* phase space is partitioned—where the solutions are *periodic* but with a period that, while being an *integer multiple* (or, occasionally, a *rational multiple*) of the basic period, is generally *different* in each of these *different* (*fully dimensional*) regions.

Our presentation does not aim at presenting *all* the *isochronous* ODEs involving a single dependent variable obtainable via our technique: we rather exhibit several examples—in some cases entire classes of ODEs, in some cases specific ODEs—in order to provide the reader with a fair idea of the potentialities of this approach. The alert practitioner will then manufacture other examples, just for the fun of it or in view of some envisaged application. Similarly, we do not discuss systematically the behavior of the solutions of the ω-modified equations exhibited below: only in some cases, which we deem particularly interesting, such information is provided. Our main goal is to convince the receptive reader that the simple approach based on the trick does, indeed, provide a useful tool for the investigation of (certain classes of) ODEs.

In the following Section 3.1 a fairly general class of *isochronous* ODEs of *arbitrary* order is exhibited. In the subsequent Section 3.2 several specific examples are presented, including both ODEs that are specific instances of those treated in Section 3.1, and ODEs that do not belong to that class. As usual the last Section 3.N provides the original references for the findings tersely reported in this chapter as well as bibliographic information on analogous results not reported herein.

3.1 A class of autonomous ODEs

In this section certain (classes of) single *autonomous* nonlinear evolution ODEs of *arbitrary* order are identified that can be modified via the trick of

Section 2.1—reviewed below—so as to generate a one-parameter family of ω-modified, also *autonomous*, ODEs with the standard properties characterizing *isochronous* ODEs: for all *positive* values of the parameter ω, these ω-modified ODEs possess *periodic* solutions with fixed period $T = 2\pi / \omega$, emerging—in the context of the *initial-value* problem—from open domains of initial data the natural measure of which, in the space of such initial data, depends on the parameter ω but is always *positive* (i.e. *nonvanishing*).

We write as follows the class of (unmodified, *autonomous*) ODEs (of *arbitrary* order N) that we take as starting point:

$$G\left({}^{(N)}\zeta,\, {}^{(N-1)}\zeta, \ldots,\, {}^{(j)}\zeta, \ldots,\, {}^{(1)}\zeta,\, \zeta\right) = 0. \tag{3.1}$$

Here and throughout we use the short-hand notation

$${}^{(j)}\zeta \equiv {}^{(j)}\zeta\,(\tau) \equiv \frac{d^j\,\zeta\,(\tau)}{d\,\tau^{\,j}}, \quad j = 0, 1, \ldots, N \tag{3.2}$$

(entailing, of course ${}^{(0)}\zeta \equiv \zeta$).

We assume the function $G \equiv G\,(w_N,\, w_{N-1}, \ldots,\, w_j, \ldots,\, w_1,\, w_0)$ to depend *analytically* on *all* its $N+1$ arguments w_j, $j = 0, 1, 2, \ldots, N$, and to be such that (3.1) indeed qualifies as an "evolution equation", namely it can be "solved" for the highest-order derivative ${}^{(N)}\zeta$, see (3.2). We moreover assume the dependence of the function G on its $N+1$ arguments to satisfy the following *grading* property:

$$\begin{aligned} &G\left(c^{\lambda+N}\, w_N,\, c^{\lambda+N-1}\, w_{N-1}, \ldots,\, c^{\lambda+j}\, w_j, \ldots,\, c^{\lambda+1}\, w_1,\, c^{\lambda}\, w_0\right) \\ &= c^S\, G\,(w_N,\, w_{N-1}, \ldots,\, w_j, \ldots,\, w_1,\, w_0)\,, \end{aligned} \tag{3.3a}$$

where λ is a *rational* number,

$$\lambda = \frac{p}{q} \tag{3.3b}$$

with p, q two *arbitrary* coprime *integers* (hereafter, for definiteness, we assume q to be *positive*, $q > 0$, and $q = 1$ if $p = \lambda = 0$). The constant S is *arbitrary*, and it plays, generally, only a secondary role.

We now introduce the new (generally *complex*) dependent variable $z \equiv z\,(t)$ via the following version of the trick:

$$z(t) = \exp\,(i\,p\,\tilde{\omega}\,t)\;\zeta\,(\tau)\,, \tag{3.4a}$$

$$\tau = \frac{\exp\,(i\,q\,\tilde{\omega}\,t) - 1}{i\,q\,\tilde{\omega}}, \tag{3.4b}$$

which actually coincides with the version of Section 2.1, see (2.3), via (3.3b) (which coincides with (2.6)) and the relation

$$\omega = q\,\tilde{\omega}. \tag{3.5}$$

Here and below, as usual, the new independent variable t is *real*, to be interpreted as *time*. We will use in the following one and/or the other of the two *positive* constants ω and $\tilde{\omega}$ at our convenience, and similarly for the *rational* number λ and/or the two integers p and q.

The ω-modified ODE satisfied by the new dependent variable $z(t)$ reads

$$G\left({}^{[N]}z,\ {}^{[N-1]}z, \ldots, {}^{[j]}z, \ldots, {}^{[1]}z,\ z\right) = 0, \tag{3.6}$$

where the notation ${}^{[j]}z$ denotes the following linear combination of the derivatives (of degree j and less) of z:

$${}^{[j]}z = {}^{(j)}z + \sum_{k=1}^{j} \left[{}^{(j-k)}z\, b_{jk}(p,\ q)\ (i\,\tilde{\omega})^{k}\right]. \tag{3.7a}$$

Here, in analogy to (3.2), we introduced the short-hand notation

$${}^{(j)}z \equiv {}^{(j)}z(t) \equiv \frac{d^j\, z(t)}{d\,t^{\,j}}, \quad j = 0, 1, \ldots, N, \tag{3.7b}$$

while the numerical coefficients $b_{jk}(p,\ q)$ can be easily obtained, by recursion, from the trick relation (3.4). For instance, the first few relations (3.7a) read

$$\begin{aligned}
{}^{[0]}z &= z,\\
{}^{[1]}z &= \dot{z} - i\,p\,\tilde{\omega}\,z,\\
{}^{[2]}z &= \ddot{z} - i\,(2\,p + q)\,\tilde{\omega}\,\dot{z} - p\,(p + q)\,\tilde{\omega}^{\,2}\,z,\\
{}^{[3]}z &= \dddot{z} - 3\,i\,(p + q)\,\tilde{\omega}\,\ddot{z} - \left(3\,p^2 + 6\,p\,q + 2\,q^2\right)\tilde{\omega}^{\,2}\,\dot{z} + i\,p\,(p + q)\,(p + 2\,q)\,\tilde{\omega}^{\,3}\,z,\\
{}^{[4]}z &= \ddddot{z} - 2\,i\,(2\,p + 3\,q)\,\tilde{\omega}\,\dddot{z} + (6\,p^2 + 18\,p\,q + 11\,q^2)\,\tilde{\omega}^{\,2}\,\ddot{z}\\
&\quad + i\,\left(4\,p^3 + 18\,p^2\,q + 21\,p\,q^2 + 3\,q^3\right)\tilde{\omega}^{\,3}\,\dot{z}\\
&\quad + p\,(p + q)\,(p + 2\,q)\,(p + 3\,q)\,\tilde{\omega}^{\,4}\,z.
\end{aligned} \tag{3.7c}$$

Here of course superimposed dots denote time differentiations.

The mechanism that causes the ω-modified N-th order ODE (3.6) with (3.7) to be *isochronous* has been explained in the preceding chapter. We refer to the literature (see Section 3.N) for a detailed proof of the existence of an open (*fully dimensional*) region in phase space where the solutions $z(t)$ are *periodic*,

$$z\left(t + \tilde{T}\right) = z(t), \tag{3.8a}$$

with a period $\tilde{T}$ which is an *integer* multiple of the basic period

$$T = \frac{2\,\pi}{\omega} = \frac{2\,\pi}{q\,\tilde{\omega}}, \tag{3.8b}$$

$$\tilde{T} = \frac{2\,\pi}{\tilde{\omega}} = q\,T. \tag{3.8c}$$

The boundary of the region of *isochrony* generally yields—in the context of the initial-value problem—solutions that become *singular* in a finite time

$t_S < T$. Of much interest is as well the behavior of the solutions living *outside* the *isochrony* region. For some (special ones of these) ODEs of type (3.6) with (3.7) (and (3.3a)), *all* their *nonsingular* solutions are *periodic* with primitive periods which are *all* integer multiples of the basic period T (*entire isochrony*); for others of these ODEs there are certain domains of initial data that yield *periodic* solutions with such periods, but also domains that yield *aperiodic*, possibly "chaotic," solutions. This general phenomenology—including the "transition to chaos"—of the solutions of this class of evolution ODEs, (3.6) with (3.7) (and (3.3a)), is now under active study (as mentioned in Section 2.N); but its analysis for the ODEs introduced herein exceeds the scope of this monograph, as it requires a case-by-case treatment; we do however exhibit below (see the *Remark 3.1.3* and the following Section 3.2) specific solutions possessed by the entire class of ODEs (3.6) with (3.7) (and (3.3a)) that display some aspects of this phenomenology (*not* including the transition to chaos).

Several remarks are now appropriate.

Remark 3.1-1.. In this section we focus on the (class of) unmodified ODEs that are themselves *autonomous* and yield deformed ODEs which are as well *autonomous*—because *autonomous* ODEs are more likely to be of (theoretical or applicative) interest. But this approach is as well applicable if the unmodified ODEs, or the deformed ODEs, or both, are *not autonomous.* Then the λ-grading property (3.3a) might have to be appropriately modified, or possibly just forsaken. ⊡

Remark 3.1-2.. The ω-modified ODE (3.6) with (3.7) is, generally, *complex* (and in any case *only* in the *complex* context is the *isochrony* property guaranteed); but it can of course be transformed into a system of *two real* coupled ODEs for the *two real* dependent variables $x \equiv x(t), y \equiv y(t)$ by setting

$$z \equiv z(t) = x + i\,y \equiv x(t) + i\,y(t)\,, \tag{3.9}$$

and by considering the two ODEs that obtain from (3.6) with (3.7) via this formula, (3.9). An alternative possibility is to use the amplitude-phase representation of the complex number z, $z(t) = r(t)\exp[i\,\theta(t)]$. ⊡

Remark 3.1-3.. The λ-grading property (3.3a) entails (as can be verified by a straightforward computation) that the ODE (3.1) admits the special "similarity solution(s)"

$$\zeta(\tau) = c\,(\tau - \tau_b)^{-\lambda}\,, \tag{3.10}$$

where τ_b is an *arbitrary*, generally *complex*, constant—the suffix b serving as reminder that this function, (3.10), generally has an (algebraic, see (3.3b)) *branch point* at $\tau = \tau_b$—while the value(s) of the constant c are required to satisfy the nondifferential equation

$$G\Big(c\,(\lambda)_N\,,\,c\,(\lambda)_{N-1}\,,\ldots,\,c\,(\lambda)_j\,,\ldots,\,c\,\lambda,\,c\Big) = 0. \tag{3.11}$$

Here and throughout we use the standard notation

$$(\lambda)_0 \equiv 1, \quad (\lambda)_m \equiv \lambda\,(\lambda-1)\cdots(\lambda-m+1) = \frac{\Gamma(\lambda+1)}{\Gamma(\lambda+1-m)}. \tag{3.12}$$

It is, moreover, easy to verify that if one linearizes (3.1) around the solution (3.10) by setting

$$\zeta(\tau) = c\,(\tau-\tau_b)^{-\lambda} + \varepsilon\,\varphi(\tau), \tag{3.13}$$

with ε a (small) expansion parameter, one obtains for $\varphi \equiv \varphi(\tau)$ a *linear* N-th order ODE *all* solutions of which are, generically, also (sums of) simple powers,

$$\varphi(\tau) = \sum_{n=1}^{N} C_n\,(\tau-\tau_b)^{\alpha_m}, \tag{3.14}$$

where the N constants C_n are of course *arbitrary* and the N exponents α_m are the N roots of the N-th degree polynomial equation in α (see (3.12))

$$\sum_{j=0}^{N} g_j\,(\alpha)_j = 0, \tag{3.15}$$

where the coefficients g_j are just the partial derivatives of the function $G(z_N, z_{N-1}, \ldots, z_j, \ldots, z_1, z_0)$, evaluated at the same values of its $N+1$ arguments as featured in the left-hand side of (3.11):

$$g_j = \frac{\partial G}{\partial z_j}\left(c\,(\lambda)_N, c(\lambda)_{N-1}, \ldots, c\,(\lambda)_j, \ldots, c\lambda, c\right), \tag{3.16}$$

where of course the constant c is the same one appearing in (3.13) and satisfying (3.11). $\boxdot$

Remark 3.1-4.. In correspondence with the special "similarity solution(s)" (3.10) of the unmodified ODE (3.1), the ω-modified ODE (3.6) with (3.7) (and (3.3b)) possesses the special solution

$$z(t) = c\,\exp(i\,p\,\tilde{\omega}\,t)\left[-\tau_b + \frac{\exp(i\,q\,\tilde{\omega}\,t)-1}{i\,q\,\tilde{\omega}}\right]^{-p/q}, \tag{3.17a}$$

$$z(t) = c\left[-\tau_b\,\exp(-i\,q\,\tilde{\omega}\,t) + \frac{1-\exp(-i\,q\,\tilde{\omega}\,t)}{i\,q\,\tilde{\omega}}\right]^{-p/q}, \tag{3.17b}$$

where the constant c is of course required to be a solution of the nondifferential equation (3.11), while the (generally *complex*) constant τ_b is *arbitrary*. And it is clear that this solution is *periodic* with period $\tilde{T}$, see (3.8c), if

$$\left|\tau_b - \frac{i}{q\,\tilde{\omega}}\right| > \frac{1}{q\,\tilde{\omega}}, \quad \text{i.e.,} \quad |\tau_b|^2 > \frac{2\,\mathrm{Im}\,\tau_b}{q\,\tilde{\omega}}, \tag{3.18a}$$

that it is, as well, *periodic* but with the (smaller or equal) period T, see (3.8b), if instead

$$\left|\tau_b - \frac{i}{q\,\tilde{\omega}}\right| < \frac{1}{q\,\tilde{\omega}}, \quad \text{i.e.,} \quad |\tau_b|^2 < \frac{2\,\mathrm{Im}\,\tau_b}{q\,\tilde{\omega}}, \tag{3.18b}$$

while it becomes *singular* in the intermediate case

$$\left|\tau_b - \frac{i}{q\,\tilde{\omega}}\right| = \frac{1}{q\,\tilde{\omega}}, \quad \text{i.e.,} \quad |\tau_b|^2 = \frac{2\,\mathrm{Im}\,\tau_b}{q\,\tilde{\omega}}, \tag{3.18c}$$

at the finite time t_b defined mod (T) as follows:

$$\tau_b = \frac{\exp\left(i\,q\,\tilde{\omega}\,t_b\right) - 1}{i\,q\,\tilde{\omega}}. \tag{3.19}$$

Note that (3.18c) indeed entails that t_b, as defined by this formula, is *real.* ⊡

Remark 3.1-5.. The degree of the N-th order ODE (3.1) can be generally lowered to order $N-1$ by taking advantage of its *autonomous* character, namely by setting

$$\zeta' \equiv^{(1)} \zeta = u\left(\zeta\right), \tag{3.20}$$

since via this definition the ODE (3.1) becomes a *nonautonomous* ODE of order $N-1$ featuring the dependent variable $u \equiv u\left(\zeta\right)$ and the independent variable ζ; and then any solution $u \equiv u\left(\zeta\right)$ of this ODE, when inserted in (3.20), yields via a quadrature a solution of (3.1). An additional reduction by one order is moreover generally possible by taking advantage of the λ-grading property (3.3a), and the corresponding symmetry of the ODE (3.1). Hence the generic possibility to reduce to quadratures the solution of the ODE (3.1) not only for $N = 1$ (when this is indeed obvious from (3.1) itself), but as well for $N = 2$; and this carries of course over to the ω-modified versions, see (3.6), of these ODEs, obtained merely via the change of variables (3.4). This might seem to decrease the interest of the main finding reported in this section, concerning the *periodicity* properties of the solutions of these ω-modified equations, for ODEs of order less than *three* ($N = 1, 2$); but in fact, even when solutions are obtainable by quadratures, the recognition of their *periodicity* properties is often far from trivial [95] [101]. ⊡

Remark 3.1-6.. Examples of N-th order ODEs that satisfy *automatically* the λ-grading property (3.3a) are

$$F\left({}^{(N)}\zeta\,\zeta^{-(\lambda+N)/\lambda},\ {}^{(N-1)}\zeta\,\zeta^{-(\lambda+N-1)/\lambda},\dots,\ {}^{(N-j)}\zeta\,\zeta^{-(\lambda+N-j)/\lambda},\dots,\right.$$
$$\left.\zeta''\,\zeta^{-(\lambda+2)/\lambda},\zeta'\,\zeta^{-(\lambda+1)/\lambda}\right)=0, \tag{3.21}$$

$$F\left({}^{(N)}\zeta\,\left[\zeta'\right]^{-(\lambda+N)/(\lambda+1)},\ {}^{(N-1)}\zeta\,\left[\zeta'\right]^{-(\lambda+N-1)/(\lambda+1)},\dots,\right.$$
$${}^{(N-j)}\zeta\,\left[\zeta'\right]^{-(\lambda+N-j)/(\lambda+1)},\dots,\ \zeta''\,\left[\zeta'\right]^{-(\lambda+2)/(\lambda+1)},$$
$$\left.\zeta\,\left[\zeta'\right]^{-\lambda/(\lambda+1)}\right)=0, \tag{3.22}$$

$$F\left({}^{(N)}\zeta\,\left[\zeta'\right]^{-N},\ {}^{(N-1)}\zeta\,\left[\zeta'\right]^{-(N-1)},\dots,\ {}^{(N-j)}\zeta\,\left[\zeta'\right]^{-(N-j)},\dots,\right.$$
$$\left.\zeta''\,\left[\zeta'\right]^{-2},\zeta\right)=0, \tag{3.23}$$

where the functions F are now only required to depend *analytically* on *all* their N arguments, but do not have to satisfy any grading property. It is, indeed, plain that these ODEs satisfy the grading property (3.3a), with $S = 0$. Note that (3.23) is merely the special case of (3.22) with $\lambda = 0$ (corresponding to $p = 0$ and $q = 1$, see (3.3b)). □

Let us end this section by considering the class of ODEs of type (3.1) which are *linear* in the highest derivative and *polynomial* in *all* the others:

$${}^{(N)}\zeta=\sum_{\underline{m}}\left\{c\left(\underline{m}\right)\prod_{\ell=0}^{N-1}\left[{}^{(\ell)}\zeta\right]^{m_\ell}\right\}, \tag{3.24a}$$

where we denote by $\underline{m}\equiv(m_0,\,m_1,\dots,\,m_{N-1})$ the N-vector that has as its N components the N *nonnegative integer* exponents m_ℓ, $\ell=0,1,\dots,N-1$. To guarantee that this ODE satisfy the grading property (3.3a) the sum in the right-hand side of this evolution ODE, (3.24a), is restricted to the sets of values of these N *nonnegative integer* exponents m_ℓ that satisfy the restriction

$$p\sum_{\ell=0}^{N-1}m_\ell+q\sum_{\ell=1}^{N-1}\ell\,m_\ell=p+q\,N \tag{3.24b}$$

or equivalently (see (3.3b))

$$\sum_{\ell=0}^{N-1}\left[(\lambda+\ell)\,m_\ell\right]=\lambda+N. \tag{3.24c}$$

The constants $c\left(\underline{m}\right)$ in (3.24a) are *arbitrary* (possibly *complex*), and, here and below, we are of course using the notation (3.2) as well as (3.3b) with p and $q>0$ two coprime *integers*.

Verification that the class of evolution equations (3.24) belongs to the class (3.1), and, in particular, that it satisfies the λ-grading condition (3.3a) (with $S = N + \lambda$) is a trivial task. And it is also plain that the condition (3.24c) can be satisfied by setting

$$\lambda = \frac{-N + \sum_{\ell=1}^{N-1} \ell \, m_\ell}{1 - \sum_{\ell=0}^{N-1} m_\ell} \tag{3.25}$$

if *only one* of the coefficients $c\,(\underline{m})$ appearing in the right-hand side of (3.24a) does *not* vanish (then of course the numbers m_ℓ in the above formula, (3.25), are those for which $c\,(\underline{m})$ does *not* vanish); namely, the ODE

$$^{(N)}\zeta = c\,(\underline{m}) \prod_{\ell=0}^{N-1} \left[{}^{(\ell)}\zeta\right]^{m_\ell}, \tag{3.26}$$

for any *arbitrary* assignment of the nonnegative-integer-valued N-vector $\underline{m} \equiv (m_0, m_1, \ldots, m_{N-1})$, belongs to the class (3.1), as it clearly satisfies the λ-grading condition (3.3a) with $S = N + \lambda$ and the *rational* number λ defined by (3.25) (excluding of course the trivial case in which one of the N nonnegative integers m_ℓ equals unity and all the others vanish, causing the denominator in the right-hand side of this formula to vanish).

The corresponding ω -modified (hence *isochronous*) ODE obtains of course from the ODE satisfied by $\zeta\,(\tau)$ via the replacement

$$^{(j)}\zeta\,(\tau) \longmapsto {}^{[j]}z\,(t)\,, \quad j = 0, 1, \ldots, N, \tag{3.27}$$

see (3.7).

We end this section by listing the evolution ODEs of type (3.24) of order $N = 2$, $N = 3$ and $N = 4$ (but systematically omitting from this list ODEs of the simple type (3.26)). Hereafter the coefficients $c\,(\underline{m})$ are *arbitrary* (possible *complex*) constants, and by the notation f (possibly embellished by some label when we need to distinguish several such functions) we generally denote a function depending *analytically*, but otherwise *arbitrarily*, on all its arguments, this requirement being sufficient to guarantee that, after the replacement (3.27), the ODEs displayed below indeed possess a lot of *periodic* solutions. The additional restrictions that the coefficients $c\,(\underline{m})$ or the functions f must satisfy in each case in order to guarantee the polynomial character of the right-hand side of the ODEs displayed below will not be specified, since they are, in each case, evident by inspection; a *necessary* condition is of course that the functions f be themselves polynomial in all their arguments; in several cases this is also *sufficient*. For each ODE reported below we also write the corresponding restrictions on the values of the two coprime *integers* p and $q > 0$ characterizing the trick transformation (3.4) that turns these unmodified ODEs into their ω-modified *isochronous* counterparts (as directly given by the replacement (3.27)). The selection of the ODEs to be displayed, including their ordering, is mainly motivated by "aesthetic" considerations, or equivalently by the wish to maximize the "user-friendly" character of this list.

Second-order ODEs:

$$\zeta'' = (\zeta')^2 \, \zeta^{-1} \, f\left([\zeta']^{-p} \, \zeta^{q+p}\right), \quad (0 < -p \leq q), \tag{3.28}$$

$$\zeta'' = (\zeta')^2 \, f(\zeta), \quad (p = 0, \; q = 1), \tag{3.29}$$

$$\zeta'' = c(q,1) \, \zeta' \, \zeta^q + c(2q+1,0) \, \zeta^{2q+1}, \quad (p = 1). \tag{3.30}$$

Third-order ODEs:

$$\zeta''' = \zeta'' \, \zeta' \, \zeta^{-1} \, f\left([\zeta'']^{-p} \, \zeta^{2q+p}, \, [\zeta']^{-p} \, \zeta^{q-p}\right), \quad (0 < -p \leq q), \tag{3.31}$$

$$\zeta''' = \zeta'' \, \zeta' \, f_1(\zeta) + (\zeta')^3 \, f_2(\zeta), \quad (p = 0, \; q = 1), \tag{3.32}$$

$$\zeta''' = c(q,0,1) \, \zeta'' \, \zeta^q + c(q-1,2,0) \, (\zeta')^2 + c(2q,1,0) \, \zeta' \, \zeta^{2q} + c(3q+1,0,0) \, \zeta^{3q+1}, \quad p = 1. \tag{3.33}$$

Fourth-order ODEs:

$$\zeta'''' = \zeta''' \, \zeta' \, \zeta^{-1} \, f\left([\zeta''']^{-p} \, \zeta^{3q+p}, \, [\zeta'']^{-p} \, \zeta^{2q+p}, \, [(\zeta')]^{-p} \, \zeta^{q+p}\right), \quad (0 < -p \leq q), \tag{3.34}$$

$$\zeta'''' = \zeta''' \, \zeta' \, f_1(\zeta) + (\zeta'')^3 \, f_2(\zeta) + \zeta' \, \zeta^2 \, f_3(\zeta) + \zeta^4 \, f_4(\zeta), \quad (p = 0, \; q = 1), \tag{3.35}$$

$$\zeta'''' = c(q,0,0,1) \, \zeta''' \, \zeta^q + c(q-1,1,1,0) \, \zeta'' \, \zeta' \, \zeta^{q-1} + c(2q,0,1,0) \, \zeta'' \, \zeta^{2q} + c(q-3,3,0,0) \, (\zeta')^3 \, \zeta^{q-3} + c(2q-2,2,0,0) \, (\zeta')^2 \, \zeta^{2q-2} + c(3q-1,1,0,0) \, \zeta' \, \zeta^{3q-1} + c(4q,0,0,0) \, \zeta^{4q}, \quad (p = 1). \tag{3.36}$$

Let us repeat that the ω-modified, hence *isochronous*, versions of these equations obtain via the replacement (3.27), see (3.2) and (3.7c). Some specific examples of these *isochronous* ODEs are discussed in the following Section 3.2, as well as some other *isochronous* ODEs *not* belonging to this class.

3.2 Examples

In this section we provide a representative sample of ω-modified ODEs, almost all of which are therefore *isochronous* (the *only* exception is (3.61) and its *real* version (3.65), see below). We organize the presentation in several subsections: in each of these we exhibit firstly (or at least pretty soon) the ω-modified ODEs and in each case we indicate the origin—and in some cases the extent—of their *isochronous* character, by identifying the unmodified equations they originated from (via the trick, as explained above); in some *solvable* cases we also exhibit the general solution of these ODEs. Our presentation below is generally terse: readers interested in additional details may find them in the literature, as identified in Section 3.N.

Notation. The dependent variable is generally a single *complex* function of time, $z \equiv z(t)$, but in some cases we also display the (*real*) system of two coupled ODEs in two (*real*) dependent variables $x \equiv x(t)$, $y \equiv y(t)$ (with $z = x + i\,y$) obtained by separating every *complex* number into its *real* and *imaginary* parts. Superimposed dots indicate differentiations with respect to the *real* variable t ("time"), while appended primes indicate differentiations with respect to the argument of the function they are appended to (generally the *complex* independent variable τ). The other symbols indicate generally *arbitrary constants*: we generally use Greek letters for *complex* constants, Latin letters at the beginning of the alphabet for *real* constants; letters such as n, m, ℓ, p, q stand instead for *integers*, while the constants Ω (as defined below) and ω are *real* (and ω is hereafter assumed, without loss of generality, to be *positive*, $\omega > 0$), with the following relation among them:

$$\Omega = \lambda\,\omega, \quad \lambda = \frac{p}{q} \tag{3.37}$$

with p and q two *coprime integers* ($q > 0$, and $q = 1$ if $p = \lambda = 0$). Hence, as above, the constant λ is a *real rational* number. Note the difference of the constant Ω, as introduced here, from the constant $\tilde{\omega}$ used above, see (3.5), entailing that the trick formula can now be written in the form

$$z\,(t) = \exp\,(i\,\Omega\,t)\;\zeta\,(\tau) = \exp\,(i\,\lambda\,\omega\,t)\;\zeta\,(\tau)\,, \quad \tau = \frac{\exp\,(i\,\omega\,t) - 1}{i\,\omega}. \tag{3.38}$$

In order to simplify the look of some formulas we hereafter use both constants, ω and Ω, as well as the two (primitive) periods corresponding to these two circular frequencies,

$$T = \frac{2\,\pi}{\omega}, \quad \tilde{T} = q\,T. \tag{3.39}$$

Finally let us emphasize that neither we systematically eliminate every constant which could be gotten rid of by trivial transformations such as rescalings or shifts of the dependent or independent variables, nor on the contrary we introduce as many such constants as possible.

3.2.1 *First-order algebraic complex ODE*

This *isochronous* ODE reads

$$\dot{z} - i\,\Omega\,z = \alpha\,z^{p/q}. \tag{3.40a}$$

It is the ω-modified version of the ODE

$$\zeta' = \alpha\,\zeta^{p/q}, \tag{3.40b}$$

and it is obtained from it via the trick (3.38) with

$$\lambda = \frac{q}{p-q}. \tag{3.41}$$

(Beware of the difference of this formula from (3.37)).

The *general* solution of this ODE, (3.40a), reads

$$z(t) = z(0)\,\exp(i\,\Omega\,t)\left\{1 - \alpha\,[z(0)]^{1/\lambda}\,\frac{\exp(i\,\omega\,t) - 1}{i\,\Omega}\right\}^{-\lambda}, \tag{3.42}$$

hence *all* its *nonsingular* solutions are *periodic* with a period that is an *integer* multiple of the basic period T, whose value depends on the initial datum $z(0)$ via the sign of the quantity

$$u = \left|1 + \frac{i\,\Omega}{\alpha\,[z(0)]^{1/\lambda}}\right| - 1: \tag{3.43}$$

if u is *positive*, the period is $q{\cdot}T$, if u is *negative* the period is T; while if u vanishes the solution (3.42) becomes singular at a finite (*real*) value of t.

3.2.2 *Polynomial vector field in the plane*

The *complex* version of this ω-modified first-order ODE is just the special case of the preceding ODE (3.40a) with $q = 1$ and p larger than unity, $p = 2, 3, \ldots$; the title of this section refers to its *real* avatar, which reads (with $\alpha = a_1 + i\,a_2$)

$$\dot{x} + \Omega\,y = a_1\,X - a_2\,Y, \qquad \dot{y} - \Omega\,x = a_1\,Y + a_2\,X, \tag{3.44a}$$

$$X \equiv \sum_{m=0}^{\lfloor p/2 \rfloor} (-1)^m \binom{p}{2m} x^{p-2m}\,y^{2m}, \tag{3.44b}$$

$$Y \equiv \sum_{m=0}^{\lfloor (p-1)/2 \rfloor} (-1)^m \binom{p}{2m+1} x^{p-2m-1}\,y^{2m+1}. \tag{3.44c}$$

Here the symbol $\lfloor x \rfloor$ denotes the *integer* part of x: $\lfloor p/2 \rfloor = p/2$ if p is *even*, $\lfloor p/2 \rfloor = (p-1)/2$ if p is *odd*, and the symbol $\binom{n}{m}$ denotes the standard binomial coefficient,

$$\binom{n}{m} = \frac{n!}{m!\,(n-m)!}. \tag{3.45}$$

In the special case with $p = 2$ this *real* system reads

$$\dot{x} + \Omega\,y = a_1\,(x^2 - y^2) - 2\,a_2\,x\,y, \quad \dot{y} - \Omega\,x = a_2\,(x^2 - y^2) + 2\,a_1\,x\,y. \tag{3.46}$$

3.2.3 *Oscillator with additional inverse-cube force*

This well-known *isochronous Newtonian* ODE,

$$\ddot{z} + \left(\frac{\Omega}{2}\right)^2 z = a^2 z^{-3}, \tag{3.47}$$

may be seen, in this context, as the avatar of (3.46) with $a_1 = 0$, $a_2 = a \neq 0$, obtained by time-differentiating the first of the (3.46), by using the second of the (3.46) to eliminate $\dot{y}$, by then using the first of the (3.46) to eliminate y, and finally by setting $x(t) = [z(t)]^{-2} - \Omega \,/\, (2a)$. The fact that *all real* solutions of this ODE are *periodic* with period $\tilde{T}$ (see (3.37) and (3.39)) is of course a well-known result; in fact, *all complex* solutions are as well *periodic*, except for those that go through the origin in the *complex* z-plane, which are clearly *singular* (this cannot happen in the *real* case due to the *repulsive* force, divergent at the origin: see the right-hand side of this *Newtonian* ODE).

3.2.4 *Isochronous versions of the first and second Painlevé ODEs (complex and real versions)*

The *complex* versions of these (*nonautonomous*) *entirely isochronous* ODEs read

$$\ddot{z} + \Omega^2 z = \left(\alpha z^2 + \gamma\right) \exp(5\, i\, \Omega\, t); \tag{3.48}$$

$$\ddot{z} - 5\, i\, \omega\, \dot{z} - 6\, \omega^2 z = \alpha z^2 + \gamma \exp(5\, i\, \omega\, t); \tag{3.49}$$

$$\ddot{z} + 5\, i\, \omega\, \dot{z} - 6\, \omega^2 z = \alpha z^2 \exp(5\, i\, \omega\, t) + \gamma; \tag{3.50}$$

$$\ddot{z} - 3\, i\, \Omega\, \dot{z} - 2\, \Omega^2 z = \alpha z^3 + (\gamma z + \delta) \exp(3\, i\, \Omega\, t); \tag{3.51}$$

and clearly their *real* versions read

$$\ddot{x} + \Omega^2 x = \cos(5\,\Omega\, t)\, \left[a_1 \left(x^2 - y^2\right) - 2\, a_2\, x\, y + c_1\right] \\ - \sin(5\,\Omega\, t)\, \left[a_2 \left(x^2 - y^2\right) + 2\, a_1\, x\, y + c_2\right], \tag{3.52a}$$

$$\ddot{y} + \Omega^2 y = \sin(5\,\Omega\, t)\, \left[a_1 \left(x^2 - y^2\right) - 2\, a_2\, x\, y + c_1\right] \\ + \cos(5\,\Omega\, t)\, \left[a_2 \left(x^2 - y^2\right) + 2\, a_1\, x\, y + c_2\right]; \tag{3.52b}$$

$$\ddot{x} + 5\,\omega\,\dot{y} - 6\,\omega^2 x = a_1 \left(x^2 - y^2\right) - 2\, a_2\, x\, y + c_1 \cos(5\,\omega\, t) - c_2 \sin(5\,\omega\, t), \tag{3.53a}$$

$$\ddot{y} - 5\,\omega\,\dot{x} - 6\,\omega^2 y = a_2 \left(x^2 - y^2\right) + 2\, a_1\, x\, y + c_2 \cos(5\,\omega\, t) + c_1 \sin(5\,\omega\, t); \tag{3.53b}$$

$$\ddot{x} - 5\,\omega\,\dot{y} - 6\,\omega^2 x = \left[a_1 \left(x^2 - y^2\right) - 2\, a_2\, x\, y\right] \cos(5\,\omega\, t) \\ - \left[a_2 \left(x^2 - y^2\right) + 2\, a_1\, x\, y\right] \sin(5\,\omega\, t) + c_1, \tag{3.54a}$$

$$\ddot{y} + 5\,\omega\,\dot{x} - 6\,\omega^2 y = \left[a_1 \left(x^2 - y^2\right) - 2\, a_2\, x\, y\right] \sin(5\,\omega\, t) \\ + \left[a_2 \left(x^2 - y^2\right) + 2\, a_1\, x\, y\right] \cos(5\,\omega\, t) + c_2; \tag{3.54b}$$

$$\ddot{x} + 3\,\Omega\,\dot{y} - 2\,\Omega^2\,x = a_1\,x\,\left(x^2 - 3\,y^2\right) - a_2\,y\,\left(3\,x^2 - y^2\right) + \left(c_1\,x - c_2\,y + d_1\right)\,\cos(3\,\Omega\,t) - (c_2\,x + c_1\,y + d_2)\,\sin(3\,\Omega\,t), \tag{3.55a}$$

$$\ddot{y} - 3\,\Omega\,\dot{x} - 2\,\Omega^2\,y = a_2\,x\,\left(x^2 - 3\,y^2\right) + a_1\,y\,\left(3\,x^2 - y^2\right) + \left(c_1\,x - c_2\,y + d_1\right)\,\sin(3\,\Omega\,t) + (c_2\,x + c_1\,y + d_2)\,\cos(3\,\Omega\,t). \tag{3.55b}$$

All solutions—except for a *lower dimensional* set—of these ODEs are *periodic* with a period which is a simple *rational* multiple (whose value the alert reader will easily figure out, see below) of the basic period T. This is demonstrated as follows.

All solutions $\zeta\,(\tau)$ of the first Painlevé ODE,

$$\zeta'' = \alpha\,\zeta^2 + \beta\,\tau, \tag{3.56}$$

are *meromorphic* functions of the independent variable τ (in the entire *complex* τ-plane), this being the very property that characterizes the Painlevé class of second-order ODEs. We now use a variant of the trick, replacing the second of the two definitions (3.38) by

$$\tau = \frac{\exp\,(i\,\omega\,t)}{i\,\omega}. \tag{3.57}$$

We thereby get from (3.56)

$$\ddot{z} - i\,(2\,\lambda + 1)\,\omega\,\dot{z} - \lambda\,(\lambda + 1)\,\omega^2\,z = \alpha\,\exp[i\,(2 - \lambda)\,\omega\,t]\,z^2 + \gamma\,\exp[i\,(\lambda + 3)\,\omega\,t], \tag{3.58a}$$

where we set

$$\beta = i\,\omega\,\gamma. \tag{3.58b}$$

Three choices of λ appear of special interest: $\lambda = -1/2$, $\lambda = 2$, $\lambda = -3$. And they yield the three ODEs (3.48), (3.49) and (3.50), whose *entire isochrony* is thereby ascertained. The demonstration that (3.51) is *entirely isochronous* is completely analogous, except that one now starts from the second Painlevé equation,

$$\zeta'' = \alpha\,\zeta^3 + \beta\,\tau\,\zeta + \delta, \tag{3.59}$$

rather than the first, (3.56), and sets $\lambda = 1$.

3.2.5 *Autonomous second-order ODEs (complex and real versions)*

In this Section 3.2.5 we begin by listing several ODEs in the class identified by its title, and we then demonstrate their *isochronous* character, generally by displaying their *general* solutions.

The *complex* versions of a first group of these *autonomous* ODEs read

$$\ddot{z} - \frac{5}{2}\, i\,\Omega\, \dot{z} - \frac{3}{2}\,\Omega^2 z = \alpha\, z^2; \tag{3.60}$$

$$\ddot{z} - i\,\Omega\, \dot{z} - 2\,\Omega^2 = \alpha\, \exp(z); \tag{3.61}$$

$$\ddot{z} - i\,\Omega\, \dot{z} + 2\,\Omega^2 z = (\dot{z} - i\,\Omega\, z)\, z; \tag{3.62}$$

$$\ddot{z} - 3\, i\,\Omega\, \dot{z} - 2\,\Omega^2 z = (\dot{z} - i\,\Omega\, z)\, z. \tag{3.63}$$

The general solutions of these four *autonomous* ODEs are exhibited below: it is thereby seen that *all* solutions of the first, third, and fourth are *periodic* with (primitive) period (at most) T (*entire isochrony*), while *all* solutions of the second are *periodic* up to a constant shift (see below). Note the striking similarity yet substantial difference (see below) among the last two of these ODEs (incidentally: an *arbitrary* constant α multiplying the right-hand sides of these two ODEs could of course be introduced by rescaling the dependent variable). The corresponding *real* versions of these four ODEs clearly read

$$\ddot{x} + \frac{5}{2}\,\Omega\, \dot{y} - \frac{3}{2}\,\Omega^2 x = a_1\, \left(x^2 - y^2\right) - 2\, a_2\, x\, y, \tag{3.64a}$$

$$\ddot{y} - \frac{5}{2}\,\Omega\, \dot{x} - \frac{3}{2}\,\Omega^2 y = a_2\, \left(x^2 - y^2\right) + 2\, a_1\, x\, y; \tag{3.64b}$$

$$\ddot{x} + \Omega\, \dot{y} - 2\,\Omega^2 = \exp(x)\, [a_1\, \cos(y) - a_2\, \sin(y)], \tag{3.65a}$$

$$\ddot{y} - \Omega\, \dot{x} = \exp(x)\, [a_2\, \cos(y) + a_1\, \sin(y)]; \tag{3.65b}$$

$$\ddot{x} + \Omega\, \dot{y} + 2\,\Omega^2 x = \dot{x}\, x - \dot{y}\, y + 2\,\Omega\, x\, y, \tag{3.66a}$$

$$\ddot{y} - \Omega\, \dot{x} + 2\,\Omega^2 y = \dot{x}\, y + x\, \dot{y} - \Omega\, \left(x^2 - y^2\right); \tag{3.66b}$$

$$\ddot{x} + 3\,\Omega\, \dot{y} - 2\,\Omega^2 x = \dot{x}\, x - \dot{y}\, y + 2\,\Omega\, x\, y, \tag{3.67a}$$

$$\ddot{y} - 3\,\Omega\, \dot{x} - 2\,\Omega^2 y = \dot{x}\, y + x\, \dot{y} - \Omega\, \left(x^2 - y^2\right). \tag{3.67b}$$

A second group of *autonomous isochronous* ODEs read

$$\ddot{z} - 3\, i\,\Omega\, \dot{z} - 2\,\Omega^2 z = \alpha\, z\, (\dot{z} - i\,\Omega\, z) + \beta\, z^3; \tag{3.68a}$$

$$\ddot{z} - 3\, i\,\Omega\, \dot{z} - 2\,\Omega^2 z = \alpha\, z\, (\dot{z} - i\,\Omega\, z) - \left(\frac{\alpha}{3}\right)^2 z^3. \tag{3.68b}$$

A third group of *autonomous isochronous* ODEs read

$$\ddot{z} + \frac{3\, q_1\, q_2 + p_1\, q_2 - p_2\, q_1}{q_1\, (p_2 - 2\, q_2)}\, i\,\Omega\, \dot{z} + \frac{q_2\, (p_1 + q_1)}{q_1\, (p_2 - 2 q_2)}\,\Omega^2 z = \alpha\, (\dot{z} - i\,\Omega\, z)^{p_2/q_2}\, z^{p_1/q_1}; \tag{3.69a}$$

$$\ddot{z} - 3\, i\,\Omega\, \dot{z} - 2\,\Omega^2 z = \alpha\, (\dot{z} - i\,\Omega\, z)^3\, z^{-3}; \tag{3.69b}$$

$$\ddot{z} + (n\, m + n - 1)\, i\,\Omega\, \dot{z} + n\, (m + 1)\,\Omega^2 z = \alpha\, (\dot{z} - i\,\Omega\, z)^{(2n+1)/n}\, z^m; \tag{3.69c}$$

$$\ddot{z} - (2\, n\, m + 1)\, i\,\Omega\, \dot{z} - 2\, n\, m\,\Omega^2 z = \alpha\, (\dot{z} - i\,\Omega\, z)^{(2n+1)/n}\, z^{-(2m+1)}. \tag{3.69d}$$

The ω-modified character of the *complex* evolution ODE (3.69a), as shown below, entails its *isochronous* character. This ODE features the four *integers* p_1, q_1, p_2, q_2: but only the two rational numbers p_1/q_1, p_2/q_2 actually enter in it. Only the three special cases of this ODE corresponding to (3.69b), (3.69c) and (3.69d), with n a *positive* integer in (3.69c) and (3.69d) and m a *nonnegative* integer in (3.69c) and a *positive* integer in (3.69d), are treated in any detail below. In all three cases, (3.69b)–(3.69d), *all nonsingular* solutions are *periodic*. In the case of (3.69b), the *general* solution is exhibited below: in this case the *nonsingular* solutions split into two (*fully dimensional*) sets, both however *periodic* with the *same* period T; and these two sets are separated by a *lower dimensional* set of initial data, to which there correspond *singular* solutions, namely solutions such that $\dot{z}\,(t)$ diverges at a *real* time $t = t_b$. In the other two cases, (3.69c) and (3.69d), the situation is analogous, but richer. Depending on the values of the integers n and m, many more periodicities are possible: and again these different periodicities correspond to different (*fully dimensional*) sets of initial data, $z(0)$, $\dot{z}\,(0)$, separated by *lower dimensional* sets of such data yielding *singular* solutions (for which $\dot{z}\,(t)$ diverges at some *real* time).

Finally a *complex autonomous isochronous* second-order ODE, featuring the *arbitrary* (but *analytic*) function $f\,(z)$, reads

$$\ddot{z} - i\,\omega\,\dot{z} = \dot{z}^{\,2}\,f(z). \tag{3.70}$$

Let us now discuss all these ODEs (the following treatment is rather terse, but the reader not interested in the demonstration of the *isochronous* character of the ODEs reported above—in this Section 3.2.5—is advised to proceed directly to the following Section 3.2.6).

To obtain (3.60) one takes as starting point the ODE

$$\zeta'' = \alpha\,\zeta^{\,2} \tag{3.71}$$

and applies the trick (3.38) with $\lambda = 2$. Recalling that the Weierstrass doubly-periodic elliptic function $\wp\,(\tau) \equiv \wp\,(\tau;\, g_2,\, g_3) \equiv \wp\,(\tau\,|\,\omega_1,\, \omega_2)$ (which is a *meromorphic* function of τ) satisfies the ODE

$$\wp''\,(\tau) = 6\,\wp^{\,2}\,(\tau) - \frac{g_2}{2}, \tag{3.72}$$

we conclude that the *general* solution of the ODE (3.60) is

$$z\,(t) = \frac{6}{\alpha}\,\exp\,(i\,\Omega\,t)\;\wp\left(\frac{\exp\,(i\,\omega\,t)}{i\,\omega} + \beta;\; 0,\; g_3\right) \tag{3.73}$$

with β and g_3 *arbitrary* constants. It is thus clear that *all* solutions of this ODE, (3.60), are *periodic* with period T, except the *lower dimensional* set of *singular* solutions (characterized by the conditions $1\,/\,\omega = |\beta + n_1\,\tau_1 + n_2\,\tau_2|$ with τ_1, τ_2 the two periods of the Weierstrass function $\wp\,(\tau;\, 0,\, g_3)$ and n_1, n_2 two arbitrary *integers*). There is moreover a (*lower dimensional*) set of solutions which are *periodic* with primitive period $\tilde{T} = T/2$, due to the periodic character of the Weierstrass function.

To obtain (3.61) one takes as starting point the (*solvable*, Liouville) ODE

$$\zeta'' = \alpha \exp(\zeta) \tag{3.74}$$

and uses the following modified version of the trick:

$$z(t) = 2\,i\,\omega\,t + \zeta(\tau)\,, \quad \tau = \frac{\exp(i\,\omega\,t) - 1}{i\,\omega}\,. \tag{3.75}$$

This yields indeed (3.61) (with $\omega = \Omega$). On the other hand this version of the trick, (3.75), shows that the solutions $z \equiv z(t)$ of (3.61) are generally *not* periodic in t (they are *periodic* with period T up to a shift); of course their time-derivatives, $\dot{z}(t)$, are therefore *periodic* with period T. This is confirmed by the following expression of the *general* solution of (3.61),

$$z(t) = a\,\exp(i\,\omega\,t) + b + 2\,i\,\omega\,t$$
$$- 2\,\log\left\{\frac{\alpha}{2\,(a\,\omega)^2}\,\exp(b) + \exp[a\exp(i\,\omega\,t)]\right\}, \tag{3.76a}$$

which features the two *arbitrary* constants a, b, and which entails

$$\dot{z}(t) = i\,\omega\,a\exp(i\,\omega\,t) + 2\,i\,\omega\,\frac{1 - a\exp(i\,\omega\,t)\exp[a\,\exp(i\,\omega\,t)]}{\frac{\alpha}{2\,(a\,\omega)^2}\,\exp(b) + \exp[a\exp(i\,\omega\,t)]}\,. \tag{3.76b}$$

Clearly *all nonsingular* functions $\dot{z}(t)$ are periodic in t with period T, and it is also easily seen that a condition sufficient to guarantee that the solution $z(t)$, see (3.76a), as well of course as its time-derivative $\dot{z}(t)$, see (3.76b), be *nonsingular* for *all* (*real*) times, is validity of the inequality

$$|a| \neq \left|b - \log\left(-\frac{2\,a^2\,\omega^2}{\alpha}\right)\right|\,. \tag{3.77}$$

To obtain (3.62), time-differentiate (3.61), eliminate $\alpha\,\exp(z)$ using (3.61), and then replace formally (as a notational change) $\dot{z}(t)$ with $z(t)$. This derivation entails of course that the *general* solution, $z = z(t)$, of (3.62) is provided by the right-hand side of (3.76b); it is clearly *nonsingular* if the inequality (3.77) holds, and *periodic* in t with period T.

The derivation of (3.63) is provided just below, since this ODE—as well as (3.68b)—is a special case of the next ODE we consider, (3.68a).

To obtain this ODE, (3.68a), we take as starting point the evolution ODE

$$\zeta'' = \alpha\,\zeta'\,\zeta + \beta\,\zeta^3, \tag{3.78}$$

and apply the trick (3.38). This yields

$$\ddot{z} - i\,(2\,\lambda + 1)\,\omega\,\dot{z} - \lambda\,(\lambda + 1)\,\omega^2\,z = \alpha\exp[i\,(1 - \lambda)\,\omega\,t]\,z\,(\dot{z} - i\,\lambda\,\omega\,z)$$
$$+ \beta\,\exp[2\,i\,(1 - \lambda)\,\omega\,t]\,z^3, \tag{3.79}$$

which indeed, for $\lambda = 1$, coincides with (3.68a).

Clearly (3.63), respectively (3.68b), are the two special cases of (3.68a) corresponding to $\beta = 0$ (and $\alpha = 1$) respectively $\beta = -(\alpha/3)^2$. The motivation for singling out these two cases is because, by setting

$$\zeta(\tau) = \frac{\gamma\,\psi'(\tau)}{\psi(\tau)}, \quad \gamma = \frac{\alpha - \left(\alpha^2 + 8\,\beta\right)^{1/2}}{2\,\beta}, \tag{3.80}$$

one transforms (3.78) into

$$\psi'''\,\psi = \eta\,\psi'\,\psi'', \quad \eta = \frac{\alpha^2\left\{1 + \frac{6\,\beta}{\alpha^2} - \left[1 + \frac{8\,\beta}{\alpha^2}\right]^{1/2}\right\}}{2\beta}, \tag{3.81}$$

and for $\beta = 0$ respectively $\beta = -(\alpha/3)^2$ one gets $\eta = 1$ respectively $\eta = 0$, two values for which this ODE is particularly easy to integrate.

Indeed in the first case, $\beta = 0$, the ODE (3.78) is itself easily integrated, yielding

$$\zeta(\tau) = \frac{a\,[b + \exp(a\,\tau)]}{\alpha\,[b - \exp(a\tau)]}, \tag{3.82}$$

with a and b arbitrary constants. Via (3.38) (with $\lambda = 1$) this yields

$$z(t) = i\,\omega\,\exp(i\,\omega\,t)\,\frac{A\,\{B + \exp[A\exp(i\,\omega\,t)]\}}{\alpha\,\{B - \exp[A\exp(i\,\omega t)]\}} \tag{3.83a}$$

with A and B,

$$A = -\frac{i\,a}{\omega}, \quad B = b\,\exp(A), \tag{3.83b}$$

two *arbitrary* constants. This formula (3.83) (with $\alpha = 1$) provides the *general* solution of (3.63); it is clearly *periodic* in t with period T, and clearly it is *nonsingular* provided the following *inequality* holds:

$$|A| \neq |\log(B)|. \tag{3.84}$$

In the second case, $\beta = -(\alpha/3)^2$, $\eta = 0$ (which entails $\gamma = -3/\alpha$, see (3.80)) one gets from (3.81)

$$\psi(\tau) = a + b\tau + \tau^2 \tag{3.85}$$

hence, via (3.80),

$$\zeta(\tau) = -\frac{3\,(b + 2\,\tau)}{\alpha\,(a + b\,\tau + \tau^2)} \tag{3.86}$$

with a and b two *arbitrary* constants. Via the trick formula (3.38) (with $\lambda = 1$) this yields

$$z(t) = \frac{3\,i\,\omega\,[(2 - B)\,\exp(i\,\omega\,t) - 2\,\exp(2\,i\,\omega\,t)]}{\alpha\,[1 - A - B + (B - 2)\,\exp(i\,\omega\,t) + \exp(2\,i\,\omega\,t)]} \tag{3.87a}$$

with A and B,

$$A = a\,\omega^2, \quad B = i\,\omega\,b, \tag{3.87b}$$

two *arbitrary* constants. This is the general solution of (3.68b); it is clearly *periodic* in t with period T, and it is *nonsingular* for *all (real)* values of t provided the following *inequality* holds:

$$\left|2 - B \pm \left[B^2 + 4\,A\right]^{1/2}\right| \neq 2. \tag{3.88}$$

To discuss the third group of ODEs, (3.69), we take as starting point the ODE

$$\zeta'' = \alpha\,(\zeta')^{p_2/q_2}\,\zeta^{p_1/q_1}, \tag{3.89}$$

and we apply the trick (3.38) with

$$\lambda = -\frac{q_1\,(2\,q_2 - p_2)}{(q_1\,q_2 - p_1\,q_2 - p_2\,q_1)}. \tag{3.90}$$

We thereby get for $z \equiv z(t)$ the (*autonomous*!) ODE (3.69a). Hence the solutions of this ODE, (3.69a), can be obtained, via (3.38), from the *general* solution of (3.89), which is yielded by the quadrature formula

$$\int^{\zeta} dx\,\left[x^{(p_1+q_1)/q_1} + a^2\right]^{q_2/(p_2-2q_2)} = b + \left[-\frac{\alpha\,q_1\,(p_2 - 2q_2)}{q_2\,(p_1 + q_1)}\right]^{-q_2/(p_2-2q_2)} \tau, \tag{3.91}$$

where a^2 and b are two *arbitrary* constants.

We forsake here a discussion of this formula for an arbitrary choice of the two rational numbers $p_1\,/\,q_1$ and $p_2\,/\,q_2$, and we limit our consideration to the three examples corresponding to the three ODEs (3.69b)–(3.69d).

The first obtains by setting $p_1\,/\,q_1 = -3$, $p_2\,/\,q_2 = 3$, entailing $\lambda = 1$; thereby (3.69a) becomes (3.69b). In this case (3.91) reads

$$-\zeta^{-1} + a^2\,\zeta = b + \frac{2\,\tau}{\alpha}, \tag{3.92a}$$

entailing, say,

$$\zeta\,(\tau) = \frac{b + \frac{2\,\tau}{\alpha} - \left[\left(b + \frac{2\,\tau}{\alpha}\right)^2 + 4\,a^2\right]^{1/2}}{2\,a^2}, \tag{3.92b}$$

so that

$$\zeta(0) = z(0) = \frac{b - \left(b^2 + 4\,a^2\right)^{1/2}}{2\,a^2}, \tag{3.92c}$$

$$\zeta'(0) = \dot{z}\,(0) - i\,\omega\,z(0) = \frac{\left(b^2 + 4\,a^2\right)^{1/2} - b}{\alpha\,a^2\,\left(b^2 + 4\,a^2\right)^{1/2}}; \tag{3.92d}$$

and it is easily seen that the initial data, $z(0)$ and $\dot{z}\,(0)$, of (3.69b) are then split into two sets, all of which however yield *nonsingular* solutions *periodic* with period T. These two sets are separated by a (topologically nontrivial, *lower dimensional*) set of initial data, characterized by the *equalities*

$$|2+\alpha\,\omega\,(\pm 2\,a - i\,b)| = 2; \tag{3.93a}$$

and clearly these special initial data yield solutions which become *singular* (namely, which are such that $\dot{z}$ diverges) at the *real* times t_b characterized by the relation

$$\exp(i\,\omega\,t_b) = 1+\alpha\,\omega\,\left(\pm a - \frac{i\,b}{2}\right) \tag{3.93b}$$

(the fact that the values of t_b, as defined mod(T) by this equation, are *real* is of course implied by the previous formula).

The second example, see (3.69b), obtains by setting

$$\frac{p_1}{q_1} = m, \quad \frac{p_2}{q_2} = \frac{2\,n+1}{n}, \tag{3.94a}$$

entailing (see (3.90))

$$\lambda = -\frac{1}{n\,m+n+1}, \tag{3.94b}$$

with m a *nonnegative* integer and n a *positive* integer; thereby (3.69a) becomes (3.69c). In this case (3.91) reads

$$P_{n\,m+n+1}(\zeta) = b + \left[-\frac{n\,(m+1)}{\alpha}\right]^{n}\tau, \tag{3.95}$$

where $P_{n\,m+n+1}(\zeta)$ is a polynomial of degree $n\,m+n+1$ in ζ, whose coefficients are time-independent (for given n and m, they only depend on the constant a). Since τ, see (3.38), is a *periodic* function of t with period T, clearly the set of the $n\,m+n+1$ roots of this algebraic equation is also *periodic* with the same period; but each root, if followed continuously as function of t, need only be periodic with period $T\,|n\,m+n+1|$. Hence we may conclude, via (3.38), that *all nonsingular* solutions $z(t)$ of (3.69c) are *periodic* with a period which is a *rational multiple* of T falling in the (closed) interval between the Least Common Multiple among T and $\tilde{T} = T/\,|n\,m-n+1|$, and the Least Common Multiple among T and $T\,|n\,m+n+1|$. The *singular* solutions correspond to *lower dimensional* sets of initial data $z(0)$, $\dot{z}\,(0)$, such that, for some *real* time, (at least) two roots of the polynomial (3.95) coincide, at which time $\dot{z}\,(t)$ diverges.

The third example, see (3.69c), obtains by setting

$$\frac{p_1}{q_1} = -(2\,m+1), \quad \frac{p_2}{q_2} = \frac{2\,n+1}{n}, \tag{3.96}$$

entailing (see (3.90))

$$\lambda = \frac{1}{2\,n\,m-1}, \tag{3.97}$$

with m and n *positive* integers; thereby (3.69a) becomes (3.69d). In this case (3.91) reads

$$P_{2\,m\,n-1}(\zeta) = \left[b + (2\,m\,n)^n\,\alpha^{-n}\,\tau\right]\,\zeta^{2\,m\,n-1}. \tag{3.98}$$

Hence, by an analysis closely analogous to that given above, in this case we also conclude that *all nonsingular* solutions $z(t)$ of (3.69d) are *periodic* with a period which is a *rational multiple* of T falling in the (closed) interval between the Least Common Multiple among T and $\tilde{T} = T/\,|2\,n\,m - 1|$, and the Least Common Multiple among T and $T\,|2\,n\,m - 1|$.

Finally, let us discuss the ODEs (3.70), featuring the largely *arbitrary* function $f\,(z)$. It is plain that via the trick (3.38) with $\lambda = 0$ this ODE is related to the ODE

$$\zeta'' = (\zeta')^2\,f\,(\zeta) \tag{3.99}$$

(which in fact coincides with (3.29)). This entails its *isochronous* character. A specific ODE of this type is discussed below (see Section 3.2.7).

3.2.6 *Autonomous third-order ODEs (complex and real versions)*

The following third-order *autonomous* ODEs,

$$\dddot{z} - 10\,i\,\omega\,\ddot{z} - 31\,\omega^2\dot{z} + 30\,i\,\omega^3\,z = \alpha\,(2\,\dot{z} - 5\,i\,\omega\,z)\,z, \tag{3.100}$$

$$\dddot{z} - 10\,i\,\omega\,\ddot{z} - 19\,\omega^2\dot{z} - 30\,i\,\omega^3\,z = \alpha\,(2\,\dot{z} - 5\,i\,\omega\,z)\,z, \tag{3.101}$$

are merely two *avatars* of the second-order *nonautonomous* ODE (3.49), and clearly their *real* versions read as follows:

$$\dddot{x} + 10\,\omega\,\ddot{y} - 31\,\omega^2\,\dot{x} - 30\,\omega^3\,y = 2\,a_1\,(\dot{x}\,x - \dot{y}\,y + 5\,\omega\,x\,y) \\ - a_2\,\left[2\,(\dot{x}\,y + \dot{y}\,x) - \omega\,\left(x^2 - y^2\right)\right], \tag{3.102a}$$

$$\dddot{y} - 10\,\omega\,\ddot{x} - 31\,\omega^2\,\dot{y} + 30\,\omega^3\,x = 2\,a_2\,(\dot{x}\,x - \dot{y}\,y + 5\,\omega\,x\,y) \\ + a_1\,\left[2\,(\dot{x}\,y + \dot{y}\,x) - 5\,\omega\,\left(x^2 - y^2\right)\right]; \tag{3.102b}$$

$$\dddot{x} + 10\,\omega\,\ddot{y} - 19\,\omega^2\,\dot{x} + 30\,\omega^3\,y = 2\,a_1\,(\dot{x}\,x - \dot{y}\,y + 5\,\omega\,x\,y) \\ - a_2\,\left[2\,(\dot{x}\,y + \dot{y}\,x) - 5\,\omega\,\left(x^2 - y^2\right)\right], \tag{3.103a}$$

$$\dddot{y} - 10\,\omega\,\ddot{x} - 19\,\omega^2\,\dot{y} - 30\,\omega^3\,x = 2\,a_2\,(\dot{x}\,x - \dot{y}\,y + 5\,\omega\,x\,y) - a_1\,\left[2\,(\dot{x}\,y + \dot{y}\,x) \right. \\ \left. - 5\,\omega\,\left(x^2 - y^2\right)\right]. \tag{3.103b}$$

Likewise, the following two third-order *autonomous* ODEs,

$$\dddot{z} + 31\,\omega^2\,\dot{z} + 30\,i\,\omega^3 z + 5\,i\,\omega\,\gamma = 2\,(\ddot{z} + 5\,i\,\omega\,\dot{z} - \gamma)\frac{\dot{z}}{z}, \tag{3.104}$$

$$\dddot{z} - 5\,i\,\omega\,\ddot{z} + \omega^2 \dot{z} - 5\,i\,\omega^3 z = \frac{2\,z\,\dot{z}\,\left(\ddot{z} + \omega^2 z\right)}{z^2 + \eta}, \tag{3.105}$$

are merely two *avatars* of the second-order *nonautonomous* ODEs (3.50) and (3.48). Hence *all* the *nonsingular* solutions of *all* these third-order ODEs are *periodic* with period $\tilde{T}$, see (3.39) and (3.37).

Indeed, time-differentiation of (3.49) yields

$$\dddot{z} - 5\,i\,\omega\,\ddot{z} - 6\,\omega^2\,\dot{z} = 2\,\alpha\,z\,\dot{z} + 5\,i\,\omega\,\gamma\,\exp(5\,i\,\omega\,t), \tag{3.106}$$

and using again (3.49) to eliminate the last term in the right-hand side of this equation, one gets (3.100). Note that the constant γ, see (3.49), does not appear in this ODE, (3.100). Hence, *all* solutions of (3.100) also satisfy (3.49) (for some appropriate value of γ); or, equivalently, the solution of the *initial-value* problem for (3.100) (namely, of the problem to evaluate the solution $z(t)$ of (3.100) which corresponds to given initial data $z(0)$, $\dot{z}\,(0)$, $\ddot{z}(0)$) is provided by the solution of the initial-value problem for (3.49), with the same data $z(0)$, $\dot{z}\,(0)$, and with $\gamma = z(0) - 5i\,\omega\,\dot{z}\,(0) - 6\,\omega^2\,z(0)$. Hence the solutions of (3.100) have the same periodicity properties as the solutions of (3.49).

As for (3.101), it is merely another *avatar* of (3.49). Indeed if, in this ODE, (3.49), we replace $z(t)$ with $z(t) + c$, $\dot{c} = 0$ before going through the procedure described just above, we get

$$\dddot{z} - 10\,i\,\omega\,\ddot{z} - \left(31\,\omega^2 + 2\,\alpha\,c\right)\,\dot{z} + 10\,i\,\left(3\,\omega^2 + \alpha\,c\right)\,\omega\,z$$
$$+ 5\,i\,\left(6\,\omega^2 + \alpha\,c\right)\,\omega\,c = \alpha\,(2\,\dot{z} - 5i\,\omega\,z)\,z, \tag{3.107}$$

which is a more general *avatar* of (3.49) than (3.100), and reduces to (3.101) for $\alpha\,c = -6\,\omega^2$.

The derivation of (3.104) from (3.50) is completely analogous to the derivation given above of (3.100) from (3.49); and the comments given above on the possibility to obtain *all* the solutions of (3.100) from those of (3.49) are as well applicable now, except that the role previously played by the constant γ (which appears in (3.49) but not in (3.100)) is now played by the constant α (which appears in (3.50) but not in (3.104)).

Finally, the starting point to obtain (3.105) is (3.48), which we now write as follows (by setting $\gamma = \alpha\,\eta$):

$$\ddot{z} + \Omega^2\,z = \alpha\,\left(z^2 + \eta\right)\,\exp(5\,i\,\Omega\,t). \tag{3.108}$$

We now time-differentiate this ODE, and then eliminate the explicitly time-dependent term using again this same ODE. This yields (3.105). Again we note that the constant α, which appears in (3.108), has dropped out of (3.105); hence we may again conclude that *all* solutions of (3.105) also satisfy (3.108) (with an appropriate value of α, possibly including $\alpha = 0$), and that the solution of the *initial-value* problem for (3.105) can be obtained from the solution of the *initial-value* problem for (3.108) (or equivalently (3.48)).

3.2.7 *Isochronous version of a solvable second-order ODE due to Painlevé*

At the beginning of last century Paul Painlevé introduced the nonlinear *autonomous* second-order ODE

$$\zeta'' = (\zeta')^2 \left[\left(1-\zeta^2\right)\left(1-k^2\zeta^2\right)\right]^{-1/2} \cdot\left\{-\alpha^{-1} + \left[\left(1-\zeta^2\right)\left(1-k^2\zeta^2\right)\right]^{-1/2} \zeta\left[2k^2\zeta^2 - k^2 - 1\right]\right\}, \quad (3.109)$$

and pointed out that it features the *general* solution

$$\zeta(\tau) = \mathrm{sn}\left(\alpha \log\left[A\,\tau - B\right], k\right) \quad (3.110)$$

where $\mathrm{sn}(u, k)$ is the first Jacobian elliptic function, A and B are two *arbitrary* constants and the rest of the notation is, we trust, self-explanatory.

This ODE, (3.109), is a classical, much quoted, example of nonlinear ODE a simple local analysis of which only evidences poles as singularities in the finite part of the complex τ-plane, but which in fact features—at the point $\tau = B\,/\,A$, see (3.110)—an essential singularity where the value of the solution itself is undefined; hence it is an example of (*autonomous*!) nonlinear ODE that, contrary to the expectation suggested by such a local analysis, does *not* possess the "Painlevé property" to feature only *meromorphic* solutions, namely solutions the singularities of which in the (finite part of the) *complex* τ-plane are *only* poles.

This *autonomous* ODE, (3.109), belongs to the class (3.99): via the trick (3.38) with $\lambda = 0$ it gets transformed into the following, also *autonomous*, *isochronous* ODE:

$$\ddot{z} - i\,\omega\,\dot{z} = (\dot{z})^2 \left[\left(1-z^2\right)\left(1-k^2z^2\right)\right]^{-1/2} \cdot \left\{-\alpha^{-1} + \left[\left(1-z^2\right)\left(1-k^2z^2\right)\right]^{-1/2} z\left[2k^2z^2 - k^2 - 1\right]\right\}, \quad (3.111)$$

which therefore features the *general* solution

$$z(t) = \mathrm{sn}\left(\alpha \log\left[a \exp(i\,\omega\,t) - b\right], k\right) \quad (3.112a)$$

where

$$a = -\frac{i\,A}{\omega}, \quad b = B - \frac{i\,A}{\omega} \quad (3.112b)$$

are again two *arbitrary* constants. The *isochronous* character of the ω-modified ODE (3.111) is confirmed by the fact, see (3.112a), that, provided there holds the *inequality*

$$|a| < |b| , \tag{3.113}$$

all the *nonsingular* solutions (3.112a) are evidently *periodic* with period T; while the solution (3.112a) (with (3.113)) is *singular* (for some *real* t) only if there holds the *equality*

$$|a| = \left| b + \exp\left[\frac{2\,m\,K\,(k) + (2\,n+1)\;i\,K'\,(k)}{\alpha}\right]\right| \tag{3.114}$$

for some *integer* values of m and n (because the first Jacobian elliptic function $\mathrm{sn}(u,\,k)$ has a pole whenever $u = 2\,m\,K\,(k) + (2\,n+1)\;i\,K'\,(k)$ with m and n integers). Here $K\,(k)$ and $K'\,(k)$ denote the complete elliptic integrals.

Let us end this Section 3.2.7 by mentioning two special cases of the *isochronous* ODE (3.111).

If $k = 1$ the ODE (3.111) becomes

$$\ddot{z} - i\,\omega\,\dot{z} = (\dot{z})^2\;\frac{2\,z + \alpha^{-1}}{z^2 - 1}. \tag{3.115}$$

This *isochronous* ODE belongs, of course again, to the class (3.99), and it possesses the *general* solution

$$z\,(t) = \frac{u\,(t) - 1}{u\,(t) + 1}, \quad u\,(t) = [\beta\;\exp\,(i\,\omega\,t) - \gamma]^{2\,\alpha}, \tag{3.116}$$

with β and γ two *arbitrary* constants. Clearly this solution, unless it is *singular,* is *periodic* with period T if $|\beta| < |\gamma|$, and it is also *periodic* with period $T/\;|2\,\alpha|$ if $|\beta| > |\gamma|$ but only provided α is *real;* it is *singular* if $|\beta| = |\gamma|$ (unless α is a *positive* integer) or if

$$|\beta| = \left|\gamma + \exp\left(\frac{i\;(2\,n+1)\;\pi}{2\,\alpha}\right)\right|, \quad n = \textit{arbitrary}\text{ integer.} \tag{3.117}$$

Likewise, if $k = 0$ the ODE (3.111) becomes

$$\ddot{z} - i\,\omega\,\dot{z} = -\,(\dot{z})^2\;\frac{z + \left(1 - z^2\right)^{1/2}}{\alpha\;(1 - z^2)}, \tag{3.118}$$

and its *general* solution reads

$$z\,(t) = \frac{[v\,(t)]^2 - 1}{2\,i\,v\,(t)}, \quad v\,(t) = [\beta\;\exp\,(i\,\omega\,t) - \gamma]^{i\,\alpha}, \tag{3.119}$$

with β and γ two *arbitrary* constants. This solution, unless it is *singular,* is *periodic* with period T if $|\beta| < |\gamma|$, and it is also *periodic* with period $T/\;|2\,\alpha|$ if $|\beta| > |\gamma|$ but only provided α is *imaginary*; it is *singular* if $|\beta| = |\gamma|$.

3.2.8 *Isochronous versions of five solvable ODEs due to Chazy*

Almost a century ago Jean Chazy introduced the four (*autonomous*) ODEs

$$\zeta'' = \zeta' \zeta \left[\left(\zeta^4 + 4 \zeta' \right)^{1/2} - \zeta^2 \right], \tag{3.120a}$$

$$\zeta'' = 3 \zeta^2 \left\{ \left[\zeta^6 - (\zeta')^2 \right]^{1/2} + \zeta^3 \right\}, \tag{3.120b}$$

$$\zeta'' = 2 \zeta' \left[\left(c^2 + \zeta^2 - \zeta' \right)^{1/2} + \zeta \right], \tag{3.120c}$$

$$\zeta''' = 2 \left[\zeta'' \zeta + (\zeta')^2 \right] = (\zeta^2)''; \tag{3.120d}$$

and (in a separate paper) the ODE

$$\zeta''' = 2 \zeta'' \zeta - 3 (\zeta')^2 . \tag{3.121}$$

In this section we review tersely Chazy's motivations for introducing and studying these ODEs, and we then tersely discuss the following *isochronous* versions of these equations,

$$\ddot{z} - 5 i \omega \dot{z} - 4 \omega^2 z = z (\dot{z} - i \omega z) \left\{ \left[z^4 + 4 (\dot{z} - i \omega z) \right]^{1/2} - z^2 \right\}, \tag{3.122a}$$

$$\ddot{z} - 4 i \omega \dot{z} - 3 \omega^2 z = 3 z^2 \left\{ \left[z^6 - (\dot{z} - i \omega z)^2 \right]^{1/2} + z^3 \right\}, \tag{3.122b}$$

$$\ddot{z} - 3 i \omega \dot{z} - 2 \omega^2 z = 2 (\dot{z} - i \omega z) \left[\left(c^2 \exp(-2 i \omega t) + z^2 - \dot{z} + i \omega z \right)^{1/2} + z \right], \tag{3.122c}$$

$$\dddot{z} - 6 i \omega \ddot{z} - 11 \omega^2 \dot{z} + 6 i \omega^3 z = 2 \left[z \ddot{z} + (\dot{z})^2 \right] - 10 i \omega z \dot{z} - 6 \omega^2 z^2, \tag{3.122d}$$

$$\dddot{z} - 6 i \omega \ddot{z} - 11 \omega^2 \dot{z} + 6 i \omega^3 z = 2 \ddot{z} z - 3 (\dot{z})^2 - \omega^2 z^2, \tag{3.123}$$

obtained—as indicated below, with appropriate assignments of the two integers p and q—via the following version of the trick,

$$z(t) = \exp(i p \omega t) \, \zeta(\tau), \quad \tau = \frac{\exp(i q \omega t) - 1}{i q \omega}, \tag{3.124}$$

which differs from (3.38) only notationally (corresponding to the replacement of ω with $q\omega$, and then setting as above $\lambda = p/q$). Note that four of these five ω-modified ODEs are *autonomous*; one of them, (3.122c), is *not* autonomous, unless the—*a priori arbitrary*—constant c vanishes.

The four ODEs (3.120) were considered remarkable by Chazy for the following reasons.

The first one of these four ODEs, (3.120a), possesses, in addition to the general solution

$$\zeta(\tau) = A \tan\left(A^3 \tau + B\right), \qquad (3.125a)$$

the *special* solution

$$\zeta(\tau) = \left[\frac{4}{3(\tau - \tau_b)}\right]^{1/3}. \qquad (3.125b)$$

Here A, B and τ_b are *arbitrary* constants. Note that the special solution (3.125b) cannot be obtained as a limiting case of the general solution (3.125a) and, in contrast to (3.125a) which is *meromorphic* in the entire *complex* τ-plane (it clearly has simple poles at $\tau = \tau_n = [-B + (2n+1)\pi/2]/A^3$, with n an *arbitrary* integer), (3.125b) features a branch point of order one-third at $\tau = \tau_b$; hence this ODE, (3.120a)—in contrast to what might be naively inferred from knowledge of its *general* solution (3.125a)—does *not* feature only solutions the only *movable* singularities of which are *poles*, namely it does *not* possess the "Painlevé property" (reminder: the *movable* singularities are those occurring at values of the independent variable that cannot be predicted *a priori*, namely at values that, in the context of the *initial-value* problem, *do* depend on the initial data; in the case of *autonomous* ODEs, *all* singularities are of course of this type). Note moreover that the existence of the two different solutions (3.125a) and (3.125b) demonstrates the lack of uniqueness of the *initial-value* problem for (3.120a) whenever the initial data, $\zeta(0)$, $\zeta'(0)$, satisfy the condition $[\zeta(0)]^4 + 4\zeta'(0) = 0$, namely whenever they entail the vanishing of the square root in the right-hand side of (3.120a).

Likewise, the second one of these four ODEs, (3.120b), possesses, in addition to the general solution

$$\zeta(\tau) = A\,\wp\left(A^2 \tau + B;\, 0,\, 4\right), \qquad (3.126a)$$

the *special* solutions

$$\zeta(\tau) = [\pm 2(\tau - \tau_b)]^{-1/2}. \qquad (3.126b)$$

Here A, B and τ_b are again *arbitrary* constants, and $\wp(u;\, 0,\, 4)$ is the Weierstrass elliptic function. Clearly the same remarks made above apply here, up to obvious adjustments; hence we do not repeat them.

The third one of these four ODEs, (3.120c), has been written above in a slightly more general form than used by Chazy, who wrote unity in place of the constant c. Actually it is easily seen that, for $c \neq 0$, the "cosmetic" rescaling

$$\zeta(\tau) = c\,\tilde{\zeta}(\tilde{\tau}), \quad \tilde{\tau} = c\,\tau, \qquad (3.127)$$

entails that $\tilde{\zeta}(\tilde{\tau})$ satisfies an ODE analogous to (3.120c) but with the constant c replaced by unity, namely the Chazy version. Our motivation for using the more general form (3.120c) is because we shall also be interested below in the case $c = 0$.

As the two ODEs discussed just above, (3.120c) features both a *general* solution and a *special* solution, which can both be written in explicit form. The former reads

$$\zeta(\tau) = \exp(A\,\tau + B) + \frac{A^2 - 4\,c^2}{4\,A}, \tag{3.128a}$$

and the latter reads

$$\zeta(\tau) = -c\,\cot\left[c\,(\tau - \tau_0)\right]. \tag{3.128b}$$

Here A, B and τ_0 are again *arbitrary* constants. Some of the comments given above (after (3.125b)) are clearly as well applicable to this case, hence they are not repeated. We note however one difference: the *general* solution is now *entire* (see (3.128a); in the previous two cases it was *meromorphic*, see (3.125a) and (3.126a)), the *special* solution is now *meromorphic* (see (3.128b); in the two previous cases it had a branch point, see (3.125b) and (3.126b)). Also note that both the *general* solution (3.128a), and the *special* solution (3.128b), are as well valid in the $c = 0$ case, when of course the ODE (3.120c) reads

$$\zeta'' = 2\,\zeta'\,\left[\left(\zeta^2 - \zeta'\right)^{1/2} + \zeta\right], \tag{3.128c}$$

and its (*general* and *special*) solutions read

$$\zeta(\tau) = \exp(A\,\tau + B) + \frac{A}{4}, \tag{3.129a}$$

$$\zeta(\tau) = -\frac{1}{\tau - \tau_0}. \tag{3.129b}$$

Finally (3.120d) admits the general meromorphic solution

$$\zeta(\tau) = -\frac{1}{2\,(\tau - \tau_0)} - \frac{3\,A}{2}\,(\tau - \tau_0)^{1/2} \times \frac{J'_{1/3}\left[A\,(\tau - \tau_0)^{3/2}\right] + B\,J'_{-1/3}\left[A\,(\tau - \tau_0)^{3/2}\right]}{J_{1/3}\left[A\,(\tau - \tau_0)^{3/2}\right] + B\,J_{-1/3}\left[A\,(\tau - \tau_0)^{3/2}\right]}. \tag{3.130}$$

Here A, B and τ_0 are again three *arbitrary* constants, $J_{\pm 1/3}(u)$ are the standard Bessel functions of order $\pm 1/3$, and the primes appended to these functions (see the numerator in the right-hand side) denote of course differentiation with respect to their arguments. Note that this solution is a *meromorphic* function of the independent variable τ: its poles occur at the zeros of the denominator in the right-hand side, while there is no singularity at $\tau = \tau_0$ (except in the special case $B = 0$).

It is now a matter of trivial algebra to verify that the four ω-modified ODEs (3.122) obtain from the four Chazy equations (3.120) via the trick (3.124), with the following assignments of the parameters p and q: to go from (3.120a) to

(3.122a), $p = 1$, $q = 3$; to go from (3.120b) to (3.122b), $p = 1$, $q = 2$; to go from (3.120c) to (3.122c), and as well from (3.120d) to (3.122d), $p = 1$, $q = 1$.

Taking advantage of the explicit solvability of the four Chazy ODEs (3.120)—as described above—one can as well exhibit—via the trick formula (3.124), with the appropriate assignments of p and q, see above—*all* the solutions of the four ω-modified ODEs, (3.122). We leave this easy task to the diligent reader, who will thereby verify that *all* the *nonsingular* solutions (namely, those that do not blow up in a *finite* time) of these four ODEs, (3.122), are *periodic* with period T (there exist also some solutions whose *primitive* periods are a *rational multiple* of T), and who will as well identify, case-by-case, the *singular* solutions, that constitute in each case only a *lower dimensional* subset (namely, they feature only *one* arbitrary constant, rather than two). (The lazy reader may find all these results in the literature, see Section 3.N).

We end this section by discussing the remaining, fifth ODE (3.121), and by indicating how the *isochronous* character of its ω-modified version (3.123) can be demonstrated. The third-order ODE (3.121)—which, incidentally, is just the special case of (3.33) with

$$q = 1, \quad c(1,0,1) = 2, \quad c(0,2,0) = -3, \quad c(2,1,0) = c(4,0,0) = 0 \qquad (3.131)$$

—was originally introduced by Chazy as an example of ODE whose solutions feature a *natural boundary* when considered as analytic functions of the independent variable. The ω-modified version (3.123) of this ODE is obtained via the trick (3.124) with $p = q = 1$. The *general* solution of (3.121), hence as well of its ω-modified version (3.123), is known, but only in a rather *implicit* form; and it is not so easy to infer from such an implicit form information on the periodicity properties of the solutions of (3.123). On the other hand it is not difficult to prove that the requirement that the initial data of the original Chazy ODE (3.121) be all *small*,

$$|\zeta(0)| << \omega, \quad |\zeta'(0)| << \omega^2, \quad |\zeta''(0)| << \omega^3, \qquad (3.132)$$

(for more precise versions of these conditions see the literature, as identified in Section 3.N), is *sufficient* to guarantee that its solutions be *holomorphic* in a closed disk D centered, in the complex τ-plane, at $\tau = 0$, and having a radius larger than $2/\omega$. As we now know (see Section 2.1) this fact is sufficient to guarantee that the corresponding solutions of the ω-modified ODE (3.123) be *completely periodic* with period T; and it is then a matter of trivial algebra to transform these conditions, (3.132), into corresponding conditions on the initial data for the ω-modified ODE (3.123). And the fact that the initial data for the unmodified Chazy ODE (3.121) which satisfy the restrictions (3.132) occupy an open, *fully dimensional*, domain in phase space entails of course that the same conclusion also holds for the corresponding initial data for the ω-modified ODE (3.123), implying the *isochronous* character of this ODE.

In view of the special character of the analyticity structure of the solutions of the original Chazy ODE (3.121) an interesting *open* problem is the behavior of the solutions of the ω-modified ODE (3.123) evolving from initial data *not* satisfying the restrictions (3.132); but such problems exceed the scope of this monograph.

3.N Notes to Chapter 3

For the notation and properties of the elliptic functions used in this chapter see, for instance, [91].

The findings reported in Section 3.1 are based on [63], where the diligent reader will find more material than has been covered herein. Beware of notational changes, and of the fact that eq. (1.4) of that paper contains a misprint: in the right-hand side q should appear in the denominator rather than in the numerator (also, the Reference 9 listed in that paper does not exist; and the elimination of many articles in that paper is due to an editorial intervention I consider unwarranted...).

The findings reported in Section 3.2 are mainly based on [62]: again, beware of notational changes. For the *isochronous* character of (3.70) see [61].

The material of Sections 3.2.7, respectively 3.2.8, is based—up to notational changes—on [39], respectively [41] (for the first four ODEs) and [63] (for the fifth ODE); the interested reader will also find in these papers references to the original articles by Painlevé and by Chazy.

4

SYSTEMS OF ODEs: MANY-BODY PROBLEMS, NONLINEAR HARMONIC OSCILLATORS

In this chapter we review *systems* of evolution ODEs—involving *several* dependent variables—that are *isochronous*, namely feature in their phase space an open, *fully dimensional*, region where *all* solutions are *completely periodic*—meaning that *all* the dependent variables are *periodic*—with the *same* fixed period, independently of their initial data provided they fall within the *isochrony* region. Many of the systems we consider are interpretable as many-body problems characterized by *Newtonian* equations of motions ("acceleration equal force"). These *isochronous* systems are generally ω-modified versions obtained from unmodified systems via the trick described and used in the two preceding chapters.

In the following Section 4.1 we consider a large class of *isochronous* dynamical systems: in Section 4.1.1 we provide a *lemma* identifying a large class of *isochronous* systems characterized by equations of motion of Newtonian type, and in Section 4.1.2 we identify six representative (classes of) many-body problems whose *isochrony* is implied by that *lemma*.

The subsequent sections—in which we revisit systems identified in Section 4.1.2, as well as other *isochronous* systems—are organized according to the dimensionality of the space in which live the dependent variables—identified as the coordinates of point particles whenever the system under consideration is interpretable as a many-body problem. But there is in this respect an element of ambiguity requiring some clarification, which we now provide.

As emphasized earlier, generally when we write the dependent variables of the *isochronous* system under consideration as z_n we understand these coordinates to be *complex* numbers: quite often, this is evident because the equations of motion that determine their time evolution are themselves *complex*, but even when the evolution equations are *real* (and therefore admit *real* solutions), often the *isochrony* property—namely, the *existence* in phase space of an open, *fully dimensional*, region where *all* solutions are *completely periodic* with the *same* fixed period—is only true in the more general context of *complex* solutions. Hence the time evolution of the, say, N *complex* quantities $z_n(t)$ takes place in the (*two-dimensional*) *complex z-plane*, and is completely equivalent, via the identification of the *real* and *imaginary* parts of z_n,

$$z_n = x_n + i\, y_n, \tag{4.1}$$

to the evolution of the $2\,N$ *real* coordinates $x_n(t)$, $y_n(t)$, each of which evolves, of course, on the (*one-dimensional*) *real* axis: should we categorize such a system as *two-dimensional* or as *one-dimensional*?

We conventionally solve this ambiguity by categorizing generally as *one-dimensional* any system whose dependent variables are *scalar* coordinates z_n, even though we will consider these coordinates as *complex*—but, generally, restricting attention to systems characterized by *analytic* ODEs. We will instead categorize as *two-dimensional* any system whose dependent variables can be represented as *two-vector* coordinates $\vec{r}_n \equiv (x_n, y_n)$—but only provided its equations of motion have a *covariant* look, namely they are by inspection *rotation-invariant* (in the plane: perhaps up to the presence of some *constant* two-vectors identifying certain preferred directions, thereby breaking *rotation invariance,* unless these *constant* two-vectors are themselves considered as dependent variables, albeit *not* evolving in time: see below). Of course, these *two-dimensional* systems—when their dependent variables are *real* two-vectors $\vec{r}_n$, suitable to represent "physical" point particles moving in the *real* plane—generally correspond to *complex one-dimensional* systems, to which they are related via (4.1)—as shall be explained below (see Section 4.3).

Likewise we will categorize as *three-dimensional* (or, more generally, as *S-dimensional*) any system whose dependent variables are *three-vector* coordinates $\vec{r}_n \equiv (x_n, y_n, z_n)$ (or *S-dimensional* vectors $\underline{r}_n \equiv (x_{1n}, x_{2n}, \ldots, x_{Sn})$), even though we will generally consider these coordinates as *complex;* but we will generally restrict attention to systems characterized by ODEs that are both *analytic* and have a *covariant* look, namely are by inspection *rotation-invariant* (in *three-dimensional* space, or in *S-dimensional* space, as the case may be—again, up to the possible presence of some constant vectors: see below).

Needless to say, a certain ambiguity remains: it could be accommodated by treating *repeatedly,* in different contexts, essentially the *same* system, but we shall generally refrain from doing so. We shall instead try and provide as much cross-referencing as possible (in the presentation, and also in Section 4.N); but let us nevertheless warn the hasty reader who might be looking for some specific model, that some systems treated later in the context of a *higher-dimensional* space also have obvious analogous counterparts—not necessarily explicitly highlighted—in a *lower-dimensional* context, and possibly viceversa.

Let us end this introduction by re-emphasizing that the *isochronous* systems reported below are merely a representative sample, selected so as to give the reader a flair of the universe of *isochronous* systems that can be manufactured via the trick. The alert reader will surely try and manufacture other such models, just for the fun of it or to use them in applicative contexts.

4.1 A class of isochronous dynamical systems

In this section we treat a *large* class of *isochronous* dynamical systems. In Section 4.1.1 we formulate and prove a *lemma* that allows to identify such a class. The generality of this *lemma* might mask its actual relevance to identify *interesting* cases. Hence in the subsequent Section 4.1.2 we review several such instances.

4.1.1 A lemma

In this section we state and prove a *lemma* identifying a *large* class of *isochronous* dynamical systems characterized by equations of motion of *Newtonian* type, hence generally interpretable as many-body problems.

Lemma 4.1.1-1. Let N be an arbitrary *positive* integer; let λ be an arbitrary *rational* number (positive, negative or zero),

$$\lambda = \frac{p}{q}, \tag{4.2}$$

where p and q are two *arbitrary* coprime *integers* (and we hereafter assume for definiteness that q is *positive*, $q > 0$); let K be an arbitrary *positive* integer; let the $2\,K$ (possibly *complex*) numbers a_k and b_k be restricted by the K conditions

$$\lambda\ (a_k + b_k - 1) + b_k \equiv \lambda\, a_k + (\lambda + 1)\, b_k - \lambda = k + 1, \quad k = 1, \ldots, K; \tag{4.3}$$

let, moreover, these $2\,K$ numbers a_k, b_k be such that, for some appropriate *real* constant c, the K *real* constants δ_k defined as follows,

$$\delta_k = \text{Re}\left[c\ (a_k + b_k - 1) + b_k - 2\right], \quad k = 1, \ldots, K, \tag{4.4a}$$

are *all nonnegative*,

$$\delta_k \geq 0, \quad k = 1, \ldots, K. \tag{4.4b}$$

[The following four facts should be noted: (i) these last conditions, (4.4), are independent of λ; (ii) if for some value of k there holds the condition

$$\text{Re}\,(a_k + b_k - 1) = 0, \tag{4.5a}$$

then the conditions (4.4), together with (4.3), imply that for that value of k

$$a_k = -k + i\ (\lambda + 1)\ \eta_k, \quad b_k = 1 + k - i\,\lambda\,\eta_k, \tag{4.5b}$$

with η_k arbitrary but *real*; (iii) if $K = 1$ the conditions (4.4) reduce to a single inequality that can *always* be satisfied by an appropriate choice of c (except in the special case $\text{Re}\,(a_1 + b_1 - 1) = 0$, when (4.5) applies); (iv) the following conditions on a_k, b_k are *sufficient* to allow that (4.4b) be satisfied by an appropriate assignment of c (indicated in each case below within square brackets)]:

$$\text{Re}(b_k) \geq 2, \quad k = 1, \ldots, K, \quad [c = 0]\,, \tag{4.6a}$$

$$\text{Re}(a_k + b_k) > 1, \quad k = 1, \ldots, K, \quad [c \rightsquigarrow +\infty]\,, \tag{4.6b}$$

$$\text{Re}(a_k + b_k) < 1, \quad k = 1, \ldots, K, \quad [c \rightsquigarrow -\infty]\,. \tag{4.6c}$$

Hence, validity of any one of these three inequalities (4.6) (which must, of course, hold for *all* values of the index $k, k = 1, \ldots, K$) for the quantities a_k, b_k—the role of which is specified immediately below, see (4.8)—is *sufficient* for the validity of (4.4b). But note that the conditions (4.3), which are moreover required to hold for the validity of this *lemma*, are more stringent, being equalities rather than inequalities]. Now introduce K (N-vector-valued) functions of the two N-vectors $\underline{z} \equiv (z_1, \ldots, z_N)$ and $\underline{\tilde{z}} \equiv (\tilde{z}_1, \ldots, \tilde{z}_N)$,

$$\underline{F}^{(a_k,\, b_k)}\left(\underline{z}, \underline{\tilde{z}}\right) \equiv \left(F_1^{(a_k,\, b_k)}\left(\underline{z}, \underline{\tilde{z}}\right), \ldots, F_N^{(a_k,\, b_k)}\left(\underline{z}, \underline{\tilde{z}}\right)\right), \quad k = 1, \ldots, K, \tag{4.7}$$

that are required to be *analytic* (but not necessarily *meromorphic*) in the $2N$ variables z_n, $\tilde{z}_n$, $n = 1, \ldots, N$, and are moreover required to satisfy the scaling properties

$$\underline{F}^{(a_k,\, b_k)}\left(\alpha\, \underline{z}, \beta\, \underline{\tilde{z}}\right) = \alpha^{a_k}\, \beta^{b_k}\, \underline{F}^{(a_k,\, b_k)}\left(\underline{z}, \underline{\tilde{z}}\right), \quad k = 1, \ldots, K. \tag{4.8}$$

Then the dynamical system of *Newtonian* type defined by the (N-vector) equation of motion

$$\ddot{\underline{z}} - i\,(2\,\lambda + 1)\,\omega\,\dot{\underline{z}} - \lambda\,(\lambda + 1)\,\omega^2\,\underline{z} = \sum_{k=1}^{K} \underline{F}^{(a_k,\, b_k)}(\underline{z}, \dot{\underline{z}} - i\,\lambda\,\omega\,\underline{z}), \tag{4.9}$$

is *isochronous*, namely there exists an *open* (hence *fully dimensional*) domain of initial data $\underline{z}(0)$, $\dot{\underline{z}}(0)$ such that *all* solutions of (4.9) characterized by such initial data are *completely λ-periodic* with period

$$T = \frac{2\,\pi}{\omega}, \tag{4.10}$$

namely,

$$\underline{z}(t + T) = \exp\left(2\,\pi\, i\, \lambda\right)\, \underline{z}(t), \tag{4.11a}$$

entailing of course (see (4.2)) that they are *completely periodic* with period $q\,T$,

$$\underline{z}(t + q\,T) = \underline{z}(t). \tag{4.11b}$$

[*Notational reminder*: here and throughout, the N-vector $\underline{z} \equiv \underline{z}(t)$ is the dependent variable; superimposed dots denote of course differentiations with respect to the independent variable t, which we interpret as *time* hence we assume to be *real*; and of course we assume the constant ω to be *real* as well (indeed, without significant loss of generality, *positive*)]. ⊡

Remark 4.1.1-2. The *Newtonian* equation of motion (4.9) is generally *complex*, unless $\lambda = -1\,/\,2$; and even if $\lambda = -1\,/\,2$, the open (*fully dimensional*) domain of initial data $\underline{z}(0)$, $\dot{\underline{z}}(0)$ yielding *isochronous* solutions might *not* include *only real* data (see below). Indeed the proof of *Lemma 4.1.1-1* (see below, in particular, the formulas (4.24d), (4.24b), (4.26), (4.27) and (4.15)) entails that the *open* domain of initial data $\underline{z}(0)$, $\underline{\dot{z}}(0)$ yielding *isochronous* motions are generally characterized by the condition:

$$\underline{\dot{z}}(0) - i\,\lambda\,\omega\,\underline{z}(0) \approx 0, \tag{4.12}$$

with the components of the N-vector $\underline{z}(0)$ being *quite small* (in modulus) if the constant c, see (4.4), is *positive*, $c > 0$, being instead *quite large* (in modulus) if the constant c is *negative*, $c < 0$, and having *intermediate* values if the constant c vanishes. [In rough terms, these restrictions on the initial data $\underline{z}\,(0)$ guarantee that, in the neighborhood of the initial data, the right-hand side of the *Newtonian* equation of motion (4.9) be *small*, while the restriction (4.12) corresponds to the requirement that the modulus of $\underline{\zeta}'\,(0)$ be *small*: see below (4.13), (4.15) and (4.24d), (4.24b), (4.26), (4.27)]. ⊡

Before turning to the proof of *Lemma 4.1.1-1*, let us emphasize that the conditions stated above for its validity are *sufficient* to guarantee the existence of, and indeed also to largely *characterize* (see below the proof of this *lemma*), an open (*fully dimensional*) domain of initial data out of which only *isochronous* motions originate, but they are not *necessary*, it is indeed possible that *isochronous* motions originate also from initial data that are *outside* that domain: see for instance some of the examples reported in the following Section 4.1.2, for which the *isochronous* region corresponds to the *entire* phase space, possibly up to a *lower dimensional* set of initial data yielding *singular* solutions.

The proof of *Lemma 4.1.1-1* hinges on the, by now standard, trick (see (2.3) and (3.4)), that we now re-write in the following guise:

$$\underline{z}(t) = \exp\left(i\,\lambda\,\omega\,t\right)\,\underline{\zeta}(\tau), \tag{4.13a}$$

$$\tau \equiv \tau(t) = \frac{\exp\left(i\,\omega\,t\right) - 1}{i\,\omega}. \tag{4.13b}$$

Here, as in the formulation of the *lemma*, the underlined notation identifies N-vectors: $\underline{z} \equiv (z_1, \ldots, z_N)$, $\underline{\zeta} \equiv (\zeta_1, \ldots, \zeta_N)$; and from the preceding discussions of the trick in Chapters 2 and 3 we know that a condition *sufficient* to guarantee that the (N-vector-valued) function $\underline{z}(t)$ satisfy the periodicity property (4.11) is that the (N-vector-valued) function $\underline{\zeta}(\tau)$ be *holomorphic* in τ in the closed circular disk

$$|\tau| \leq \frac{2}{\omega}. \tag{4.14}$$

Hence, to prove *Lemma 4.1.1-1*, we first obtain the ODE (see (4.16) below) in the independent variable τ entailed, via the trick (4.13), for $\underline{\zeta}(\tau)$ by the evolution equation (4.9) satisfied by $\underline{z}(t)$, and we then show that, under the hypotheses required for the validity of the *lemma*, this ODE (4.16) satisfied by $\underline{\zeta}(\tau)$ entails that there exists an open, *fully dimensional*, domain of initial data $\underline{\zeta}(0)$, $\underline{\zeta}'(0)$ guaranteeing the *holomorphy* of $\underline{\zeta}(\tau)$ in the closed circular disk (4.14). (Here and throughout appended primes denote of course differentiation with respect to the complex variable τ). Since clearly (4.13) entails

$$\underline{z}(0) = \underline{\zeta}(0), \quad \dot{\underline{z}}(0) = \underline{\zeta}'(0) + i\,\lambda\,\omega\,\underline{\zeta}(0), \tag{4.15}$$

the existence of such an *open* domain of initial data $\underline{\zeta}(0)$, $\underline{\zeta}'(0)$ guaranteeing the *holomorphy* of $\underline{\zeta}(\tau)$ in the closed circular disk (4.14) entails the existence of an *open* domain of initial data $\underline{z}(0)$, $\dot{\underline{z}}(0)$ entailing the *periodicity* property (4.11), thereby proving the *lemma.*

It is easily seen, by taking advantage of the relation (4.13) among $\underline{z}(t)$ and $\underline{\zeta}(\tau)$, of the evolution equation (4.9) satisfied by $\underline{z}(t)$, and as well of the scaling properties (4.8) with the conditions (4.3), that the ODE satisfied by $\underline{\zeta}(\tau)$ reads as follows:

$$\underline{\zeta}'' = \sum_{k=1}^{K} (1 + i\,\omega\,\tau)^{k-1}\; \underline{F}^{(a_k, b_k)}(\underline{\zeta}, \underline{\zeta}'). \tag{4.16}$$

Note that (the right-hand side of) this evolution equation is *holomorphic* (in fact, *polynomial*) in the independent variable τ.

Let us now prove that, under the hypotheses of the *lemma*, there is an *open* domain of initial data $\underline{\zeta}(0)$, $\underline{\zeta}'(0)$ such that the (N-vector-valued) solution $\underline{\zeta}(\tau)$ of this ODE, (4.16), is *holomorphic* in a circular disk centered at the origin of the complex τ-plane and having an *arbitrarily* large radius, hence in particular in the disk (4.14). To make contact with the standard notation, we now set

$$\varphi_n(\tau) = \frac{\zeta_n(\tau) - \zeta_n(0)}{\alpha}, \quad \varphi_{N+n}(\tau) = \frac{\zeta_n'(\tau) - \zeta_n'(0)}{\beta}, \quad n = 1, \ldots, N, \tag{4.17}$$

where α and β are two *positive* constants (that we shall conveniently assign below). Note that this definition of the $2\,N$ quantities $\varphi_\ell(\tau)$ entails that they *all* vanish at the origin,

$$\varphi_\ell(0) = 0, \quad \ell = 1, \ldots, 2\,N. \tag{4.18}$$

By this definition (4.17), the *second-order* (N-vector) ODE (4.16) gets reformulated as the following *first-order* system of $2\,N$ coupled ODEs:

$$\varphi_n' = f_n, \quad \varphi_{N+n}' = f_{N+n}, \quad n = 1, \ldots, N, \tag{4.19}$$

with

$$f_n\,(\tau) = \frac{\zeta_n'(0) + \beta\,\varphi_{N+n}\,(\tau)}{\alpha}, \quad n = 1, \ldots, N, \tag{4.20a}$$

$$f_{N+n}(\tau) = \frac{1}{\beta} \sum_{k=1}^{K} (1 + i\,\omega\,\tau)^{k-1}$$
$$\times\, \underline{F}^{(a_k, b_k)} \left[\zeta_m(0) + \alpha\,\varphi_m(\tau)\,,\, \zeta'_m(0) + \beta\,\varphi_{N+m}(\tau)\right],$$
$$n = 1, \dots, N. \tag{4.20b}$$

Note that, for notational convenience, in the right-hand side of (4.20b) we replaced some N-vectors with their components (on the understanding that the index m always ranges from 1 to N).

We now use the standard result [105] that, taking advantage of the *holomorphy* of the *explicit* dependence upon τ of the $2\,N$ functions f_ℓ, $\ell = 1, \dots, 2\,N$ (see (4.20b)), states that the $2\,N$ functions $\varphi_\ell(\tau)$—the solutions of (4.19)—are *holomorphic* in τ in a circular disk centered, in the complex τ-plane, at the origin ($\tau = 0$, where the initial conditions (4.18) are now assigned), the radius ρ of which is bounded below by the formula

$$\rho \geq \vartheta \left\{1 - \exp\left[-\frac{\varphi}{(2\,N + 1)\,\vartheta\,M(\vartheta, \varphi)}\right]\right\}, \tag{4.21}$$

where the *positive* constant ϑ is *arbitrarily large*, the *positive* quantity φ is characterized by the requirement that the $2\,N$ functions f_ℓ, $\ell = 1, \dots, 2\,N$, be *holomorphic* in the $2\,N$ variables φ_j, $j = 1, \dots, 2\,N$ provided

$$\left|\varphi_j\right| < \varphi, \quad j = 1, \dots, 2\,N, \tag{4.22}$$

and the *positive* quantity $M(\vartheta, \varphi)$ is defined by the formula

$$M(\vartheta, \varphi) = \max_{|\tau| < \vartheta;\ \left|\varphi_j\right| < \varphi, j=1,\dots,2\,N;\ \ell=1,\dots,2\,N} |f_\ell|\,. \tag{4.23}$$

Note that the *lower bound* (4.21) holds *a fortiori* if we *overestimate* the quantity $M(\vartheta, \varphi)$, as we shall indeed do in the following.

Using the explicit expressions (4.20) of the quantities f_ℓ, $\ell = 1, \dots, 2\,N$, we therefore can now write

$$M(\vartheta, \varphi) \leq \max_{|\tau| < \vartheta;\ \left|\varphi_j\right| < \varphi,\ j=1,\dots,2\,N;\ n=1,\dots,N} \left[\frac{\beta}{\alpha}\,(v + \varphi),\right.$$
$$\left.\frac{1}{\beta} \sum_{k=1}^{K} (1 + \omega\vartheta)^{k-1} \left|F_n^{(a_k, b_k)}\left(\alpha\left[r_m + \varphi_m\right],\, \beta\left[v_m + \varphi_{N+m}\right]\right)\right|\right] \tag{4.24a}$$

having set

$$\zeta'_n(0) = \beta\,v_n, \quad n = 1, \dots, N, \tag{4.24b}$$
$$v = \max_{n=1,\dots,N} |v_n|\,, \tag{4.24c}$$
$$\zeta_n(0) = \alpha\,r_n, \quad n = 1, \dots, N. \tag{4.24d}$$

Next we take advantage of the rescaling property (4.8) to re-write (4.24a) as follows:

$$M(\vartheta,\varphi) \leq \varepsilon \max_{|\tau|<\vartheta;\, |\varphi_j|<\varphi,\; j=1,\ldots,2N;\; n=1,\ldots,N} \Bigg[v+\varphi, \sum_{k=1}^{K} (1+\omega\,\vartheta)^{k-1}\,\varepsilon^{\delta_k} \left| F_n^{(a_k,b_k)}\left(r_m+\varphi_m, v_m+\varphi_{N+m}\right)\right| \Bigg], \tag{4.25}$$

where we introduced the *positive* constant ε (the choice of which still remains our privilege) by setting

$$\alpha = \varepsilon^{c}, \qquad \beta = \varepsilon^{c+1}, \tag{4.26}$$

where the constant c is of course the same that appears in (4.4a) (and we indeed used this equation, (4.4a), to write (4.25): see the exponent of ε in the summand in the right-hand side of this equation (4.25)).

We now consider this expression (4.25) of $M(\vartheta,\varphi)$ for *small but finite* ε,

$$\varepsilon \rightsquigarrow 0, \tag{4.27}$$

and, by taking advantage of (4.4b), we thereby get

$$M(\vartheta,\varphi) \leq C\,\varepsilon, \tag{4.28}$$

where the *positive* constant C is *finite*. Indeed, if the inequality (4.4b) did hold strictly, namely with $\geq$ replaced by $>$, for *all* values of k, then clearly the assignment $C = v+\varphi$ would be appropriate. Otherwise the expression of C is a bit more complicated, but always *finite*, provided values of the quantities r_n and v_n, as well as the value of the positive quantity φ, are chosen so that, for all values of φ_j satisfying the inequality $|\varphi_j| \leq \varphi$, $j=1,\ldots,2N$, see (4.22), the functions $F_n^{(a_k,b_k)}\left(r_m+\varphi_m,\ v_m+\varphi_{N+m}\right)$ are *finite*. This can always be guaranteed by restricting the quantities r_n and v_n, hence the initial data $\zeta_n(0)$, $\zeta_n'(0)$ (see (4.24d) and (4.24b)), to be within an open (*fully dimensional*) domain of such data (note that in this second case the *positive* constant C shall also depend on ϑ; but again all we need to know, see below, is that C is *finite* for *any finite* value of ϑ, which is certainly the case).

We now insert this upper bound to $M(\vartheta,\varphi)$, which is valid for small ε ($\varepsilon \rightsquigarrow 0$), in the expression (4.21) of ρ, getting thereby the following lower bound to ρ:

$$\rho \geq \vartheta \left\{ 1-\exp\left[-\frac{\varphi}{(2N+1)\,\vartheta\,C\,\varepsilon}\right]\right\}. \tag{4.29}$$

It is then evident that, by choosing ε *sufficiently small*, see (4.27), one can guarantee that ρ is *larger* than a quantity which is arbitrarily close to ϑ, and since ϑ is itself an *arbitrarily large* constant, we conclude that the restriction of the initial data $\underline{\zeta}(0)$, $\underline{\zeta}'(0)$ to an appropriate open *(fully dimensional)* domain guarantees that ρ is *arbitrarily large*, and in particular *larger* than $\frac{2}{\omega}$,

$$\rho > \frac{2}{\omega}. \tag{4.30}$$

This entails the *holomorphy* of $\underline{\varphi}(\tau)$ (hence as well of $\underline{\zeta}(\tau)$; see (4.17)) in the closed circular disk (4.14), hence, via (4.13), the *complete* λ*-periodicity*, indeed the *isochrony*, of $\underline{z}(t)$, see (4.11). Q. E. D.

Let us end this Section 4.1.1 with the following observation.

Hamiltonian systems yielding second-order equations of motion of *Newtonian* type for N scalar dependent variables (such as the system of ODEs considered above, see (4.9)) are, loosely speaking, *completely integrable* if they feature N functionally independent and globally defined constants of motion (including the Hamiltonian itself) that Poisson-commute among themselves; they are *superintegrable* if they feature $N-1$ *additional* functionally independent and globally defined constants of motion. *All* the *confined* motions of such *superintegrable* Hamiltonian systems are *completely periodic*, since their evolution then becomes essentially slave to a single variable; but they need *not* be *isochronous*, since their periods generally do depend on the initial data. And let us emphasize that, for a many-degree-of-freedom system, this phenomenology of *completely periodic* (rather than *multiply periodic*, or *aperiodic*, possibly "chaotic") motions is quite a nontrivial phenomenon (resulting from *superintegrability*); and of course even more so *isochrony*.

The notions of *integrability* and *superintegrability* can be loosely extended also to dynamical systems that are *not* Hamiltonian (as several of those considered herein), on the basis of an analogous simple comparison of the number of available (functionally independent) constants of motion and the degrees of freedom of the system—at least for systems with a finite number of degrees of freedom, to which our consideration has been so far restricted. It is moreover convenient to introduce the notion of *partially integrable* (and *partially superintegrable*) dynamical systems, to include the possibility that a system feature these properties only in a "part" of its (natural) phase space. Of course for this notion to be reasonable it is required (at least as long as we restrict attention to *autonomous* dynamical systems, as we indeed did above) that this part of phase space remain *invariant* throughout the evolution, namely that *all* motions originating from it *always* remain in it. But then this distinction among *integrable* (or *superintegrable*) and *partially integrable* (or *partially superintegrable*) dynamical systems may be made redundant if one includes in the very definition of a dynamical system the domain in phase space in which it is supposed to live, rather than relying on the *natural* definition of its phase space.

As noted above, a *superintegrable* system, as long as it only produces *confined* motions, always yields *completely periodic* evolutions, which, however, need not be *isochronous*. Hence, *superintegrability* does not entail *isochrony*; while the converse, at least loosely speaking, is clearly the case, namely *isochrony* does entail *superintegrability*. Hence, most of the systems discussed in this book, and in particular *all* the dynamical systems to which the *Lemma 4.1.1-1* is applicable, should be considered *superintegrable*, or at least *partially superintegrable* to the extent these two notions are distinct (see previous discussion). In view of the generality of the class of dynamical systems covered by this *Lemma 4.1.1-1*—as also demonstrated by the examples treated in the following Section 4.1.2 and beyond—this observation, trivial as it might be, underlines the remarkable nature of the results reported herein.

4.1.2 *Examples*

First of all let us display in *Table 4.1* various (*interesting*: see below) assignments of the constant λ and the corresponding values of $-(2\,\lambda+1)$ and $-\lambda\,(\lambda+1)$ (see (4.9)), of the constants a_k and b_k (all of them guaranteeing the validity of the condition (4.3)) and of the corresponding values of δ_k (see (4.4a)); note that *all* these assignments are compatible, via an appropriate choice of the constant c, with the condition (4.4b), see the last column of *Table 4.1*:

TABLE 4.1.

λ	$-(2\lambda+1)$	$-\lambda(\lambda+1)$	a_k	b_k	δ_k	$\delta_k \geq 0$ if
$-\frac{1}{2}$	0	$\frac{1}{4}$	$-1-2k$	0	$-2c(1+k)$	$c \leq -\frac{1}{1+k}$
0	-1	0	a_k	$1+k$	$c(k+a)+k-1$	$c=0$
-1	1	0	$-k$	b_k	$c(b-k-1)+b-2$	$c=-1$
$-\frac{2}{3}$	$\frac{1}{3}$	$\frac{2}{9}$	$-\frac{1+3\,k}{2}$	0	$-\frac{3}{2}c(1+k)-2$	$c \leq -\frac{2}{3}$
1	-3	-2	$2+k$	0	$c(1+k)-2$	$c \geq 1$
2	-5	-6	$\frac{3+k}{2}$	0	$\frac{1}{2}\left[c(k+1)-4\right]$	$c \geq 2$

Note that in the second row of *Table 4.1* the constants a_k are arbitrary, and in the third row the constants b_k are as well arbitrary.

Let us now consider six examples corresponding to the six rows of this *Table 4.1*.

Example 4.1.2-1 The assignment $\lambda = -1/2$ with $b_k = 0$, corresponding to the *first* row of *Table 4.1*, is interesting because it yields the *Newtonian* (N-vector) equation of motion

$$\ddot{\underline{z}} + \frac{1}{4}\,\omega^2\,\underline{z} = \sum_{k=1}^{K} \underline{F}^{(-1-2\,k)}(\underline{z}), \tag{4.31a}$$

that can be both *real* and *Hamiltonian.* Note that here, consistently with (even though not necessarily required by) the vanishing of the constants b_k in the first row of *Table 4.1*, the N-vector-valued functions $\underline{F}^{(-1-2k)}(\underline{z})$ are chosen to depend only on the (N-vector-valued) argument $\underline{z}$ and (in addition to being *analytic* in all their arguments, namely in the N components z_n of this N-vector) they are required to satisfy the scaling property (see (4.8))

$$\underline{F}^{(-1-2k)}(\alpha\,\underline{z}) = \alpha^{-1-2k}\,\underline{F}^{(-1-2k)}(\underline{z}), \quad k = 1, 2, \ldots, K. \tag{4.31b}$$

This (N-vector) Newtonian equation of motion, (4.31a), obtains clearly from the Hamiltonian

$$H\left(\underline{p}, \underline{z}\right) = \frac{1}{2}\sum_{n=1}^{N}\left(p_n^2 + \frac{1}{4}\omega^2\, z_n^2\right) + \sum_{k=1}^{K} V^{(-2k)}(\underline{z}) \tag{4.31c}$$

provided

$$F_n^{(-1-2\,k)}(\underline{z}) = -\frac{\partial V^{(-2\,k)}(\underline{z})}{\partial\, z_n}, \quad k = 1, \ldots, K, \tag{4.31d}$$

where the functions $V^{(-2k)}(\underline{z})$ (in addition to being *analytic* in all their arguments) must obviously, see (4.31b) and (4.31d), obey the scaling property

$$V^{(-2\,k)}(\alpha\,\underline{z}) = \alpha^{-2k}\,V^{(-2\,k)}(\underline{z}), \quad k = 1, \ldots, K, \tag{4.31e}$$

but are otherwise *arbitrary.* It seems remarkable that such a general (class of) Hamiltonian system(s), see (4.31), is *isochronous*—although it must be remembered that this property holds in the *complex*, namely that the open (*fully dimensional*) domain of initial data out of which only solutions that are *isochronous* emerge need *not* be *real*—entailing that the corresponding *isochronous* solutions are as well *not real*—even if the functions $V^{(-2\,k)}(\underline{z})$ are all *real*, hence the *Newtonian* equations of motion (4.31a) with (4.31d) are as well *real.* As entailed by *Remark 4.1.1-1*, and by the last entry of the first row of *Table 4.1*, the open (*fully dimensional*) domain of initial data yielding *isochronous* motions whose existence is guaranteed by *Lemma 4.1.1-1*—motions *completely antiperiodic* with period T, hence *completely periodic* with period $2\,T$: see (4.10) and (4.11) (with $\lambda = -1\,/\,2$ hence $q = 2$, see (4.2))—is characterized by N-vectors $\underline{z}(0)$ the components of which are *quite large* (in modulus), while the components of the N-vectors $\dot{\underline{z}}(0)$ are as well *quite large* (in modulus) and consistent with the (approximate) condition (4.12) (of course with $\lambda = -1\,/\,2$).

A special case, particularly interesting because it yields Newtonian equations of motion that are clearly interpretable as those of an N-body problem with *translation-invariant* one- and two-body forces *only*, obtains from the assignment

$$V^{(-2\,k)}(\underline{z}) = \frac{1}{4}\sum_{m,n=1;m\neq n}^{N}\frac{g_{nm}^{(k)}}{k\,(z_n - z_m)^{2\,k}}, \tag{4.32a}$$

where the constants $g_{nm}^{(k)}$ are *arbitrary* (possibly *complex* but, of course, symmetrical, $g_{nm}^{(k)} = g_{mn}^{(k)}$). The corresponding Newtonian equations of motion read as follows:

$$\ddot{z}_n + \frac{1}{4}\omega^2 z_n = \sum_{m=1;m\neq n}^{N} \sum_{k=1}^{K} \frac{g_{nm}^{(k)}}{(z_n - z_m)^{2k+1}}. \tag{4.32b}$$

For $K = 1$ and $g_{nm}^{(1)} = g$ (namely, the *same* coupling constant acting among every interacting pair, entailing equality of the N particles) this is a well-known *completely integrable* (indeed, *superintegrable* and *solvable*) system. In this case *all* motions are *isochronous*, except for those, emerging from a *lower dimensional* domain of initial data, that become *singular* at a finite time due to the *collision* of two or more particles (moving in the *complex* z-plane). Of course no such outcome is possible in the *real* case with $g > 0$, and indeed in this case *all* motions without exception are *isochronous* with period $2T$, see (4.10). [Note that this *isochronous* N-body problem is a generalization of the model described in Chapter 1, indeed for $K = 1$ (4.32b) coincides with (1.4b)].

Example 4.1.2-2 The assignment $\lambda = 0$ with $b_k = 2$, corresponding to the *second* row of *Table 4.1*, yields the *Newtonian* (N-vector) equation of motion

$$\ddot{\underline{z}} - i\,\omega\,\dot{\underline{z}} = \underline{F}^{(\cdot,2)}(\underline{z}, \dot{\underline{z}}), \tag{4.33a}$$

where, consistently with the arbitrariness of the constants a_k in the second row of *Table 4.1*, the N-vector-valued function $\underline{F}^{(\cdot,2)}(\underline{z}, \dot{\underline{z}})$ is required (in addition to being *analytic* in all its arguments) to satisfy the scaling property (4.8) *only* in its second (N-vector-valued) argument,

$$\underline{F}^{(\cdot,2)}(\underline{z}, \beta\,\dot{\underline{z}}) = \beta^2\, \underline{F}^{(\cdot,2)}(\underline{z}, \dot{\underline{z}}). \tag{4.33b}$$

Note the remarkable generality of this class of functions $\underline{F}^{(\cdot,2)}(\underline{z}, \dot{\underline{z}})$, hence of the *isochronous* equation of motion (4.33a).

This *Newtonian* equation of motion, (4.33a), is *Hamiltonian* provided the functions $\underline{F}^{(\cdot,2)}(\underline{z}, \dot{\underline{z}})$ take (componentwise) the special form

$$F_n^{(\cdot,2)}(\underline{z}, \dot{\underline{z}}) = \dot{z}_n \sum_{m=1,m\neq n}^{N} \dot{z}_m\, f_{nm}(\underline{z}), \quad n = 1, \ldots, N, \tag{4.34a}$$

with the functions $f_{nm}(\underline{z})$ expressed as follows:

$$f_{nm}(\underline{z}) = \frac{\partial\,\Phi_n(\underline{z})}{\partial\,z_m} - \frac{\partial\,\Phi_m(\underline{z})}{\partial\,z_n}, \quad n, m = 1, \ldots, N, \tag{4.34b}$$

for some set of N (*arbitrary* but *analytic*) functions $\Phi_n(\underline{z})$. Indeed the corresponding *Newtonian* equations of motion,

$$\ddot{z}_n - i\,\omega\,\dot{z}_n = \dot{z}_n \sum_{m=1,\, m\neq n}^{N} \dot{z}_m \left[\frac{\partial\,\Phi_n(\underline{z})}{\partial\, z_m} - \frac{\partial\,\Phi_m(\underline{z})}{\partial\, z_n}\right], \quad n = 1, \ldots, N, \quad (4.34c)$$

obtain, as can be easily verified, from the *Hamiltonian*

$$H\left(\underline{p}, \underline{z}\right) = \sum_{n=1}^{N} \left\{-\frac{i\,\omega\, z_n}{c} + \exp\left[c\, p_n + \Phi_n(\underline{z})\right]\right\}, \quad (4.34d)$$

where c is an *arbitrary* (of course *nonvanishing*) constant (that does not appear in the *Newtonian* equations of motion (4.34c)). Again, it seems remarkable that such a general (class of) *Hamiltonian* system(s), see (4.34d), is *isochronous.*

Let us also display the *real* version of the *complex* equations of motion (4.34c), obtained, of course, by setting

$$\underline{z} = \underline{x} + i\,\underline{y}, \quad \Phi_n(\underline{z}) = u_n(\underline{x},\, \underline{y}) + i\, v_n(\underline{x},\, \underline{y}),\ n = 1, \ldots, N. \quad (4.35a)$$

These *real* Newtonian equations of motion read

$$\ddot{x}_n + \omega\,\dot{y}_n = \sum_{m=1,\, m\neq n}^{N} \left\{ (\dot{x}_n\,\dot{x}_m - \dot{y}_n\,\dot{y}_m) \left[\frac{\partial\, u_n\left(\underline{x},\, \underline{y}\right)}{\partial\, x_m} - \frac{\partial\, u_m\left(\underline{x},\, \underline{y}\right)}{\partial\, x_n}\right] + (\dot{x}_n\,\dot{y}_m + \dot{y}_n\,\dot{x}_m) \left[\frac{\partial\, u_n\left(\underline{x},\, \underline{y}\right)}{\partial\, y_m} - \frac{\partial\, u_m\left(\underline{x},\, \underline{y}\right)}{\partial\, y_n}\right]\right\}, \quad n = 1, \ldots, N, \quad (4.35b)$$

$$\ddot{y}_n - \omega\,\dot{x}_n = \sum_{m=1,\, m\neq n}^{N} \left\{ (\dot{x}_n\,\dot{y}_m + \dot{y}_n\,\dot{x}_m) \left[\frac{\partial\, u_n\left(\underline{x},\, \underline{y}\right)}{\partial\, x_m} - \frac{\partial\, u_m\left(\underline{x},\, \underline{y}\right)}{\partial\, x_n}\right] - (\dot{x}_n\,\dot{x}_m - \dot{y}_n\,\dot{y}_m) \left[\frac{\partial\, u_n\left(\underline{x},\, \underline{y}\right)}{\partial\, y_m} - \frac{\partial\, u_m\left(\underline{x},\, \underline{y}\right)}{\partial\, y_n}\right]\right\}, \quad n = 1, \ldots, N. \quad (4.35c)$$

Note that, in writing these *real* Newtonian equations of motion, we only used the (*real*) functions $u_n\left(\underline{x},\, \underline{y}\right)$, which of course, being the *real* part of an *analytic* function, depend *harmonically* on the coordinates x_m and y_m,

$$\left[\frac{\partial^2}{\partial\, x_m^2} + \frac{\partial^2}{\partial\, y_m^2}\right] u_n\left(\underline{x},\, \underline{y}\right) = 0, \quad n, m = 1, \ldots, N, \quad (4.35d)$$

but are otherwise *arbitrary.* As it can be verified, this *isochronous* system, (4.35), of $2\,N$ *real* Newtonian ODEs satisfied by the $2\,N$ *real* coordinates x_n, y_n is obtained in the standard manner from the Hamiltonian

$$H(\underline{\xi},\underline{\eta};\underline{x},\underline{y})=\sum_{n=1}^{N}\left\{\frac{\omega\, y_n}{c}+\exp\left[c\,\xi_n+u_n(\underline{x},\underline{y})\right]\,\cos\left[c\,\eta_n-v_n(\underline{x},\underline{y})\right]\right\},\tag{4.35e}$$

where the functions $v_n(\underline{x},\underline{y})$, being the *imaginary* parts of the *same* analytic functions of which the functions $u_n\left(\underline{x},\,\underline{y}\right)$ are the *real* parts, are related to these functions as follows:

$$\frac{\partial\, u_n\left(\underline{x},\,\underline{y}\right)}{\partial\, x_n}=\frac{\partial\, v_n\left(\underline{x},\,\underline{y}\right)}{\partial\, y_n},\quad \frac{\partial\, u_n\left(\underline{x},\,\underline{y}\right)}{\partial\, y_n}=-\frac{\partial\, v_n\left(\underline{x},\,\underline{y}\right)}{\partial\, x_n},\qquad n=1,\ldots,N.\tag{4.35f}$$

Particularly interesting is the case characterized by the assignment

$$\Phi_n(\underline{z})=\sum_{m=1,\,m\neq n}^{N}\left\{g_{nm}^{(0)}\,\log\left(z_m-z_n\right)+\sum_{\ell=1}^{L}\frac{g_{nm}^{(\ell)}\,\left(z_m-z_n\right)^{\gamma_\ell+1}}{\gamma_\ell+1}\right\},\qquad n=1,\ldots,N,\tag{4.36a}$$

because it yields *Newtonian* equations of motion that are clearly interpretable as those of an N-body problem with *translation-invariant* one-body and two-body forces *only*:

$$\ddot{z}_n-i\,\omega\,\dot{z}_n=2\,\dot{z}_n\sum_{m=1,\,m\neq n}^{N}\dot{z}_m\left\{\frac{g_{nm}^{(0)}}{z_n-z_m}+\sum_{\ell=1}^{L}g_{nm}^{(\ell)}\,\left(z_n-z_m\right)^{\gamma_\ell}\right\},\qquad n=1,\ldots,N.\tag{4.36b}$$

Here L is an arbitrary *nonnegative* integer (for $L=0$ it is understood that the sum over ℓ be set to zero), the coupling constants $g_{nm}^{(j)}$ (where, of course, $j=0,1,\ldots,L;\ n,m=1,\ldots,N$) are *arbitrary* (possibly *complex*) except for the symmetry requirement $g_{nm}^{(j)}=g_{mn}^{(j)}$ (that comes from the Hamiltonian structure, see (4.34d) and (4.36a)), and also *arbitrary* (possibly *complex*) are the exponents γ_ℓ (except for the obvious restriction $\gamma_\ell\neq-1$, see (4.36a)). Particularly, interesting is the $L=0$ case,

$$\ddot{z}_n-i\,\omega\,\dot{z}_n=2\,\dot{z}_n\sum_{m=1,\,m\neq n}^{N}\frac{g_{nm}^{(0)}\,\dot{z}_m}{z_n-z_m},\qquad n=1,\ldots,N,\tag{4.37a}$$

because via the (we trust) self-evident identifications

$$z_n\equiv x_n+i\,y_n,\quad \vec{r}_n\equiv\left(x_n,y_n,0\right),\quad \hat{k}\equiv(0,0,1),\quad g_{nm}^{(0)}\equiv g_{nm}+i\,\tilde{g}_{nm},\tag{4.37b}$$

where the "Cartesian coordinates" $x_n,\ y_n,$ as well as the new "coupling constants" $g_{nm},\ \tilde{g}_{nm},$ are, of course, now *all real*, the equations of motion (4.37a) can be re-interpreted as describing N "physical" point particles moving in a plane orthogonal, in three-dimensional space, to the unit vector $\hat{k}$. In fact, these equations of motion then take the following (*translation-* and *rotation-invariant*!) form:

$$\ddot{\vec{r}}_n = \omega \hat{k} \wedge \dot{\vec{r}}_n + 2 \sum_{m=1, m \neq n}^{N} \left\{ r_{nm}^{-2} \left(g_{nm} + \tilde{g}_{nm} \hat{k} \wedge \right) \right.$$
$$\left. \left[\dot{\vec{r}}_n (\dot{\vec{r}}_m \cdot \vec{r}_{nm}) + \dot{\vec{r}}_m (\dot{\vec{r}}_n \cdot \vec{r}_{nm}) - \vec{r}_{nm} (\dot{\vec{r}}_m \cdot \dot{\vec{r}}_m) \right] \right\},$$
$$n = 1, \ldots, N, \tag{4.37c}$$

where the symbols "$\cdot$" respectively "$\wedge$" identify, of course, the three-dimensional scalar respectively vector products, so that for instance if $\vec{r} \equiv (x, y, 0)$, then $\hat{k} \wedge \vec{r} \equiv (-y, x, 0)$. Note that, when *all two-body* forces are switched off by setting *all* the (two-body) coupling constants g_{nm}, $\tilde{g}_{nm}$ to zero, this model has a direct physical interpretation: it describes the motion of N (equal) charged particles moving in a plane under the action of a constant magnetic field orthogonal to that plane (i.e., a *cyclotron*; in the approximation in which the mutual interaction among the charged particles can be neglected, and the motion takes indeed place in a plane).

The alert reader has presumably recognized that—up to trivial notational differences—this N-body model coincides with the second example of Chapter 1: in particular, as discussed there, for $\tilde{g}_{nm} = 0$ and $g_{nm} = 1$ (entailing $g_{nm}^{(0)} = 1$) the system (4.37) is a well-known *completely integrable* (indeed, *superintegrable* and *solvable*) system; in this case *all* motions are *isochronous*, except for those, emerging from a *lower-dimensional* domain of initial data, that become *singular* in a finite time due to the collision of two or more particles. Also *completely integrable* (indeed, *superintegrable* and *solvable*) is the system (4.37) with $\tilde{g}_{nm} = 0$ and the interaction acting only among "nearest neighbors" (in terms of their labels) with *all* coupling constants g_{nm} equal to minus one half, $g_{nm}^{(0)} = g_{nm} = -\frac{1}{2} \left(\delta_{n,m+1} + \delta_{n,m-1} \right)$; in this case *all* motions are *isochronous*.

Example 4.1.2-3 The case $\lambda = -1$ corresponding to the *third* row of *Table 4.1* is remarkable because of its generality. Indeed, the corresponding *Newtonian* (N-vector) equation of motion reads

$$\ddot{\underline{z}} + i\,\omega\,\dot{\underline{z}} = \sum_{k=1}^{K} \underline{F}^{(-k,\cdot)}(\underline{z},\, \dot{\underline{z}} + i\,\omega\,\underline{z}), \tag{4.38a}$$

where, consistently with the arbitrariness of the constants b_k in the third row of *Table 4.1*, the N-vector-valued functions $\underline{F}^{(-k,\cdot)}(\underline{z}, \tilde{\underline{z}})$ are required (in addition to being *analytic* in *all* their arguments) to satisfy the scaling property (4.8) *only* in their *first* (N-vector-valued) argument,

$$\underline{F}^{(-k,\cdot)}(\alpha\,\underline{z}, \tilde{\underline{z}}) = \alpha^{-k}\,\underline{F}^{(-k,\cdot)}(\underline{z}, \tilde{\underline{z}}), \quad k = 1, \ldots, K. \tag{4.38b}$$

As entailed by *Remark 4.1.1-1* and by the last entry of the third row of *Table 4.1*, the open *(fully dimensional)* domain of initial data that yield *isochronous* motions (with period T, see (4.10)) whose existence is guaranteed by *Lemma 4.1.1-1*, is characterized by N-vectors $\underline{z}(0)$ the components of which are *quite large* (in modulus), while the components of the N-vectors $\dot{\underline{z}}(0)$ are as well *quite large* (in modulus) and consistent with the (approximate) condition (4.12) (of course with $\lambda = -1$).

For instance, an *isochronous* many-body problem of this kind is characterized by the equations of motion

$$\ddot{z}_n = iB\omega + (2C - 1)\, i\omega \dot{z}_n - C\omega^2 z_n + \frac{A + B\dot{z}_n + C\dot{z}_n^2}{z_n}$$
$$+ D \sum_{m=1,\, m \neq n}^{N} \frac{(\dot{z}_n + i\omega z_n)(\dot{z}_m + i\omega z_m)}{z_n - z_m}, \tag{4.39}$$

where the four constants A, B, C, D are *arbitrary*, and an *entirely isochronous* many-body problem belonging to this class (4.38) is characterized—see Section 4.2.2 below—by the equations of motion (4.48).

Example 4.1.2-4 The case $\lambda = -2\,/\,3$ corresponding to the *fourth* row of *Table 4.1* yields the *Newtonian* (N-vector) equation of motion

$$\ddot{\underline{z}} + i\,\Omega\,\dot{\underline{z}} + 2\,\Omega^2\,\underline{z} = \sum_{k=1}^{K} \underline{F}^{(-\frac{1+3\,k}{2})}(\underline{z}), \tag{4.40a}$$

where we introduced for notational convenience the constant

$$\Omega = \frac{\omega}{3}, \tag{4.40b}$$

and, consistently with (even though not necessarily required by) the vanishing of the constants b_k in the fourth row of the Table, the N-vector-valued functions $\underline{F}^{(-\frac{1+3\,k}{2})}(\underline{z})$ are chosen to depend only on the (N-vector-valued) argument $\underline{z}$ and (in addition to being *analytic* in all their arguments) are, of course, required to satisfy the scaling property (see (4.8))

$$\underline{F}^{(-\frac{1+3\,k}{2})}(\alpha\,\underline{z}) = \alpha^{-\frac{1+3\,k}{2}}\,\underline{F}^{(-\frac{1+3\,k}{2})}(\underline{z}), \quad k = 1, \ldots, K. \tag{4.40c}$$

As entailed by *Remark 4.1.1-1* and by the last entry of the fourth row of *Table 4.1*, the open (*fully dimensional*) domain of initial data that yield *isochronous* motions (with period $3\,T$, see (4.10)) the existence of which is guaranteed by *Lemma 4.1.1-1* is characterized by N-vectors $\underline{z}(0)$ the components of which are *quite large* (in modulus), while the components of the N-vectors $\dot{\underline{z}}(0)$ are as well *quite large* (in modulus) and consistent with the (approximate) condition (4.12) (of course with $\lambda = -2\,/\,3$).

This example is of interest inasmuch as it includes a (*complex*) deformation of the equations of motion of the classical gravitational many-body problem. Indeed clearly a special case of (4.40a) with $K=1$ (and $N=3J$) reads as follows:

$$\ddot{\vec{r}}_j + i\,\Omega\,\dot{\vec{r}}_j + 2\,\Omega^2\,\vec{r}_j = \sum_{\ell=1,\ell\neq j}^{J} M_\ell\,\frac{\vec{r}_{\ell j}}{r_{\ell j}^3}, \quad j=1,\ldots,J; \tag{4.41}$$

and obviously, for $\Omega = 0$, these are just the Newtonian equations of motion describing J point-like (or spherical, with constant density) bodies having masses M_j and interacting according to the Newtonian law of gravitation (here, to make contact with the real world, we are of course assuming the vectors $\vec{r}_j \equiv \vec{r}_j(t)$ to be three-dimensional; and we use the short-hand notation $\vec{r}_{\ell j} \equiv \vec{r}_\ell - \vec{r}_j$).

Example 4.1.2-5 The case $\lambda = 1$ corresponding to the *fifth* row of *Table 4.1* yields the *Newtonian* (N-vector) equation of motion

$$\ddot{\underline{z}} - 3\,i\,\omega\,\dot{\underline{z}} - 2\,\omega^2\,\underline{z} = \sum_{k=1}^{K} \underline{F}^{(2+k)}(\underline{z}), \tag{4.42a}$$

where, consistently with (even though not necessarily required by) the vanishing of the constants b_k in the fifth row of *Table 4.1*, the N-vector-valued functions $\underline{F}^{(2+k)}(\underline{z})$ are chosen to depend only on the (N-vector-valued) argument $\underline{z}$ and (in addition to being *analytic* in *all* their arguments) are of course required to satisfy the scaling property (see (4.8))

$$\underline{F}^{(2+k)}(\alpha\,\underline{z}) = \alpha^{2+k}\,\underline{F}^{(2+k)}(\underline{z}), \quad k=1,\ldots,K. \tag{4.42b}$$

As entailed by *Remark 4.1.1-1* and by the last entry of the fifth row of *Table 4.1*, the *open* domain of initial data that yield *isochronous* motions (with period T, see (4.10)) the existence of which is guaranteed by *Lemma 4.1.1-1*, is characterized by N-vectors $\underline{z}(0)$ the components of which are *quite small* (in modulus), while the components of the N-vectors $\dot{\underline{z}}(0)$ are as well *quite small* (in modulus) and thereby also consistent with the (approximate) condition (4.12) (of course with $\lambda = 1$).

An interesting special case is that in which the functions $\underline{F}^{(2+k)}(\underline{z})$ are *polynomials* (of course *homogeneous*, of degree $2+k$) in the N components z_n of the N-vector $\underline{z}$; and particularly interesting is the even more special case with $K = 1$, that describes an assembly of oscillators interacting via (*one-body velocity-dependent*) *linear* forces and (*three-body velocity-independent*) *cubic* forces:

$$\ddot{z}_n - 3\,i\,\omega\,\dot{z}_n - 2\,\omega^2\,z_n = \sum_{j,k,m=1}^{N} a_{njkm}\,z_j\,z_k\,z_m, \quad n=1,\ldots,N, \tag{4.43a}$$

where the N^4 "coupling constants" a_{njkm} are *arbitrary*. An equivalent *real* version of these Newtonian equations of motion, obtained by setting $z_n \equiv x_n + i\,y_n$,

$a_{njkm} \equiv \alpha_{njkm} + i\,\beta_{njkm}$, reads

$$\ddot{x}_n + 3\,\omega\,\dot{y}_n - 2\,\omega^2\,x_n = \sum_{j,k,m=1}^{N} \left(\alpha_{njkm}\,X_{jkm} + \beta_{njkm}\,Y_{jkm}\right), \quad n = 1, \ldots, N, \tag{4.43b}$$

$$\ddot{y}_n - 3\,\omega\,\dot{x}_n - 2\,\omega^2\,y_n = \sum_{j,k,m=1}^{N} \left(-\alpha_{njkm}\,Y_{jkm} + \beta_{njkm}\,X_{jkm}\right), \quad n = 1, \ldots, N, \tag{4.43c}$$

$$X_{jkm} = x_j\,x_k\,x_m - x_j\,y_k\,y_m - y_j\,x_k\,y_m - y_j\,y_k\,x_m, \tag{4.43d}$$

$$Y_{jkm} = y_j\,y_k\,y_m - y_j\,x_k\,x_m - x_j\,y_k\,x_m - x_j\,x_k\,y_m. \tag{4.43e}$$

(The alert reader has of course recognized this system as that already treated in Section 2.2, see (2.23)).

Example 4.1.2-6 The last case we tersely discuss is characterized by $\lambda = 2$ and corresponds to the *sixth* row of *Table 4.1*. It yields the *Newtonian* (N-vector) equation of motion

$$\ddot{\underline{z}} - 5\,i\,\omega\,\dot{\underline{z}} - 6\,\omega^2\,\underline{z} = \sum_{k=1}^{K} \underline{F}^{\left(\frac{3+k}{2}\right)}(\underline{z}), \tag{4.44a}$$

where, consistently with (even though not necessarily required by) the vanishing of the constants b_k in the sixth row of *Table 4.1.2-1*, the N-vector-valued functions $\underline{F}^{\left(\frac{3+k}{2}\right)}(\underline{z})$ are chosen to depend only on the (N-vector-valued) argument $\underline{z}$ and (in addition to being analytic in all their arguments) are, of course, required to satisfy the scaling property (see (4.8))

$$\underline{F}^{\left(\frac{3+k}{2}\right)}(\alpha\,\underline{z}) = \alpha^{\frac{3+k}{2}}\,\underline{F}^{\left(\frac{3+k}{2}\right)}(\underline{z}), \quad k = 1, \ldots, K. \tag{4.44b}$$

As entailed by *Remark 4.1.1-1* and by the last entry of the sixth row of *Table 4.1*, the open (*fully dimensional*) domain of initial data that yield *isochronous* motions (with period T, see (4.10)) whose existence is guaranteed by *Lemma 4.1.1-1* is characterized by N-vectors $\underline{z}(0)$ the components of which are *quite small* (in modulus), while the components of the N-vectors $\dot{\underline{z}}(0)$ are as well *quite small* (in modulus) and thereby consistent with the (approximate) condition (4.12) (of course with $\lambda = 2$).

The reason why we singled out this case is because it includes what might be considered the simplest instance of *nonlinearity* (only *quadratic*). Indeed, setting $K = 1$ and restricting consideration to the case of *polynomial* nonlinearity, the Newtonian equation of motion (4.44a) becomes (in the *complex* version, with arbitrary *complex* "coupling constants" c_{njk})

$$\ddot{z}_n - 5\,i\,\omega\,\dot{z}_n - 6\,\omega^2\,z_n = \sum_{j,k=1}^{N} c_{njk}\,z_j\,z_k, \quad n = 1, \ldots, N, \tag{4.45a}$$

or equivalently (in the *real* version, with arbitrary *real* "coupling constants" a_{njk}, b_{njk}, where $c_{njk} = a_{njk} + i b_{njk}$)

$$\ddot{x}_n + 5\,\omega\,\dot{y}_n - 6\,\omega^2\,x_n = \sum_{j,k=1}^{N} (a_{njk}\,X_{jk} - b_{njk}\,Y_{jk}), \quad n = 1, \ldots, N, \tag{4.45b}$$

$$\ddot{y}_n - 5\,\omega\,\dot{x}_n - 6\,\omega^2\,y_n = \sum_{j,k=1}^{N} (a_{njk}\,Y_{jk} + b_{njk}\,X_{jk}), \quad n = 1, \ldots, N, \tag{4.45c}$$

$$X_{jk} = x_j\,x_k - y_j\,y_k, \quad Y_{jk} = x_j\,y_k + y_j\,x_k. \tag{4.45d}$$

In conclusion, let us re-emphasize that the findings reported above have all been obtained via the simple trick (4.13). Although some of the results (about *isochrony*) reported in this paper might also be proven by standard "Poincaré–Dulac" techniques, the effectiveness of this trick is demonstrated by the generality of the results reported above, as well as by the ease with which these findings have been proven—including moreover the possibility to obtain *explicit* bounds on the size of the open (*fully dimensional*) domain of initial data that yield *isochronous* outcomes: see the proof of *Lemma 1.1.1-1* (although we did not insist there on this aspect, being satisfied with demonstrating the *existence* of such an *open* domain of initial data; such explicit bounds are provided in some of the papers mentioned in Section 4.N).

4.2 One-dimensional systems

We consider useful to subdivide this section into several subsections—to facilitate locating the treatment of some dynamical system by looking at the Table of contents.

4.2.1 *Many-body problems with two-body velocity-independent forces*

The most interesting *isochronous* systems of this kind have already been treated in the context of **Example 4.1.2-1**, and moreover in the introductory Chapter 1 and in Section 2.2; and the simplest, *superintegrable* system of this type, see (1.1), is too well known to justify any additional elaboration here. Enough material has been provided above to justify leaving to the interested reader the task to manufacture other *isochronous* systems belonging to the class identified by the title of this section.

4.2.2 *Goldfishing*

In this, quite extended, Section 4.2.2 we review several *entirely isochronous* many-body problems, characterized by *Newtonian* equations of motion with one-body and two-body velocity-dependent forces, obtained by modifying *solvable* systems identified via a technique to which—for reasons explained below—we refer as "goldfishing." Hence, this section also provides a review of the class of *solvable* many-body problems "of goldfish type."

Terminology reminder: a dynamical system is called *solvable* whenever the solution of its initial-value problem can be reduced to purely algebraic operations, typically to finding the zeros of a polynomial the coefficients of which are explicitly known in terms of the initial data and of time, or equivalently to finding the eigenvalues of a matrix the time evolution of which is explicitly known.

Over a quarter century ago a class of *solvable* dynamical systems was introduced, naturally interpretable as classical many-body problems characterized by *Newtonian* equations of motion ("accelerations proportional to forces": with appropriate velocity-dependent one-body and two-body forces). The simplest example of this kind is characterized by the equations of motion (see (1.10))

$$\ddot{z}_n - i\,\omega\,\dot{z}_n = 2 \sum_{m=1,\,m\neq n}^{N} \frac{\dot{z}_n\,\dot{z}_m}{z_n - z_m}. \tag{4.46}$$

Notational reminder: here and hereafter superimposed dots denote differentiations with respect to the (*real*) independent variable t ("time"), i is the imaginary unit ($i^2 = -1$), the dependent variables $z_n \equiv z_n(t)$ are generally *complex*, N is a positive integer ($N \geq 2$), indices such as n, m generally range from 1 to N (unless otherwise indicated), and ω is a *real* constant (for definiteness, *nonnegative*) to which (whenever this constant does not vanish) is associated the basic period

$$T = \frac{2\,\pi}{\omega}. \tag{4.47}$$

The model (4.46) with $\omega = 0$ might be considered the simplest one; but, as already pointed out in the introductory Chapter 1, the model with $\omega > 0$ has the remarkable property to be *entirely isochronous* (see the solution, via (1.11), of its initial-value problem). The origin of this fact can be traced to the property of the equations of motion of this system, (4.46) with $\omega > 0$, to be just the special case with $\lambda = 0$ of the system of ODEs

$$\ddot{z}_n - (2\,\lambda + 1)\,i\,\omega\,\dot{z}_n - \lambda\,(\lambda + 1)\,\omega^2\,z_n = 2 \sum_{m=1,\,m\neq n}^{N} \frac{(\dot{z}_n - i\,\lambda\,\omega\,z_n)\,(\dot{z}_m - i\,\lambda\,\omega\,z_m)}{z_n - z_m}, \tag{4.48}$$

which are related via the usual trick (see for instance (4.13)) to the equations of motion of the system (4.46) with $\omega = 0$, reading

$$\zeta_n'' = 2 \sum_{m=1,\,m\neq n}^{N} \frac{\zeta_n'\,\zeta_m'}{\zeta_n - \zeta_m} \tag{4.49}$$

(here as usual, in view of the application of the trick formula (4.13), the dependent variables are denoted as $\zeta_n(\tau)$ and differentiations with respect to the *complex* independent variable τ are denoted by appended primes). This finding is in fact already discussed above, see **Example 4.1.2-2** and Chapter 1.

Remark 4.2.2-1. As implied by the discussion of **Example 4.1.2-2**, and as it can in any case be easily verified, the *Newtonian* equations of motion (4.46) are yielded in the standard manner by the *Hamiltonian*

$$H(\underline{p},\, \underline{z}) = \sum_{n=1}^{N} \left[\frac{i\,\omega\, z_n}{c} + \exp\left(c\, p_n\right) \prod_{m=1,\, m\neq n}^{N} \frac{1}{z_n - z_m} \right] \tag{4.50}$$

where c is an *arbitrary* (nonvanishing) constant. Some, but not all, of the *Newtonian* equations of motion written below (in this Section 4.2.2) are as well *Hamiltonian*, namely they can be obtained in the standard manner from an appropriate *Hamiltonian*. ⊡

The neat character of the *Newtonian* equations of motion (4.46), as well as their *Hamiltonian* character, see (4.50), and their physical version, see (1.7) (or equivalently (4.37c) with $g_{nm} = 1$ and $\tilde{g}_{nm} = 0$), suggested [36] to highlight this many-body problem by calling it a "goldfish": this name was originally attributed to the model (4.46) with $\omega = 0$, and then was as well used for this model (4.46) with *positive* ω, having the additional remarkable property to be *entirely isochronous* (anyway these two models are related by the simple trick transformation (4.13) with $\lambda = 0$). This terminology originated from the following description of the search for integrable systems given by V. E. Zakharov: "A mathematician, using the dressing method to find a new integrable system, could be compared with a fisherman, plunging his net into the sea. He does not know what a fish he will pull out. He hopes to catch a goldfish, of course. But too often his catch is something that could not be used for any known to him purpose. He invents more and more sophisticated nets and equipments and plunges all that deeper and deeper. As a result he pulls on the shore after a hard work more and more strange creatures. He should not despair, nevertheless. The strange creatures may be interesting enough if you are not too pragmatic. And who knows how deep in the sea do goldfishes live?" [143]. Subsequently, the name "goldfish" has been, more generally, used for several variants (possibly even *non-integrable* ones) of the original model (4.46) with $\omega = 0$, and for their ω-modified *isochronous* versions; it seems of course particularly appropriate for ω-modified models obtained from *integrable*, indeed *solvable*, systems, which as a consequence are generally *entirely isochronous*, namely *completely periodic* with period T—or possibly an *integer* multiple of T—in their *entire* natural phase space (except possibly for a *lower-dimensional* set characterized by *singular* solutions). Several such systems have been identified recently via a technique—described below, in this Section 4.2.2—that in our opinion justifies the title of this section inasmuch as it calls to mind the research approach poetically described by V. E. Zakharov, as quoted above.

Indeed, as explained below, a natural first outcome of this technique is to yield *solvable* N-body problems characterized by *Newtonian* equations of motion with one-body and two-body forces, the latter of which, however, feature generally multiplicative coefficients playing the role of "coupling constants" but being instead *time dependent*, their time evolution being characterized by a system of coupled ODEs that also contain the independent variables $z_n(t)$ identifying the particle positions in the N-body context. Only in special cases—each of which seems to require a "miracle" to occur—one manages to get rid of these *time-dependent* coefficients (playing the role of "auxiliary variables") via an appropriate *ansatz* expressing them in terms of the independent variables z_n (and possibly also of their time derivatives $\dot{z}_n$): in this manner a proper *solvable* N-body problem is obtained, involving *only* the particle coordinates $z_n(t)$ (and possibly also their time derivatives $\dot{z}_n(t)$) but *no* additional auxiliary variables. Hence this research strategy has so far proceeded via a sequence of contributions each of which has identified (and investigated) a new *solvable* many-body problem obtained in this manner, generally characterized by equations of motions "of goldfish type". Below—after describing the technique—we review these *solvable* many-body problems. Each of these models can then be ω-modified via the trick, obtaining thereby a many-body model which is both *solvable* and *entirely isochronous:* indeed, its *solvable* character generally entails that its *isochrony* region coincides with its *entire* natural phase space (possibly up to a *lower dimensional* region characterized by *singular* solutions).

But before proceeding with the presentation of these findings, it is appropriate to interject two general considerations.

A method to identify *solvable* many-body models—introduced almost three decades ago, just in the paper [29] that identified the *solvable* character of the original goldfish model (4.46)—is based on the nonlinear mapping among the N zeros $z_n(t)$ of a time-dependent polynomial $\psi(z, t)$, of degree N in the variable z (and, without significant loss of generality, *monic*), and its N coefficients $c_m(t)$:

$$\psi(z, t) = \prod_{n=1}^{N} [z - z_n(t)] = z^N + \sum_{m=1}^{N} c_m(t)\, z^{N-m}. \tag{4.51}$$

The idea was to select (*cum grano salis*) a *simple* (in some sense *solvable*) time evolution of the polynomial $\psi(z, t)$, generally entailing a *simple* (and no less *solvable*) time evolution of the N coefficients $c_m(t)$, and to then focus on the corresponding time evolution of the N zeros $z_n(t)$, which generally then turns out to be given by the *Newtonian* equations of motion of an N-body problem of *goldfish* type. Consider for instance the *simple* time evolution

$$\psi_{tt} - i\,\omega\,\psi_t = 0. \tag{4.52a}$$

Notation: here and hereafter subscripted variables denote partial differentiations.

This evolution PDE is clearly compatible with the assumed polynomial character of ψ—see (4.51), implying then

$$\ddot{c}_m - i\,\omega\,\dot{c}_m = 0 \tag{4.52b}$$

—and clearly the solution of its initial-value problem reads

$$\psi(z,\,t)=\psi(z,\,0)+\psi_t(z,\,0)\,\frac{\exp(i\,\omega\,t)-1}{i\,\omega}. \tag{4.52c}$$

It is on the other hand easy to verify that, for the N zeros $z_n(t)$, this time evolution (4.52) corresponds just to the *goldfish* system of ODEs (4.52). [To verify this one can take advantage—if need be—of the formulas (A.5) and (A.10) of Appendix A; note moreover that the solution formula (4.52c) yields, via (4.51) together with (A.5), the (implicit) solution formula, (1.11), of the *goldfish* model, see (1.10) or (4.46)].

It should be immediately clear how this approach can be extended so as to obtain an entire class of *solvable* N-body problems of *goldfish* type, whose signature can be inferred by the presence of terms associated with the right-hand side of (A.10). These findings are too old [29] and too well known [37] to justifying any detailed elaboration of them here. Let us merely display the *Newtonian* equations of motion of a general class of *solvable* many-body models of *goldfish* type yielded by this approach:

$$\begin{aligned}\ddot{z}_n &= a_1\,\dot{z}_n+a_2+a_3\,z_n-2\,(N-1)\,a_{10}\,z_n^2\\&\quad+\sum_{m=1,\,m\neq n}^{N}\Big\{(z_n-z_m)^{-1}\,[2\,\dot{z}_n\,\dot{z}_m+(\dot{z}_n+\dot{z}_m)\,(a_4+a_5\,z_n)\\&\quad+a_6\,(\dot{z}_n\,z_m+\dot{z}_m\,z_n)\,z_n+2\,(a_7+a_8\,z_n+a_9\,z_n^2+a_{10}\,z_n^3)]\Big\}.\end{aligned} \tag{4.53}$$

Here the ten "coupling constants" a_j are *arbitrary*, and the *solvable* character of this class of many-body problems is due to the *linear*—hence *solvable*—character of the corresponding evolution equations satisfied by the coefficients $c_m(t)$, see (4.51):

$$\begin{aligned}&\ddot{c}_m-(N+1-m)\,a_4\,\dot{c}_{m-1}-[(N-m)\,a_5+a_1]\,\dot{c}_m+m\,a_6\,\dot{c}_{m+1}\,(N+1-m)\\&\quad(N+2-m)\,a_7\,c_{m-2}+(N+1-m)\,[(N-m)\,a_8+a_2]\,c_{m-1}\\&\quad-m\,[(2\,N-1-m)\,a_9+a_3]\,c_m+m\,(m+1)\,a_{10}\,c_{m+1}=0.\end{aligned} \tag{4.54}$$

For appropriate assignments of the ten coupling constants a_j this class of many-body problems includes several models displaying interesting phenomenologies [37]. Consistently with the focus of this monograph we merely highlight the sub-classes of these models characterized by an *entirely isochronous* behavior. They are characterized by coupling constants a_j satisfying the following restrictions, that clearly entail that the solutions $c_m(t)$ of the *linear* evolution equations (4.54) are *all* periodic with period $T=2\pi/\omega$:

$$c_m(t+T)=c_m(t)\,. \tag{4.55}$$

Indeed restrictions that are *sufficient* to guarantee this outcome—hence, via (4.51), the *entirely periodic* character of the corresponding many-body model

(4.53)—clearly read as follows (as the diligent reader will easily verify by solving the *linear* evolution equations (4.54), which are triangular thanks to the first of the following set of restrictions (see (4.56))): either

$$a_6 = a_{10} = 0, \tag{4.56a}$$

or

$$a_2 = a_4 = a_7 = a_8 = 0, \tag{4.56b}$$

with, moreover,

$$a_1 = 2\left(k_1 + N k_2\right) i\omega, \quad a_3 = \left(k_2 - k_3\right)\left[k_1 + \left(N - \frac{1}{2}\right)\left(k_2 + k_3\right)\right]\omega^2, \tag{4.56c}$$

$$a_5 = -2k_2 i\omega, \quad a_9 = -\left(k_2^2 - k_3^2\right)\omega^2, \tag{4.56d}$$

where the three *integers* k_1, k_2, k_3 are *arbitrary* except for the following restrictions:

$$k_2 \neq k_3, \tag{4.57a}$$

and

$$k_1 + m k_3 \neq 0, \quad 2k_1 + m\left(k_2 + k_3\right) \neq 0, \quad m = 1, \ldots, N. \tag{4.57b}$$

Let us emphasize that the constants a_j not restricted by these conditions remain *arbitrary*. And let us also mention that a quite recent extension of this approach—which deserves a somewhat more extensive, if still terse, treatment because of its novelty—is discussed below, see **Example 4.2.2-10**.

The point we rather like to emphasize now is that there is also another way to take advantage of the possibility to establish, via the standard formula (4.51), a relationship among a system of N ODEs characterizing the time evolution of the N zeros $z_n(t)$ of a time-dependent polynomial of degree N in the variable z and a corresponding system of N ODEs characterizing the time evolution of the N coefficients $c_m(t)$ of this polynomial. Indeed in the following we again take advantage of this correspondence; but in quite a different—in some sense opposite—manner than the past practice [29] [37]. As outlined above, this approach associated to a *linear*, hence *solvable*, time evolution of the polynomial $\psi(z, t)$—hence of its N coefficients $c_m(t)$, see (4.54)—a corresponding (generally quite *nonlinear*) time evolution of its N zeros $z_n(t)$. This evolution was interpretable as an N-body problem (see (4.53))—to be categorized as *solvable* thanks to the relation (4.51), namely because its solution is reduced to the task of finding the N zeros of a polynomial $\psi(z, t)$ whose time evolution—namely, the time evolution of its coefficients $c_m(t)$—is known (by solving the *linear* evolution equations (4.54): merely an algebraic task). Below we instead identify *solvable* N-body problems of *goldfish* type via another approach—sometimes referred to as the *direct* approach [79]—resulting from identifying the particle coordinates $z_n(t)$ of these N-body problems as the N *eigenvalues* of a time-dependent $N \times N$ matrix $U(t)$ whose time evolution is described by a *solvable* matrix ODE. This of course implies that the particle coordinates $z_n(t)$—as well as the coefficients $c_m(t)$, see (4.51)—are related to the $N \times N$ matrix $U(t)$ by the formulas

$$\det[z\,\mathbf{1}-U(t)] = \psi(z,\,t) = \prod_{n=1}^{N}[z - z_n(t)] = z^N + \sum_{m=1}^{N} c_m(t)\, z^{N-m}, \quad (4.58)$$

where $\mathbf{1}$ denotes, of course, the unit $N \times N$ matrix (in the following we will often omit to indicate this matrix, when its presence—as in this case—is obvious). Then, via this correspondence—see (4.51) and (4.58)—among the N zeros of a polynomial $\psi(z,\,t)$ and its N coefficients $c_m(t)$, one identifies—via a standard procedure based on the identities reported in Appendix A—*solvable* evolutions for these coefficients $c_m(t)$, which can also be justifiably interpreted as nontrivial N-body problems, being also characterized by systems of *nonlinear* ODEs of *Newtonian* type. This possibility is not surprising, given the close relationship—see (4.51) and (4.58)—among the *eigenvalues* of a matrix and the *zeros* of a polynomial. Yet the "alternative" N-body models obtained in this manner—describing the time evolution of the N coefficients $c_m(t)$—are nontrivial, indeed in our opinion they are rather interesting; and, when *isochronous* ω-modified variants of these N-body problems are manufactured—as we shall do below—referring to these models as assemblies of *nonlinear harmonic oscillators* appears particularly appropriate. In this connection the following *Remark* is relevant.

Remark 4.2.2-2. If the matrix $U(t)$ is *periodic*,

$$U(t+T) = U(t), \quad (4.59a)$$

each of the N coefficients $c_m(t)$ is as well *periodic* with the *same* period (see (4.58)),

$$c_m(t+T) = c_m(t), \quad (4.59b)$$

while its N *eigenvalues* $z_n(t)$ are as well *periodic*, but not necessarily with the *same* period,

$$z_n(t+T_n) = z_n(t), \quad (4.60)$$

due to the possibility that through the time evolution *different* eigenvalues get exchanged. Of course, each of the periods T_n is an *integer multiple* q_n of the basic period T,

$$T_n = q_n\,T, \quad (4.61a)$$

entailing the *complete periodicity* of the N-body problem characterizing the time evolution of the entire set of N particle coordinates $z_n(t)$ with a period $\tilde{T}$ which is the Least Common Multiple (LCM) of the periods T_n,

$$z_n\left(t+\tilde{T}\right) = z_n(t), \quad n = 1,\ldots,N, \quad \tilde{T} = \mathrm{LCM}(T_n). \quad (4.61b)$$

The *integer multiples* q_n generally depend on the initial data—even if the periodicity of $U(t)$, see (4.59a), holds *universally*—but of course none of them can exceed N,

$$1 \leq q_n \leq N; \tag{4.61c}$$

hence, the overall period $\tilde{T} = \tilde{q}\,T$—which generally also depends on the initial data—is *a priori* constrained by the inequality

$$1 \leq \tilde{q} \leq N\,!; \tag{4.62}$$

but generally $\tilde{q}$ is much smaller than the upper limit in this inequality (see [100]). ⊡

The second observation we like to interject here before plunging into the core of this section is to explain how, as a side product of the results described below, *Diophantine* results and conjectures are also obtained. Take as starting point an *entirely isochronous* N-body problem *all nonsingular* solutions of which are *periodic* with a *fixed* period: several such *nonlinear harmonic oscillators* models are indeed manufactured below. Assume then that such a model possesses some stable *equilibrium* configuration which can be explicitly found, and investigate its behavior in that neighborhood. The standard "small oscillations" theory entails that in that neighborhood the motion of the system is *multiply periodic*, the frequencies characterizing this behavior being the *eigenvalues* of a matrix—let us call it E for "equilibrium"—constructed from the equations of motions, so that its entries are given in terms of the values the dependent variables assume at equilibrium. But the *isochrony* of the system in question—valid essentially for *all* its motions—must as well apply to these small oscillations. Hence the conclusion that *all* the frequencies characterizing the multiply periodic motion near equilibrium must be *integer multiples* of a single basic frequency—implying that *all* the *eigenvalues* of the matrix E, up to a common scaling factor, must be *integer numbers*. In some cases below such *Diophantine* findings merely confirm known results (often arrived at by developments originated just from this kind of approach); in other cases they seem new. They are of course confirmed, for small values of N, by explicit numerical computations; and generally these verifications suggest explicit formulas for the *integer* eigenvalues of the matrix E, whose validity for *arbitrary* N still have, in several cases, the status of *conjectures*. We confine a terse presentation of all these findings to Appendix C.

Let us finally delve into the core of this Section 4.2.2.

The starting point of this approach to obtain *solvable* many-body problems is a *solvable matrix evolution equation*, say

$$\ddot{U} = F\left(U, \dot{U}\right). \tag{4.63}$$

Henceforth, $U = U(t)$ is a time-dependent $N \times N$ matrix, and the assumed *solvability* of this matrix ODE entails the possibility to write in *explicit* form the solution $U(t)$ of its initial-value problem.

The function $F(U, \dot{U})$ is assumed not to depend on any other matrix besides U and $\dot{U}$ (the ordering of which in its definition is of course relevant, since these two matrices generally do *not* commute), so that there hold the relation

$$R\,F(U,\,\dot{U})\,R^{-1} = F(R\,U\,R^{-1},\,R\,\dot{U}\,R^{-1}), \tag{4.64}$$

for any (invertible) $N \times N$ matrix R.

The main idea is then quite simple: *to investigate the time evolution of the eigenvalues $z_n(t)$ of the matrix $U(t)$*. And let us immediately emphasize, before proceeding to exhibit the form taken by the equations that describe this time evolution, that whenever the time evolution of the $N \times N$ matrix $U(t)$ is *periodic* with a period T, the corresponding evolution of each of its N eigenvalues—unless it runs into a singularity (this may happen, but only for special, *non-generic*, initial data)—is obviously as well *periodic*, as discussed above (see *Remark 4.2.2-2*).

To investigate the time evolution of the eigenvalues $z_n(t)$ of $U(t)$ we set

$$U(t) = R(t)\,Z(t)\,[R(t)]^{-1}, \qquad Z(t) = \mathrm{diag}\,[z_n(t)]\,, \tag{4.65}$$

and we take note of the consequential identities

$$\dot{U} = R\left\{\dot{Z} + [M,\,Z]\right\}R^{-1}, \tag{4.66a}$$

$$\ddot{U} = R\left\{\ddot{Z} + \left[\dot{M},\,Z\right] + 2\left[M,\,\dot{Z}\right] + [M,\,[M,\,Z]]\right\}R^{-1}, \tag{4.66b}$$

where we set

$$M = R^{-1}\,\dot{R}. \tag{4.67}$$

Notation: here and hereafter $[A,\,B]$ denotes the commutator of the two matrices A and B,

$$[A,\,B] \equiv A\,B - B\,A. \tag{4.68}$$

Remark 4.2.2-3. The diagonalizing matrix R, see (4.65), is clearly defined up to multiplication from the right by an *arbitrary diagonal* $N \times N$ matrix D, hence the matrix M is defined up to the "gauge transformation"

$$M \mapsto D^{-1}\,M\,D + D^{-1}\,\dot{D}, \tag{4.69}$$

which, thanks to the *arbitrariness* of the *diagonal* matrix D, entails that the *diagonal* part of the matrix M can be assigned *arbitrarily*: see the second term in the right-hand side of this formula; of course the *off-diagonal* part of the matrix M must then be modified appropriately. ⊡

Via (4.65) and (4.66) the *solvable* matrix ODE (4.63) becomes

$$\ddot{Z} + \left[\dot{M},\, Z\right] + 2\,\left[M,\, \dot{Z}\right] + [M,\, [M,\, Z]] = F\left(Z,\, \dot{Z} + [M,\, Z]\right), \tag{4.70}$$

hence by separating the diagonal and off-diagonal parts of this matrix ODE one immediately arrives at the following systems of (altogether N^2 coupled scalar) ODEs:

$$\ddot{z}_n = -2 \sum_{m=1,\, m\neq n}^{N} (z_n - z_m)\, M_{nm}\, M_{mn} + \left[F\left(Z,\, \dot{Z} + [M,\, Z]\right)\right]_{nn}, \tag{4.71a}$$

$$\begin{aligned} \dot{M}_{nm} &= -2\,\frac{\dot{z}_n - \dot{z}_m}{z_n - z_m}\, M_{nm} - (\mu_n - \mu_m)\, M_{nm} \\ &\quad + \sum_{\ell=1,\, \ell\neq m,n}^{N} \frac{z_n + z_m - 2\, z_\ell}{z_n - z_m}\, M_{n\ell}\, M_{\ell m} - \frac{\left[F\left(Z,\, \dot{Z} + [M,\, Z]\right)\right]_{nm}}{z_n - z_m}, \\ &\quad n \neq m. \end{aligned} \tag{4.71b}$$

Note that we denote the *diagonal* elements of the $N \times N$ matrix M as μ_n, $M_{nn}(t) \equiv \mu_n(t)$ (we never use the convention according to which repeated indices are summed upon), and recall that, as explained above (see *Remark 4.2.2-3*), their values can be chosen at our convenience; the rest of the notation is, we trust, self-evident. Note that (4.63) (a matrix evolution equation amounting to N^2 coupled ODEs), characterizing the time evolution of the N^2 matrix elements of the $N \times N$ matrix $U(t)$, has now been replaced by the system (4.71), amounting again altogether to $N + N\,(N-1) = N^2$ coupled ODEs, and characterizing the time evolution of the N eigenvalues $z_n(t)$ and of the $N\,(N-1)$ off-diagonal matrix elements $M_{nm}(t)$, $n \neq m$.

Because of the way these systems of ODEs have been derived they are, as it were by definition, *solvable* (recall that we started from an assumedly *solvable* matrix evolution equation, (4.63)). And it is clear that the first system of N coupled ODEs, (4.71a), has indeed the structure of a *Newtonian* N-body problem characterizing the motion of N "particle coordinates" $z_n(t)$, while the second system of $N\,(N-1)$ ODEs, (4.71b), characterizes the evolution of the $N\,(N-1)$ "auxiliary variables" $M_{nm}(t)$ (with $n \neq m$). It might be possible to attribute as well to these auxiliary variables $M_{nm}(t)$ a physical meaning in terms of internal ("spin") degrees of freedom: pioneering steps in this direction were made long ago, simultaneously and independently, by John Gibbons and Theo Hermsen [98] and by Stefan Wojciechowski [138] [139]. Here a different goal is pursued: to identify cases in which, via an appropriate *ansatz* (expressing the auxiliary variables in terms of the particle coordinates, and possibly also of their derivatives: see below), thanks to some "miraculous" identities, it is possible to satisfy *identically* the second system, (4.71b), thereby getting altogether rid of the auxiliary variables $M_{nm}\,(t)$, hence obtaining from (4.71a) a genuine N-body problem characterized by *Newtonian* equations of motion involving *only* the particle coordinates $z_n(t)$. The possibility to do so depends on the specific choice of the original *solvable* matrix ODE (see (4.63); actually in some cases, see below, it might be preferable to take as starting point a *first-order* matrix ODE or *two* coupled *first-order* matrix ODEs rather than a *second-order* matrix ODE), and moreover on the identification of an appropriate *ansatz*; general rules to identify a working combination of these two elements are not known, so the approach followed so far has been a trial and error one ("goldfishing"), which has yielded over time a few successes. We now review tersely these findings, beginning from a particularly simple example which we use as prototype to illustrate how this approach works, although in this case the results obtained are far from novel.

Example 4.2.2-1 The most trivial *second-order* matrix ODE obtains from (4.63) by setting $F\left(U, \dot{U}\right) = 0$, hence it reads

$$\ddot{U} = 0, \tag{4.72a}$$

and it is of course *solvable*:

$$U\,(t) = U\,(0) + \dot{U}(0)\,t. \tag{4.72b}$$

The corresponding version of (4.71) reads

$$\ddot{z}_n = -2 \sum_{m=1,\, m\neq n}^{N} (z_n - z_m)\, M_{nm}\, M_{mn}, \tag{4.73a}$$

$$\dot{M}_{nm} = -2\, \frac{\dot{z}_n - \dot{z}_m}{z_n - z_m}\, M_{nm} - (\mu_n - \mu_m)\, M_{nm} + \sum_{\ell=1,\, \ell\neq m,n}^{N} \frac{z_n + z_m - 2\, z_\ell}{z_n - z_m}\, M_{n\ell}\, M_{\ell m}, \quad n \neq m. \tag{4.73b}$$

It is thus seen that this system of N^2 ODEs, satisfied by the N dependent variables $z_n\,(t)$ ("particle coordinates") and by the $N\,(N-1)$ dependent variables $M_{nm}\,(t)$ with $n \neq m$ ("auxiliary variables"), is generally *solvable*—even allowing for an essentially *arbitrary* (see above) assignment of the N quantities $\mu_n\,(t) \equiv M_{nn}\,(t)$. But, as we explained above, we would like to get altogether rid of the $N\,(N-1)$ auxiliary variables $M_{nm}(t)$ (with $n \neq m$), finding a way to satisfy identically the second, (4.73b), of these two systems of ODEs, which we conveniently rewrite as follows:

$$\frac{\dot{M}_{nm}}{M_{nm}} = -2\, \frac{\dot{z}_n - \dot{z}_m}{z_n - z_m} - \mu_n + \mu_m + \sum_{\ell=1,\, \ell\neq m,n}^{N} \frac{M_{n\ell}\, M_{\ell m}}{M_{nm}}\, \frac{z_n + z_m - 2\, z_\ell}{z_n - z_m}, \quad n \neq m. \tag{4.73c}$$

Two *ansatzen* turn out to be suitable to achieve this goal.

The *first ansatz* is suggested by the first term in the right-hand side of this system of ODEs, hence it reads

$$M_{nm}(t) = \frac{g_{nm}(t)}{\left[z_n(t) - z_m(t)\right]^2}. \tag{4.74}$$

Thereby the system of ODEs (4.73c) becomes

$$\frac{\dot{g}_{nm}}{g_{nm}} = -\mu_n + \mu_m$$
$$- \sum_{\ell=1,\, \ell \neq m,n}^{N} \frac{g_{n\ell}\, g_{\ell m}}{g_{nm}} \left[\frac{1}{(z_n - z_\ell)^2} - \frac{1}{(z_m - z_\ell)^2}\right], \quad n \neq m, \tag{4.75}$$

and this suggests taking advantage of the freedom to assign the quantities $\mu_n(t)$ by setting

$$\mu_n(t) = - \sum_{\ell=1,\, \ell \neq n}^{N} \frac{g_{n\ell}(t)}{\left[z_n(t) - z_m(t)\right]^2}. \tag{4.76}$$

Thereby the system of ODEs satisfied by the new auxiliary variables g_{nm} reads

$$\frac{\dot{g}_{nm}}{g_{nm}} = \frac{g_{nm} - g_{mn}}{(z_n - z_m)^2}$$
$$- \sum_{\ell=1,\, \ell \neq m,n}^{N} \frac{1}{g_{nm}} \left[\frac{g_{n\ell}\, g_{\ell m} - g_{n\ell}\, g_{nm}}{(z_n - z_\ell)^2} - \frac{g_{n\ell}\, g_{\ell m} - g_{m\ell}\, g_{nm}}{(z_m - z_\ell)^2}\right], \quad n \neq m, \tag{4.77}$$

and it is then clear that it admits the trivial solution

$$g_{nm}(t) = (-2)^{-1/2}\, g, \quad n \neq m \tag{4.78}$$

with g an *arbitrary constant.* The factor $(-2)^{-1/2}$ in the right-hand side of this formula is introduced merely for notational convenience: to ensure that, via (4.73a) and (4.74), the *Newtonian* equations of motion determining the time-evolution of the coordinates $z_n(t)$ read

$$\ddot{z}_n = \sum_{m=1,\, m \neq n}^{N} \frac{g^2}{(z_n - z_m)^3}, \tag{4.79}$$

coinciding with the ($\omega = 0$, *nonisochronous*) version of (1.1b).

To obtain the *isochronous* version, (1.1b), of this well-known system we rewrite this system of ODEs, (4.79), as follows:

$$\zeta_n'' = \sum_{m=1,\, m \neq n}^{N} \frac{g^2}{(\zeta_n - \zeta_m)^3}. \tag{4.80}$$

This differs from the preceding version merely notationally, in a self-evident manner. We now use the standard version of the trick (2.3) with $\lambda = -1/2$,

$$z_n(t) = \exp\left(-\frac{i\,\omega}{2}t\right)\zeta_n(\tau), \quad \tau = \frac{\exp(i\,\omega\,t) - 1}{i\,\omega}, \tag{4.81}$$

and thereby obtain the *Newtonian* equations of motion (1.1b),

$$\ddot{z}_n + \left(\frac{\omega}{2}\right)^2 z_n = \sum_{m=1, m\neq n}^{N} \frac{g^2}{(z_n - z_m)^3}. \tag{4.82}$$

This method to arrive at this model implies, of course, its *solvability*, in the sense of reducing the solution of the initial-value problem for this model to the determination of the eigenvalues of an $N \times N$ matrix explicitly known in terms of the initial data and of time. Indeed, for this particular model, this was achieved by M. A. Olshanetsky and A. M. Perelomov long ago, by a technique which can be justifiably considered as a precursor of the approach described above. But we do not elaborate any further on these findings, since they constitute by now standard textbook material, see for instance [37] and references therein (and note that this model coincides with the very first model discussed in the introductory Section 1, see (1.1b)).

The *second ansatz* is somewhat less obvious: it reads

$$M_{nm}(t) = \frac{[\dot{z}_n(t)\,\dot{z}_m(t)]^{1/2}\,\eta_{nm}(t)}{z_n(t) - z_m(t)}. \tag{4.83}$$

Its insertion in (4.73a) yields

$$\ddot{z}_n = 2 \sum_{m=1,\, m\neq n}^{N} \eta_{nm}\,\eta_{mn}\,\frac{\dot{z}_n\,\dot{z}_m}{z_n - z_m}, \tag{4.84}$$

while its insertion in (4.73c) yields, via (4.84),

$$\begin{aligned}\frac{\dot{\eta}_{nm}}{\eta_{nm}} &= \frac{\eta_{nm}\,\eta_{mn} - 1}{z_n - z_m} \\ &+ \sum_{\ell=1,\, \ell\neq m,n}^{N} \frac{\eta_{n\ell}\,\eta_{\ell m}}{\eta_{nm}}\,(1 + \eta_{nm})\left(\frac{1}{z_n - z_\ell} + \frac{1}{z_m - z_\ell}\right), \quad n \neq m,\end{aligned} \tag{4.85}$$

having made in this case the simple choice $\mu_n = 0$ (or equivalently $\mu_n = \mu$, see (4.73c)).

It is now clear that this system of $N\,(N-1)$ ODEs, (4.85), admits the trivial solution

$$\eta_{nm} = -1, \quad n \neq m. \tag{4.86}$$

And the insertion of this value of η_{nm} in (4.84) yields just the original *goldfish* model (see (1.10) or (4.46), with $\omega = 0$),

$$\ddot{z}_n = 2 \sum_{m=1,\, m \neq n}^{N} \frac{\dot{z}_n\, \dot{z}_m}{z_n - z_m}, \tag{4.87}$$

from which the *entirely isochronous* version (1.10) or (4.46) (with $\omega > 0$; or, more generally, the version (4.48)) can be obtained via the trick, as explained at the beginning of this Section 4.2.2.

Again, as in the case of the model (4.82) arrived at via the *first ansatz*, in this case as well the method we just described to arrive at this *goldfish* model implies of course its *solvability*, in the sense of reducing the solution of the initial-value problem for this model to the determination of the eigenvalues of an $N \times N$ matrix explicitly known in terms of the initial data and of time. But again we do not elaborate any further on these findings, since they constitute by now standard textbook material (see for instance [37], and references therein). Moreover, the results for this *goldfish* model could also be easily retrieved from those described below for the generalized versions of it, to whose presentation we now turn.

Let us emphasize that, while in the simple case we just treated, both *ansatzen* —(4.74) and (4.83)—work, this does not always happen: indeed, while both *ansatzen* (including appropriate variants, see below) have been tried in the other cases treated below, more often than not only the second type does work, yielding solvable models of *goldfish* type—which are anyway the focus of this Section 4.2.2.

From the simple example we just treated it is clear what the strategy appropriate to get such models is: to start from some *solvable* matrix equation, to proceed (more or less) as above in order to get the analog of the system of ODEs (4.73), and then to use some appropriate variant of the *second ansatz* (4.83). The results recently obtained in this manner are tersely reviewed below: readers interested in getting more details are advised to look at the relevant references, as specified in each of the cases reported below. Some proofs of these results are confined to Appendix B, and related *Diophantine* findings are collected in Appendix C.

Example 4.2.2-2 In [65] the starting point of the treatment is the *linear* (and obviously *solvable*) second-order matrix ODE

$$\ddot{U} + a\,\dot{U} + b\,U = 0, \tag{4.88}$$

where a and b are two *arbitrary* constants. The *solvable* many-body problems thereby manufactured (using mainly the *first ansatz*) are well-known classical models [37] the *solvability* of which was already known, thanks to the technique invented long ago by M. A. Olshanetsky and A. M. Perelomov (see Section 4.N), whose close relationship to that described herein has already been mentioned above.

Example 4.2.2-3 In [47] the starting point of the treatment is the *solvable* second-order matrix ODE

$$\ddot{U} = a \dot{U} U^{-1} \dot{U}, \tag{4.89}$$

whose initial-value problem is explicitly solved by the formula

$$U(t) = A \left(1 + \frac{B}{\alpha} t\right)^{\alpha}, \quad \text{if } a \neq 1, \tag{4.90a}$$

$$U(t) = A \exp(B t), \quad \text{if } a = 1, \tag{4.90b}$$

where the constant α is related to a (provided $a \neq 1$) as follows,

$$\alpha = \frac{1}{1-a}, \quad a = \frac{\alpha - 1}{\alpha}, \tag{4.90c}$$

and the two constant $N \times N$ matrices A, B are given in terms of the initial data by the following formulas:

$$A = U(0), \quad B = [U(0)]^{-1} \dot{U}(0). \tag{4.90d}$$

The *solvable* N-body problems manufactured are again mainly well-known classical models the *solvability* of which was already known, thanks to the technique invented by M. A. Olshanetsky and A. M. Perelomov (see Section 4.N), but they include in addition the following *generalized goldfish* model:

$$\ddot{z}_n = i \omega \dot{z}_n + 2 \sum_{m=1, m \neq n}^{N} \frac{\dot{z}_n \dot{z}_m}{z_n - z_m} + a \sum_{m=1}^{N} \frac{\dot{z}_n \dot{z}_m}{z_m}. \tag{4.91}$$

For $a = 0$ these *Newtonian* equations of motion (4.91) coincide with those, (1.10) or (4.46), of the basic *goldfish* model. It has moreover been shown [47] that the *solvability* of this *generalized goldfish* model, (4.91), could be directly inferred from the *solvability* of the *goldfish* many-body model (4.46). And it is plain that, via the trick (4.13), the following *isochronous* version of this model is obtained:

$$\begin{aligned} &\ddot{z}_n - (2\lambda + 1)\, i \omega \dot{z}_n - \lambda(\lambda + 1)\, \omega^2 z_n \\ &= i \omega (\dot{z}_n - i \lambda \omega z_n) + 2 \sum_{m=1, m \neq n}^{N} \frac{(\dot{z}_n - i \lambda \omega z_n)(\dot{z}_m - i \lambda \omega z_m)}{z_n - z_m} \\ &+ a \sum_{m=1}^{N} \frac{(\dot{z}_n - i \lambda \omega z_n)(\dot{z}_m - i \lambda \omega z_m)}{z_m}, \end{aligned} \tag{4.92}$$

where λ is an *arbitrary rational* number.

Example 4.2.2-4 In [48] the starting point of the treatment is the *solvable* second-order matrix ODE

$$\ddot{U} = a \left(\dot{U} U + U \dot{U}\right), \tag{4.93}$$

where a is an arbitrary constant. It is easily seen that the *general* solution of this matrix evolution equation, (4.93), reads

$$U(t) = a^{-1} \left[\cos(A\,t) - B\,A^{-1}\,\sin(A\,t)\right]^{-1} \left[A\,\sin(A\,t) + B\,\cos(A\,t)\right], \quad (4.94a)$$

where A and B are two arbitrary *constant* $N \times N$ matrices. In terms of the initial-value problem for the matrix evolution equation (4.93) clearly (4.94a) entails

$$U(0) = a^{-1}\,B, \quad \dot{U}(0) = a^{-1}\left(A^2 + B^2\right), \quad (4.94b)$$

and these two matrix equations can be inverted to yield

$$A^2 = -a^2\,[U(0)]^2 + a\,\dot{U}(0), \quad B = a\,U(0). \quad (4.94c)$$

Note that the explicit expression (4.94a) entails that the matrix $U(t)$ is actually a function of the matrix A^2 rather than A.

The *solvable* many-body problem is characterized by the *Newtonian* equations of motion

$$\ddot{z}_n = 2\,a\,\dot{z}_n\,z_n + 2 \sum_{m=1,\,m\neq n}^{N} \frac{\dot{z}_n\,\dot{z}_m}{z_n - z_m}. \quad (4.95)$$

Here a is again an *arbitrary* (of course scalar) constant (which might, of course, be rescaled away), and these equations of motion constitute again a generalization of those of the (*nonisochronous*) goldfish model with $\omega = 0$, to which they clearly reduce for $a = 0$.

Before discussing the solvability of this model, (4.95), let us mention its *Hamiltonian* character. Indeed the Hamiltonian

$$H = \sum_{n=1}^{N} \left\{ -\frac{a}{c}\,z_n^2 + \exp(c\,p_n) \prod_{m=1,m\neq n}^{N} (z_n - z_m)^{-1} \right\}, \quad (4.96a)$$

where c is an arbitrary (nonvanishing) constant, yields the *Hamiltonian* equations

$$\dot{z}_n = \frac{\partial H}{\partial p_n} = c\,\exp(c\,p_n) \prod_{m=1,m\neq n}^{N} (z_n - z_m)^{-1}, \quad (4.96b)$$

$$\dot{p}_n = -\frac{\partial H}{\partial z_n} = \frac{1}{c} \left\{ 2\,a\,z_n + \sum_{m=1,m\neq n}^{N} \frac{\dot{z}_n + \dot{z}_m}{z_n - z_m} \right\}. \quad (4.96c)$$

Note that to write more neatly the second set of these Hamiltonian equations, (4.96c), we used the first, (4.96b). It is then obvious that t-differentiation of (the logarithm of) the first set of Hamiltonian equations, (4.96b), yields, via the second set, (4.96c), just the *Newtonian* equations of motion (4.95), demonstrating thereby their *Hamiltonian* character.

The solution of the initial-value problem for this N-body model, (4.95), is given by the following *Proposition 4.2.2-4* (proven in Appendix B).

Proposition 4.2.2-4. The coordinates $z_n(t)$ providing the solution of the initial-value problem for the N-body model (4.95) are the N *eigenvalues* of the $N \times N$ matrix (4.94a) with the two $N \times N$ matrices A^2 and B defined componentwise in terms of the initial data $z_n(0)$, $\dot{z}_n(0)$ as follows:

$$\left(A^2\right)_{nm} = -\delta_{nm}\, a^2\, z_n^2(0) + a\, \left[\dot{z}_n(0)\, \dot{z}_m(0)\right]^{1/2}, \quad B_{nm} = \delta_{nm}\, a\, z_n(0). \tag{4.97}$$

Here δ_{nm} is the standard Kronecker symbol, $\delta_{nm} = 1$ if $n = m$, $\delta_{nm} = 0$ if $n \neq m$. These formulas indicate that the $N \times N$ matrix B is *diagonal*, while the $N \times N$ matrix A^2 is the sum of a *diagonal* matrix and a *dyadic* matrix. ⊡

The *entirely isochronous* variant of the model (4.95) is characterized by the following *Newtonian* equations of motion, obtained from (4.95) via the trick (4.13) with $\lambda = 1$:

$$\ddot{z}_n - 3\, i\, \omega\, \dot{z}_n - 2\, \omega^2\, z_n = 2\, a\, \left(\dot{z}_n - i\, \omega\, z_n\right)\, z_n + 2 \sum_{m=1,\, m \neq n}^{N} \frac{\left(\dot{z}_n - i\, \omega\, z_n\right)\, \left(\dot{z}_m - i\, \omega\, z_m\right)}{z_n - z_m}. \tag{4.98}$$

It is clear from the general discussion of the trick how the solution of the initial-value problem for this ω-modified model can be obtained from the solution given above, see *Proposition 4.2.2-4*. Equivalently, the solutions $z_n(t)$ of this *entirely isochronous* N-body problem is provided by the following variant of *Proposition 4.2.2-4*.

Proposition 4.2.2-5. The solutions $z_n(t)$ of the *entirely isochronous* N-body problem (4.98) are the N *eigenvalues* of the $N \times N$ matrix ODE obtained from (4.93) via the (appropriate version of the) trick (4.13) with $\lambda = 1$,

$$\ddot{U} - 3\, i\, \omega\, \dot{U} - 2\, \omega^2\, U = a\, \left(\dot{U}\, U + U\, \dot{U} - 2\, i\, \omega\, U\right), \tag{4.99}$$

the *general* solution of which reads of course (see (4.94a))

$$U(t) = \frac{\exp\left(i\, \omega\, t\right)}{a}\, \left[\cos\left(A\, \tau\right) - B\, A^{-1}\, \sin(A\, \tau)\right]^{-1}\, \left[A\, \sin\left(A\, \tau\right) + B\, \cos(A\, \tau)\right], \tag{4.100a}$$

$$\tau = \frac{\exp\left(i\, \omega\, t\right) - 1}{i\, \omega}, \tag{4.100b}$$

with A and B two *arbitrary* $N \times N$ matrices. Clearly in the *generic* case this solution is *nonsingular* and *periodic* with period T, see (1.3),

$$U\left(t + T\right) = U\left(t\right). \tag{4.101}$$

Hence, it is plain that *all nonsingular* solutions of this *isochronous* N-body problem, (4.98), are *completely periodic* with a period which is a *finite integer multiple* of the basic period T, while the *singular* solutions constitute a *lower dimensional* set. ⊡

The alternative version of the N-body problem (4.95)—obtained via the change of dependent variables (4.51)—is easily seen (via (A.10) and (A.7a)) to read

$$\ddot{c}_m - 2\,a\,\dot{c}_{m+1} + 2\,a\,\dot{c}_1\,c_m = 0, \tag{4.102a}$$

supplemented with the "boundary conditions"

$$c_0 = 1, \quad c_{N+1} = 0. \tag{4.102b}$$

Note that (4.102a) is trivially satisfied for $m = 0$ (see (4.102b)), and that it can be integrated once for $m = 1$, yielding

$$c_2 = \frac{1}{2}\,c_1^2 + \frac{1}{2\,a}\,(\dot{c}_1 + C)\,, \tag{4.103}$$

where C is an integration constant. Insertion of this expression of c_2 in (4.102a) with $m = 2$ yields

$$\dot{c}_3 = \left(\frac{1}{2\,a}\right)^2 \dddot{c}_1 + \frac{\ddot{c}_1\,c_1 + 2\,\dot{c}_1^2 + C\,\dot{c}_1}{2\,a} + \frac{1}{2}\,\dot{c}_1\,c_1^2, \tag{4.104}$$

the right-hand side of which is however *not* an exact differential. Alternatively one could start from $m = N$—which yields, see (4.102), the Schrödinger-like linear equation

$$\ddot{c}_N + 2\,a\,\dot{c}_1\,c_N = 0, \tag{4.105a}$$

with $c_N(t)$ playing the role of eigenfunction and $\dot{c}_1(t)$ playing the role of "potential"—and work all the way down by solving sequentially (but only formally!) the series of second-order, nonhomogeneous, linear ODEs for $c_m(t)$ with $m = N-1$, $N-2, \ldots$, arriving in the end, for $m = 1$, to a highly nonlinear (integrodifferential) equation for $c_1(t)$.

The fact that the system (4.102) is *solvable* is far from trivial. It is obviously implied by the results described above, since the coefficients $c_m(t)$ of the polynomial (4.51) can be explicitly written in terms of its zeros $z_n(t)$ (see, for instance, Section 2.3.1 of [37]). More directly, see (4.58), the solution of the alternative N-body model (4.102) is given by the following

Proposition 4.2.2-6. The *general* solution of the model (4.102) is given by the formula

$$\det\left[z - U\,(t)\right] = z^{\,N} + \sum_{m=1}^{N} c_m\,(t)\;z^{\,N-m}, \tag{4.106}$$

with the $N \times N$ matrix $U\,(t)$ given by (4.94a). ⊡

The *isochronous* variant of this system, (4.102), can be obtained by first rewriting it in the following guise:

$$\gamma_m'' - 2\,a\,\gamma_{m+1}' + 2\,a\,\gamma_1'\,\gamma_m = 0, \quad \gamma_0 = 1, \quad \gamma_{N+1} = 0 \tag{4.107}$$

with $\gamma_m \equiv \gamma_m\,(\tau)$, and by then setting

$$c_m(t) = \exp(i\,m\,\omega\,t)\,\gamma_m\,(\tau)\,, \quad \tau = \frac{\exp(i\,\omega\,t) - 1}{i\,\omega}. \tag{4.108}$$

It reads

$$\begin{aligned} &\ddot{c}_m - i\,(2\,m+1)\,\omega\,\dot{c}_m - m\,(m+1)\,\omega^2\,c_m \\ &\quad - 2\,a\,[\dot{c}_{m+1} - i\,(m+1)\,\omega\,c_{m+1}] + 2\,a\,[\dot{c}_1 - i\,\omega\,c_1]\,c_m = 0, \\ &\qquad c_0 = 1, \quad c_{N+1} = 0. \end{aligned} \tag{4.109}$$

Clearly its solution is given by the following variant of *Proposition 4.2.2-6.*

Proposition 4.2.2-7. The *general* solution of the N-body model (4.109) is given by the formula (4.106), but with the $N \times N$ matrix $U\,(t)$ defined by (4.100). In the *generic* case it is *completely periodic* with period T, see (4.59b) and (1.3). ⊡

Again, it seems justified to emphasize that the *isochronous* character of the *generic* (hence *nonsingular*) solution of this system is a *nontrivial* finding (up to the observation that *all* true mathematical results are indeed *trivial*!). If need be this is confirmed by the following

Example 4.2.2-5 A model similar to that considered just above has been recently investigated by David Gómez-Ullate, Andy Hone and Matteo Sommacal [99]. The equations of motions of their model, in its "N-body problem" formulation, read

$$\ddot{z}_n = 2\,(1-N)\,z_n^3 + 2\sum_{m=1,m\neq n}^{N} \frac{\dot{z}_n\,\dot{z}_m + z_n^4}{z_n - z_m}, \tag{4.110}$$

and those of the alternative model related to it via (4.51) read

$$\ddot{c}_m + (m+1)\,(m+2)\,c_{m+2} - 2\,c_2\,c_m = 0, \tag{4.111a}$$

with the "boundary conditions"

$$c_0 = 1, \quad c_{N+1} = 0. \tag{4.111b}$$

Via the simple change of dependent variables

$$z_n = b\,\tilde{z}_n, \quad b = \left(\frac{a}{1-N}\right)^{1/2} \tag{4.112}$$

the equations of motion (4.110) can be recast in the form

$$\ddot{\tilde{z}}_n = 2\,a\,\tilde{z}_n^3 + 2 \sum_{m=1,m\neq n}^{N} \frac{\dot{\tilde{z}}_n\,\dot{\tilde{z}}_m + b^2\,\tilde{z}_n^4}{\tilde{z}_n - \tilde{z}_m}, \tag{4.113}$$

demonstrating that they are indeed somewhat similar to (4.95). Likewise, via the simple change of dependent variables

$$c_{2m} = \frac{(-a)^m\,\tilde{c}_m}{(2\,m)!} \tag{4.114}$$

the evolution equations (4.111a) (with *even* index $2m$; note that they are decoupled from those for *odd* index) become

$$\ddot{\tilde{c}}_m - 2\,a\,\tilde{c}_{m+1} + 2\,a\,\tilde{c}_1\,\tilde{c}_m = 0, \tag{4.115}$$

which are remarkably similar to the evolution equations (4.102a). However, in contrast to the systems (4.95) and (4.102), which are *solvable*, as explained above, for *all* values of the positive integer N, the systems (4.110) and (4.111) have been shown to be *solvable* only for $N < 4$; moreover, for $N = 3$ their solution generally involves transcendental (more precisely: elliptic) functions of the time variable, as well as algebraic functions of such elliptic functions (typically roots of N-degree polynomials the coefficients of which evolve in time as elliptic functions), in contrast to the solution of the models (4.95) and (4.102), which clearly only involve, for all values of N, algebraic functions of elementary (more precisely: exponential, or equivalently trigonometric) functions of the time variable (again, typically, via roots of N-degree polynomials the coefficients of which evolve exponentially, or equivalently trigonometrically, in the time variable).

An *isochronous* version of the model (4.113) can be clearly obtained via the trick (4.13) with $\lambda = 1$, and likewise an *isochronous* version of the model (4.115) can be manufactured by applying a transformation analogous to that, see (4.108), which has yielded the transition from the alternative model (4.102) to the *isochronous* model (4.109). But presumably the *generic* solutions of these models need *not* be *isochronous* if $N > 3$.

Example 4.2.2-6 In [76] the starting point of the treatment is the *solvable first-order matrix* ODE

$$\dot{U} = a\,U^2 + C, \tag{4.116}$$

where a is again an *arbitrary* scalar *constant* and C is an *arbitrary constant matrix*. Note that time-differentiation of this *first-order matrix* ODE reproduces the *second-order matrix* ODE (4.93); indeed the novelty of the results reported below, relative to those described above as **Example 4.2.2-4,** is *not* due to the selection of a different *solvable matrix* ODE as starting point of the treatment, but rather to the identification of a *different ansatz* to get rid of the auxiliary variables, yielding the *solvable* N-body problem characterized by the *Newtonian* equations of motion

$$\ddot{z}_n = 2\,a\,\dot{z}_n\,z_n + 2 \sum_{m=1,\,m\neq n}^{N} \frac{\left(\dot{z}_n - a\,z_n^2\right)\,\left(\dot{z}_m - a\,z_m^2\right)}{z_n - z_m}. \tag{4.117}$$

These *Newtonian* equations of motion differ from (4.95), but they provide again a generalization of those of the (*nonisochronous*) *goldfish* model with $\omega = 0$, to which they clearly reduce for $a = 0$.

The *solvable* character of these *Newtonian* equations of motion is demonstrated by the following *Proposition 4.2.2-8* (proven in Appendix B).

Proposition 4.2.2-8. The coordinates $z_n(t)$ of the N particles moving according to the *Newtonian* equations of motion (4.117) are the N eigenvalues of the $N \times N$ matrix

$$U(t) = [1 - a\,U(0)\,t]^{-1} \left\{ U(0) + [\chi(t)]^{-1} \left\{ [q(t)\,U(0) + r(t)]\;P \right.\right.$$
$$\left.\left. + \left[1 - \cos(\Omega\,t) + \frac{\sin(\Omega\,t) - \Omega\,t}{\Omega}\,a\,U(0)\right]\;P\;[1 - a\,U(0)\,t]^{-1}\;U(0) \right\} \right\} \tag{4.118a}$$

where

$$\chi(t) = \sum_{n=1}^{N} \left[\left(\frac{\Omega_n^2}{\Omega^2}\right) \frac{\Omega\,\cos(\Omega\,t) - \sin(\Omega\,t)\;a\,z_n(0)}{\Omega\,[1 - a\,z_n(0)\,t]} \right], \tag{4.118b}$$

$$q(t) = \sum_{n=1}^{N} \left[\left(\frac{\Omega_n^2}{\Omega^2}\right) \frac{[1 - \cos(\Omega\,t)]\;[1 + a\,z_n(0)\,t] - \Omega\,t\,\sin(\Omega\,t)}{1 - a\,z_n(0)\,t} \right], \tag{4.118c}$$

$$r(t) = \left(\frac{\Omega}{a}\right) \sum_{n=1}^{N} \left[\left(\frac{\Omega_n^2}{\Omega^2}\right) \frac{\Omega\,\sin(\Omega\,t) - 2\,[1 - \cos(\Omega\,t)]\;a\,z_n(0)}{\Omega\,[1 - a\,z_n(0)\,t]} \right], \tag{4.118d}$$

the diagonal $N \times N$ matrix $U(0)$ is given in terms of the initial values of the coordinates z_n by the simple formula

$$U(0) = \operatorname{diag}\left[z_n(0)\right], \tag{4.118e}$$

and the *dyadic* $N \times N$ matrix P is defined as follows:

$$P_{nm} = \frac{\Omega_n\,\Omega_m}{\Omega^2}. \tag{4.118f}$$

In these formulas the scalars Ω_n and Ω^2 are defined in terms of the initial data as follows:

$$\Omega_n = \left\{a\,\left[\dot{z}_n(0) - a\,z_n^2(0)\right]\right\}^{1/2}, \tag{4.118g}$$

$$\Omega^2 = a \sum_{n=1}^{N} \left(\dot{z}_n - a\,z_n^2\right) = \sum_{n=1}^{N} \Omega_n^2, \tag{4.118h}$$

so that the *dyadic* $N \times N$ matrix P is actually a *projector*, $P^2 = P$. ⊡

Several comments are now appropriate. (i) The formula (4.118h) defines Ω^2 rather than Ω, consistently with the fact that the right-hand side of (4.118a) is indeed a function of Ω^2 rather than Ω. (ii) In the spirit of the initial-value problem the values of $\dot{z}_n$ and z_n^2 to be inserted in the right-hand side of the expression (4.118h) of Ω^2 are evaluated at the initial time $t = 0$, as implied, via (4.118g), by the second version of this formula (4.118h); but let us emphasize that this quantity, Ω^2 (see the first version of (4.118h)), is a constant of motion for the system under consideration, as it is apparent by summing its equations of motion, (4.117), over the index n from 1 to N and by then noticing that the double sum in the right-hand side vanishes due to the antisymmetry of the summand under exchange of the two dummy indices n and m, while the resulting equation clearly entails that the time-derivative of Ω^2 vanishes. (iii) In writing the formula (4.118a) we implied that the constant Ω does not vanish; but the formula remains valid, via an obvious limiting procedure, even if Ω does vanish. (iv) The time dependence exhibited by this solution (4.118) is remarkable, inasmuch as it features a *rational* dependence *both* on t and on circular functions of Ωt (with Ω generally *complex,* see (4.118h)); the reader interested in a terse discussion of the corresponding behavior of the particle coordinates $z_n(t)$ is referred to the original literature [76].

An *entirely isochronous* variant of the *Newtonian* equations of motion (4.117) is obtained via the trick (2.3) with $\lambda = 1$:

$$\ddot{z}_n - 3\,i\,\omega\,\dot{z}_n - 2\,\omega^2\,z_n = 2\,a\,\left(\dot{z}_n - i\,\omega\,z_n\right)\,z_n$$
$$+2\sum_{m=1,\,m\neq n}^{N}\frac{\left(\dot{z}_n - i\,\omega\,z_n - a\,z_n^2\right)\left(\dot{z}_m - i\,\omega\,z_m - a\,z_m^2\right)}{z_n - z_m}. \quad (4.119)$$

Again, it is clear from the general discussion of the trick—and the treatment given above in the context of **Example 4.2.2-4**—how the solution of the initial-value problem for this ω-modified model can be obtained from the solution given above, see *Proposition 4.2.2-8.* And it is as well plain from the *meromorphic* character of the matrix solution (4.118) that *all nonsingular* solutions of this model (4.119) are *completely periodic* with a period which is a *finite integer multiple* of the basic period T, while the *singular* solutions constitute a *lower dimensional* set.

The alternative version of the N-body problem (4.117)—obtained via the change of dependent variables (4.51)—is easily seen (via (A.10), (A.7a), (A.9e) and (A.8n)) to read (after setting, without significant loss of generality, $a = 1$)

$$\ddot{c}_m - 2\,m\,\dot{c}_{m+1} + (m+2)\,(m-1)\,c_{m+2}$$
$$= -2\left[c_1\,\left(\dot{c}_m - m\,c_{m+1}\right) + c_2\,c_m\right], \quad (4.120a)$$

supplemented by the boundary conditions

$$c_{N+1} = c_{N+2} = 0. \quad (4.120b)$$

In (4.120a) $m = 1, \ldots, N$, but it is clear that this equation is also satisfied for $m = 0$ with $c_0 = 1$.

The *solvability* of this nonlinear system of ODEs, implied as it clearly is by the *solvability* (see *Proposition 4.2.2-8*) of the system (4.117), is a *nontrivial* result (again, up to the observation that *all* correct mathematical findings are in some sense *trivial*—after their validity has been proven!). Note that the first ODE of this system, namely that corresponding to $m = 1$,

$$\ddot{c}_1 - 2\,\dot{c}_2 = -2\,c_1\,\dot{c}_1, \tag{4.121a}$$

can be integrated once, yielding

$$c_2 = \frac{1}{2}\left(\dot{c}_1 + c_1^2 + C^{\,2}\right), \tag{4.121b}$$

with $C^{\,2}$ an arbitrary constant. But, even for $N = 2$, the next equation (for $m = 2$) of this system,

$$\ddot{c}_2 = -2\,c_1\,\dot{c}_2 - 2\,c_2^2, \tag{4.122a}$$

which via (4.121b) becomes the following *third-order* ODE for the single dependent variable $c_1(t) \equiv \theta(t)$,

$$\dddot{\theta} + 4\ddot{\theta}\,\theta + 3\dot{\theta}^{\,2} + 6\,\theta^{\,2}\,\dot{\theta} + \theta^{\,4} + 2\,C^{\,2}\left(\dot{\theta} + \theta^{\,2}\right) + C^{\,4} = 0, \tag{4.122b}$$

does not look trivially integrable (even just once, let alone three times!). Indeed, the fact that this ODE admits the explicit solution

$$\begin{aligned}
\theta(t) = &\left(\left\{\theta(0) + \frac{1}{2}\left[\theta^{\,2}(0) + \dot{\theta}(0) + C^{\,2}\right]t\right\}\cos(C\,t) + \frac{1}{2}\left\{\theta^{\,2}(0) + \dot{\theta}(0) - C^{\,2}\right.\right.\\
&\left.\left.+\left[\ddot{\theta}(0) + 3\,\dot{\theta}(0)\,\theta(0) + \theta^{\,3}(0) + C^{\,2}\,\theta(0)\right]t\right\}\frac{\sin(C\,t)}{C}\right)\cdot\\
&\cdot\left(\left\{1 - \frac{1}{2\,C^2}\left[\ddot{\theta}(0) + 3\,\dot{\theta}(0)\,\theta(0) + \theta^{\,3}(0) + C^{\,2}\,\theta(0)\right]t\right\}\cos(C\,t)\right.\\
&+\frac{1}{2\,C^2}\left\{\ddot{\theta}(0) + 3\,\dot{\theta}(0)\,\theta(0) + \theta^{\,3}(0) + 3\,C^{\,2}\,\theta(0)\right.\\
&\left.\left.+\left[\theta^{\,2}(0) + \dot{\theta}(0) + C^{\,2}\right]C^{\,2}\,t\right\}\frac{\sin(C\,t)}{C}\right)^{-1}
\end{aligned} \tag{4.122c}$$

is remarkable, as the diligent reader who tries and verifies this fact will certainly note (at least if he or she tries to do so without any computer assistance).

The *entirely isochronous* variant of the system (4.120) can be obtained by rewriting formally the system (4.120) with $c_m\,(t)$ replaced by $\gamma_m\,(\tau)$ and by then applying the following version of the trick:

$$c_m(t) = \exp(i\, m\, \omega\, t)\; \gamma_m(\tau), \quad \tau = \frac{\exp(i\, \omega\, t) - 1}{i\, \omega}. \tag{4.123}$$

This yields the following *autonomous* system of ODEs

$$\ddot{c}_m - i\,(2\,m + 1)\,\omega\,\dot{c}_m - 2\,m\,\dot{c}_{m+1} - \omega^2\, m\,(m+1)\,c_m + 2\,i\,m\,(m+1)\,\omega\,c_{m+1} \\ + (m+2)\,(m-1)\,c_{m+2} = -2\,\left[c_1\,(\dot{c}_m - i\,m\,\omega\,c_m - m\,c_{m+1}) + c_2\,c_m\right], \tag{4.124a}$$

with the boundary conditions

$$c_{N+1} = c_{N+2} = 0, \tag{4.124b}$$

which may then justifiably be called a system of *nonlinear harmonic oscillators*, since—as implied by the *meromorphic* character of the solution (4.118) (see the treatment given above in the context of **Example 4.2.2-4**)—*all* its *nonsingular* solutions are *completely periodic* with the basic period T,

$$c_m(t + T) = c_m(t), \tag{4.124c}$$

while the *singular* solutions are a *lower dimensional* set. And let us end the discussion of this example by pointing again out that, while the system of ODEs (4.124a) with (4.124b) is meant to hold for $m = 1, \ldots, N$, the ODE (4.124a) with $m = 0$ is also (trivially) satisfied with $c_0 = 1$.

In the following example we limit our presentation to exhibiting the *solvable matrix* ODE (or the *solvable* system of two coupled *matrix* ODEs) that provides the point of departure of the treatment, and the *solvable* N-body problems obtained from it—including their *entirely isochronous* versions—referring to the original papers for a more detailed coverage—including analogous results to those displayed in the preceding *Propositions* presented in this Section 4.2.2.

Example 4.2.2-7 In [20] the starting point of the treatment is the *solvable* system of two coupled *first-order matrix* ODEs

$$\dot{U} = a\, U^2 + V, \quad \dot{V} = b\, V, \tag{4.125}$$

where a is again an *arbitrary* scalar *constant* and the scalar b might be an *arbitrary* function of time. Note that for $b = 0$ this matrix ODE reduces to (4.116) (with $V = C$). The main contribution of [20] is to identify (and investigate) two *solvable* many-body problems. The first is characterized by the following (*autonomous*) *Newtonian* equations of motion:

$$\ddot{z}_n = 2\, a\, \dot{z}_n\, z_n + b\, \left[\dot{z}_n - a\, z_n^2\right] \\ +2 \sum_{m=1, m \neq n}^{N} \frac{\left[\dot{z}_n - a\, z_n^2\right] \left[\dot{z}_m - a\, z_m^2\right]}{z_n - z_m}. \tag{4.126}$$

Here b is an *arbitrary* constant; clearly for $b = 0$ this model reduces to (4.117). The second is characterized by the following (*autonomous*) *Newtonian* equations of motion (obtained via the trick from an appropriate time-dependent choice of the function $b(t)$ in (4.125)):

$$\ddot{z}_n = (3+k)\, i\,\omega\, \dot{z}_n + (2+k)\,\omega^2\, z_n + a\,\left[2\,\dot{z}_n - (2+k)\, i\,\omega\, z_n\right]\, z_n$$
$$+2 \sum_{m=1, m\neq n}^{N} \frac{\left[\dot{z}_n - i\,\omega\, z_n - a\, z_n^2\right]\, \left[\dot{z}_m - i\,\omega\, z_m - a\, z_m^2\right]}{z_n - z_m}. \qquad (4.127)$$

Here k is an *arbitrary* constant, but the case with k a *positive* or *negative integer* is particularly interesting, since this model is then *entirely isochronous* (with the single exception of the case $k = -2$, when the solutions are instead *multiply periodic*). And note that this model reduces to (4.119) for $k = 0$.

Alternative formulations of these *solvable* models, (4.126) respectively (4.127), are provided by the following two *solvable* systems of ODEs:

$$\ddot{c}_m - 2\,a\,m\,\dot{c}_{m+1} + (2\,a\,c_1 - b)\,\dot{c}_m + a^2\,(m+2)\,(m-1)\,c_{m+2}$$
$$+\,a\,\left[b - m\,(2\,a\,c_1 - b)\right]\,c_{m+1} + a\,\left[2\,a\,c_2 - b\,c_1\right]\,c_m = 0, \qquad (4.128)$$

respectively

$$\ddot{c}_m - 2\,m\,a\,\dot{c}_{m+1} + \left[2\,a\,c_1 - (2\,m+k+1)\,i\,\omega\right]\,\dot{c}_m$$
$$+\,(m+2)\,(m-1)\,a^2\,c_{m+2} + \left[-2\,m\,a^2\,c_1 + (m+1)\,(2\,m+k)\,i\,\omega\,a\right]\,c_{m+1}$$
$$+\,\left[2\,a^2\,c_2 - (2\,m+k)\,i\,\omega\,a\,c_1 - m\,(m+k+1)\,\omega^2\right]\,c_m = 0. \qquad (4.129)$$

The index m ranges as usual from 1 to N, while the quantities c_{N+1} and c_{N+2} by definition vanish,

$$c_{N+1} = c_{N+2} = 0. \qquad (4.130)$$

It can be moreover noted that, for each of these systems, the equation with $m = 0$ is identically satisfied, provided one sets $c_0 = 1$. These systems look superficially *linear*; but they are in fact *nonlinear*, due to the presence of the dependent variables $c_1(t)$ and $c_2(t)$ in the equations of motion for the dependent variable $c_m(t)$.

The second of these models, (4.129), is *entirely isochronous*, in fact its *generic* solutions are *completely periodic* with period T, see (1.3),

$$c_m(t+T) = c_m(t), \qquad (4.131)$$

provided k is an integer different from -2.

In the following two examples we do report the solutions, but the reader not specifically interested in them is advised to skip their detailed display.

Example 4.2.2-8 In [21] the starting point is the following *solvable* system of two coupled first-order matrix ODEs:

$$\dot{U} = \alpha + \beta\, V\, U, \qquad (4.132a)$$

$$\dot{V} = a\, V^2 + b\, V + c. \qquad (4.132b)$$

Here α, β, a, b, c are five *scalar* quantities, and it is clear that this system is *solvable* (at least in the sense that its solution is achievable by quadratures) if the three parameters a, b, c are all *constant* or if either a or c vanishes, while α, β can be two (*arbitrarily* assigned) functions of time: indeed the solution can then be obtained by firstly solving the second, (4.132b), of these two matrix ODEs, thereby explicitly determining $V(t)$, and by then solving the first of these two matrix ODEs, (4.132a), which is linear in $U(t)$ and hence *solvable* (at least in the sense of being reducible to quadratures) once $V(t)$ is known. But, in fact, to obtain the results reported below this system of matrix ODEs must be solved only in some special cases: for instance only the case $\beta = 1$ need to be considered, hence hereafter we set $\beta = 1$, and there are additional limitations, see below.

If α is time-independent (as we hereafter assume), the system (4.132) can clearly be reformulated as a *single second-order* (and of course as well *solvable*) ODE for the matrix $U(t)$:

$$\ddot{U} = c\,U + \left(\dot{U} - \alpha\right) \left\{b - U^{-1} \left[a\,\alpha - (a+1)\,\dot{U}\right]\right\} \tag{4.133}$$

(to obtain this matrix ODE note firstly that (4.132a) with $\beta = 1$ entails

$$V = \left(\dot{U} - \alpha\right) U^{-1}, \tag{4.134}$$

then time differentiate (4.132a), use (4.132b) to eliminate $\dot{V}$, and finally use (4.134) to eliminate V; note that to make these steps we assumed α to be time-independent, but we made no assumption on the time-dependence of a, b, c).

To arrive at the generalized *goldfish* N-body models reported below one limits moreover consideration to the $c = 0$ case (this is necessary in order that an *ansatz* analogous to (4.83) work [21]). The *solvable* N-body problem thereby identified is characterized by the system of *Newtonian* equations of motion

$$\ddot{z}_n = (\dot{z}_n - \alpha)\left[b + \frac{\alpha}{z_n} + (1+a)\sum_{m=1}^{N}\frac{\dot{z}_m - \alpha}{z_m} + 2\sum_{m=1,\,m\neq n}^{N}\frac{\dot{z}_m - \alpha}{z_n - z_m}\right], \tag{4.135}$$

where b might also be time-dependent (see below). In the *autonomous* case with constant $b = i\,\omega$, when this equation reads

$$\ddot{z}_n = (\dot{z}_n - \alpha)\left[i\,\omega + \frac{\alpha}{z_n} + (1+a)\sum_{m=1}^{N}\frac{\dot{z}_m - \alpha}{z_m} + 2\sum_{m=1,\,m\neq n}^{N}\frac{\dot{z}_m - \alpha}{z_n - z_m}\right], \tag{4.136}$$

it can be clearly considered a generalized *goldfish* model: indeed, for $\alpha = 0$ and $a = -1$ this system reduces to the standard *goldfish* model (4.46)—and this reduction can also be performed (albeit less trivially so [21]) if $\alpha = 0$ but $a \neq -1$.

But when the constant α does *not* vanish (as we hereafter assume, $\alpha \neq 0$), the N-body problem of *goldfish* type characterized by the *Newtonian* equations of motion (4.136) cannot be reduced to any previously known system: it is a *solvable* model the novelty of which is demonstrated by the fact that its solution involves special functions not featured in the solution of any previously known *solvable* model [21].

Of more interest in the context of this monograph is the possibility to manufacture an, also new, *entirely isochronous* many-body problem of *goldfish* type, by starting from the following *nonautonomous* version of the system of ODEs (4.135),

$$\zeta_n'' = (\zeta_n' - \alpha)\left[b(\tau) + \frac{\alpha}{\zeta_n} + (1+a)\sum_{m=1}^{N}\frac{\zeta_m' - \alpha}{\zeta_m} + 2\sum_{m=1,\, m\neq n}^{N}\frac{\zeta_m' - \alpha}{\zeta_n - \zeta_m}\right]. \tag{4.137a}$$

Note that this system coincides with (4.135), except for a merely notational change (the dependent variables are here denoted as $\zeta_n \equiv \zeta_n(\tau)$, appended primes indicating differentiations with respect to the independent variable τ).

It is now easy to verify that, provided

$$b(\tau) = \frac{k\,i\,\omega}{1 + i\,\omega\,\tau}, \tag{4.137b}$$

via the change of dependent and independent variables (as it were, just the version of the trick (4.13) with $\lambda = -1$)

$$z_n(t) = \exp(-i\,\omega\,t)\;\zeta_n(\tau), \tag{4.138a}$$

$$\tau = \frac{\exp(i\,\omega\,t) - 1}{i\,\omega}, \qquad \exp(i\,\omega\,t) = 1 + i\,\omega\,\tau, \tag{4.138b}$$

the (*nonautonomous*) system of ODEs (4.137) yields the *solvable* many-body problem characterized by the following *autonomous Newtonian* equations of motion:

$$\ddot{z}_n + i\,\omega\,\dot{z}_n = (\dot{z}_n + i\,\omega\,z_n - \alpha)\cdot$$
$$\cdot\left[k\,i\,\omega + \frac{\alpha}{z_n} + (1+a)\sum_{m=1}^{N}\frac{\dot{z}_m + i\,\omega\,z_m - \alpha}{z_m} + 2\sum_{m=1,\, m\neq n}^{N}\frac{\dot{z}_m + i\,\omega\,z_m - \alpha}{z_n - z_m}\right]. \tag{4.139}$$

The *solvable,* and *entirely isochronous*, character of this model is evidenced by the following [21]

Proposition 4.2.2-9. The particle coordinates $z_n(t)$ providing the solution to the initial-value problem for the N-body problem characterized by the *Newtonian* equations of motion (4.139) are the N eigenvalues of the $N \times N$ matrix $U(t)$ defined in terms of the initial data $z_n(0)$, $\dot{z}_n(0)$ as follows:

$$U(t) = \exp(-i\omega t)\{1 + [\phi(t) - 1]\, P\}\, U(0) + (1 - P)\alpha \frac{1 - \exp(-i\omega t)}{i\omega} + \alpha P u(t)\phi(t), \tag{4.140a}$$

where

$$U(0) = \mathrm{diag}\,[z_n(0)]\,, \tag{4.140b}$$

$$P_{nm} = -\frac{\{[\dot{z}_n(0) + i\,\omega\, z_n(0) - \alpha]\ [\dot{z}_m(0) + i\,\omega\, z_m(0) - \alpha]\}^{1/2}}{\eta\, i\,\omega\, z_m(0)}, \qquad P^2 = P, \tag{4.140c}$$

$$\eta = -\sum_{n=1}^{N} \frac{\dot{z}_n(0) + i\,\omega\, z_n(0) - \alpha}{i\,\omega\, z_n(0)}, \tag{4.140d}$$

$$\phi(t) = \left\{1 + a\,\eta\, \left[\exp\left(\tilde{k}\, i\,\omega\, t\right) - 1\right]\right\}^{-1/a}, \tag{4.140e}$$

and the function $u(t)$ is defined as follows: if

$$|1 - a\,\eta| \le |a\,\eta|\,, \quad \text{i.e.} \quad \mathrm{Re}\,(a\,\eta) \ge \frac{1}{2}, \tag{4.140f}$$

then

$$u(t) = (a\,\eta)^{1/a} \sum_{s=0}^{\infty} \binom{1/a}{s} \left(\frac{1 - a\,\eta}{a\,\eta}\right)^s \frac{\exp(-i\,\omega\, t) - \exp\left\{\left[-\tilde{k}\, \left(s - \frac{1}{a}\right)\right]\, i\,\omega\, t\right\}}{\left[\tilde{k}\, \left(s - \frac{1}{a}\right) - 1\right]\, i\,\omega}; \tag{4.140g}$$

if instead

$$|1 - a\,\eta| \ge |a\,\eta|\,, \quad \text{i.e.} \quad \mathrm{Re}\,(a\,\tilde{\eta}) \le \frac{1}{2}, \tag{4.140h}$$

then

$$u(t) = (1 - a\,\eta)^{1/a} \left[\sum_{s=0}^{\infty} \binom{1/a}{s} \left(\frac{a\,\eta}{1 - a\,\eta}\right)^s \frac{\exp\left(\tilde{k}\, s\, i\,\omega\, t\right) - \exp(-i\,\omega\, t)}{(\tilde{k}\, s + 1)\, i\,\omega}\right]. \tag{4.140i}$$

Note that we use throughout the short-hand notation

$$\tilde{k} = k + 1. \tag{4.140j}$$

These formulas are applicable as written provided $a \ne 0$ (and note that the sums in (4.140g) and (4.140i) terminate, yielding of course the same result, if a is the *inverse* of a *positive* integer); if $a = 0$ the definition (4.140e) must be modified to read

$$\phi(t) = \exp\left\{\eta\left[1 - \exp\left(\tilde{k}\, i\, \omega\, t\right)\right]\right\}, \tag{4.140k}$$

and the definition (4.140i) of $u(t)$ must be modified to read

$$u(t) = \exp(-\eta) \sum_{s=0}^{\infty} \frac{\eta^s}{s!} \frac{\exp\left[\tilde{k}\, s\, i\, \omega\, t\right] - \exp(-i\, \omega\, t)}{(\tilde{k}\, s + 1)\, i\, \omega}. \tag{4.140l}$$

In the special case with $k = 0$, $\tilde{k} = 1$, the sums in the definition of the function $u(t)$ can be performed in closed form, and one gets

$$u(t) = (a\, \eta)^{1/a} \frac{\exp(i\, \omega\, t\, /\, a) - \exp(-i\, \omega\, t)}{i\, \omega} \quad \text{for } a \neq 0, \tag{4.140m}$$

$$u(t) = \exp(-i\, \omega\, t) \frac{\exp\{\eta\, [\exp(i\, \omega\, t) - 1]\} - 1}{i\, \omega\, \eta} \quad \text{for } a = 0. \tag{4.140n}$$

Finally, let us note that the formulas written above assume that $\eta \neq 0$ (see in particular (4.140c)). For the special case in which η instead vanishes the expression of $U(t)$ is somewhat simpler:

$$U(t) = \exp(-i\, \omega\, t)\, U(0) + \alpha \frac{1 - \exp(-i\, \omega\, t)}{i\, \omega} V(0) \tag{4.141a}$$

with

$$V(0) = \frac{[\dot{z}_n(0) + i\, \omega\, z_n(0) - \alpha]^{1/2}\, [\dot{z}_m(0) + i\, \omega\, z_m(0) - \alpha]^{1/2}}{i\, \omega\, z_m(0)}. \;\square \tag{4.141b}$$

Remark 4.2.2-10. These formulas, as written above, are valid for *arbitrary* values of the two constants a and $\tilde{k}$, provided $\tilde{k}\, \omega$ hence $\tilde{k}$ itself is *real* ($\mathrm{Im}(\tilde{k}) = 0$; this condition is required to guarantee the convergence for all *real* values of t of the infinite series in the right-hand sides of (4.140g), (4.140i) and (4.140l)). But the most interesting case is when $\tilde{k}$ hence k (see (4.140j)) are *rational* numbers,

$$\tilde{k} = \frac{\tilde{p}}{\tilde{q}}, \quad k = \frac{\tilde{p} - \tilde{q}}{\tilde{q}}, \tag{4.142}$$

of course with $\tilde{p}$ and $\tilde{q}$ *integers* (and $\tilde{q} > 0$). Then whenever the initial data entail, via (4.140d), that there holds the inequality (4.140h), and provided moreover $\tilde{k}\, s \neq -1$ for every *positive integer* s hence $\tilde{p} \neq -1$ (see (4.142)), clearly (see (4.140a), (4.140e) and (4.140i)) the matrix $U(t)$ is *periodic* with period $\tilde{T}$, see (1.3),

$$\tilde{T} = \tilde{q}\,T, \quad U(t+\tilde{T}) = U(t). \tag{4.143}$$

If the initial data entail instead that the inequality (4.140f) prevails (rather than (4.140h)) and moreover a is also *rational*, hence $\tilde{k}\,/\,a$ is as well *rational*,

$$\frac{\tilde{k}}{a} = \frac{k+1}{a} = \frac{\check{p}}{\check{q}}, \tag{4.144}$$

of course again with $\check{p}$ and $\check{q}$ *integers* (and $\check{q} > 0$), then provided $\tilde{k}\,\left(s - \frac{1}{a}\right) \neq 1$ hence $\tilde{q}\,(\check{q} + \check{p}) \neq s\,\check{q}\,\tilde{p}$ for any *nonnegative integer* s (see (4.142) and (4.144)), clearly (see (4.140a), (4.140e) and (4.140g)) $U(t)$ is again *periodic*, but now with a period $\check{T}$ (generally) larger than $\tilde{T}$,

$$\check{T} = \max(\tilde{q},\,\check{q})\,T, \quad U(t+\check{T}) = U(t).\ \boxdot \tag{4.145}$$

Remark 4.2.2-11. As entailed by the preceding *Remark 4.2.2-10* and by *Proposition 4.2.2-9*, the N-body problem characterized by the *Newtonian* equations of motion of *goldfish* type (4.139) with k *rational* is *entirely isochronous*: if a is *real* but otherwise *arbitrary*, a condition on the initial data sufficient to guarantee that its solution be *completely periodic* with a period which is an *integer multiple* of T, see (1.3), is validity of the inequality (4.140h) (provided, moreover, $\tilde{p} \neq -1$, see (4.142)); if a is moreover *rational*, then *all* its solutions are *completely periodic* with a period which is an *integer multiple* of T (provided moreover $\tilde{k}\,\left(s - \frac{1}{a}\right) \neq 1$, hence $\tilde{q}\,(\check{q}+\check{p}) \neq s\,\check{q}\,\tilde{p}$ for any *nonnegative integers*, see (4.142) and (4.144)). $\boxdot$

To describe the alternative version of this N-body problem we prefer to get rid of the unessential constants ω and α. This can be done by appropriate rescaling, or equivalently by setting

$$\omega = 1 \quad \text{entailing} \quad T = 2\,\pi, \quad \text{and} \quad \alpha = i. \tag{4.146}$$

Let us emphasize that hereafter (in the context of this **Example 4.2.2-8**) whenever we refer to previous formulas we understand that such assignments have been made in them.

An alternative formulation of the system (4.139) is then characterized by the following system of N *Newtonian* equations of motion satisfied by the N dependent variables $c_m(t)$:

$$\begin{aligned} &\ddot{c}_m - (K - 2\,m)\,i\,\dot{c}_m + (2N + 1 - 2\,m)\,i\,\dot{c}_{m-1} + m\,(K - m)\,c_m \\ &\quad + [(N + 1 - m)\,(K - 2\,m) - \gamma]\,c_{m-1} \\ &\quad - (N - m)\,(N + 2 - m)\,c_{m-2} = 0, \\ &\quad\; m = 1, \ldots, N, \qquad c_0 = 1, \qquad c_{-1} = 0, \end{aligned} \tag{4.147a}$$

where

$$K \equiv K(t) = \tilde{k} + (a+1)\,[N + \gamma\,(t)] \tag{4.147b}$$

and

$$\gamma \equiv \gamma\,(t) = \frac{c_{N-1}\,(t) - i\,\dot{c}_N\,(t)}{c_N\,(t)}. \tag{4.147c}$$

This system of N coupled ODEs is *just as solvable* as (4.139): indeed, as implied by developments analogous to those described above in the context of **Example 4.2.2-4**, its solution is provided by the following

Proposition 4.2.2-12. The dependent variables $c_m(t)$ that solve the system (4.147) are the N coefficients $c_m(t)$ of the (monic) polynomial $\psi(z,t)$ of degree N in z having the N coordinates $z_n(t)$ solutions of (4.139) as its zeros,

$$\psi(z,t) = \prod_{n=1}^{N} [z - z_n(t)] = z^N + \sum_{m=1}^{N} c_m(t)\, z^{N-m}, \tag{4.148}$$

so that, as implied by *Proposition 4.2.2-9*, this polynomial is given by the formula

$$\psi(z,t) = \det\,[z - U(t)] \tag{4.149}$$

where $U(t)$ is the $N \times N$ matrix the time-dependence of which is given by (4.140) (with (4.146)). ⊡

Remark 4.2.2-13. Provided $\tilde{k}$ is a *rational number,* $\tilde{k} = \tilde{p}\,/\,\tilde{q}$ (with $\tilde{p}$ and $\tilde{q}$ *integers* and $\tilde{q} > 0$, see (4.142), and moreover $\tilde{p} \neq -1$), *Proposition 4.2.2-12* with *Remark 4.2.2-10* entail that the solution of the nonlinear system of N *Newtonian* equations of motion (4.147) is *completely periodic* with a period which is an *integer multiple* q of $2\,\pi$,

$$c_m(t + 2\,\pi\,q) = c_m(t), \tag{4.150}$$

in the following cases. (i) If a is an arbitrary (possibly even *complex*) number and the initial data entail the inequality

$$\mathrm{Re}\,[-a\,\{N + \gamma(0)\}] < \frac{1}{2} \tag{4.151}$$

(see (4.147c)), then (4.150) holds with $q = \tilde{q}$. (ii) If a is an *arbitrary real* and *rational* number so that the number $\tilde{k}\,/\,a$ is itself *rational,* $\tilde{k}\,/\,a = \check{p}\,/\,\check{q}$ (with $\check{p}$ and $\check{q}$ *integers* and $\check{q} > 0$, see (4.144) and moreover $\tilde{q}\,(\check{q} + \check{p}) \neq s\,\check{q}\,\tilde{p}$ for any *nonnegative integer* s), then *all* solutions of the nonlinear system of ODEs (4.147) are *completely periodic*, see (4.150) with $q = \max(\tilde{q},\,\check{q})$. ⊡

Example 4.2.2-9 In [79] the starting point is the following *solvable* $N \times N$ matrix ODE:

$$\ddot{U} = 2\,U\,\left(U^{\,2} - a^2\right), \tag{4.152}$$

where a^2 is an *arbitrary* constant (we use a^2 rather than a merely for notational convenience, see below). The *solvable* character—in terms of ratios of appropriate sigma functions—of this matrix ODE was demonstrated by V. I. Inozemtsev [106]; particularly relevant for our purposes is the implication that *all* solutions of this matrix ODE are *meromorphic* functions of the independent variable, and moreover that the *generic* solution is *nonsingular* (see also below *Lemma 4.5-1*).

The *solvable* N-body problem of *goldfish* type associated with this matrix ODE (via an analogous procedure to that outlined above in the context of **Example 4.2.2-4**) is characterized by the following equations of motion of Newtonian type:

$$\ddot{z}_n = 2\, z_n \left(z_n^2 - a^2\right) + 2 \sum_{m=1, m\neq n}^{N} \frac{\left(\dot{z}_n + z_n^2 - a^2\right)\left(\dot{z}_m + z_m^2 - a^2\right)}{z_n - z_m}. \tag{4.153}$$

Remark 4.2.2-14. Trivially related models involving additional arbitrary constants could of course be obtained by rescaling the (dependent and independent) variables and by shifting by a constant amount the dependent variables; but note that while the first factor 2 in the right-hand side of (4.153) could be changed by rescaling (we put it there for notational convenience, see below), the second factor 2 (that multiplying the sum) cannot be changed—being indeed characteristic of *solvable goldfish* models. ⊡

Remark 4.2.2-15. Although for *real* a^2 and for *real* initial data $z_n(0)$, $\dot{z}_n(0)$ the time evolution (for *real* time) of this N-body model entails that the dependent variables $z_n(t)$ are as well *real*, we generally assume the time evolution to take place in the *complex* z-plane, and generally allow the constant a to be as well *complex*; indeed, such an evolution is much more interesting due to the possibility that the "particles" characterized by the *complex* coordinates $z_n(t)$ go round each other and the related fact that initial data $z_n(0)$, $\dot{z}_n(0)$ leading to particle collisions are then *exceptional* (they generally have vanishing dimensionality relative to *generic* initial data). [If attention is instead restricted to *real* motions, then the trivial change of dependent variables $z_n \to i\, y_n$ with y_n *real* might be expedient in order to deal with *confined* motions]. Note that it is possible to reformulate these *complex* equations of motions as *real* (and even *covariant*, even *rotation-invariant*) equations of motion describing the motion of *real* point particles in the *real* (say, *horizontal*) plane, but we postpone all such considerations to Section 4.3. ⊡

The *solvable* character of these equations of motion is evidenced by the following

Proposition 4.2.2-16. The solution of the initial-value problem for the equations of motion (4.153) is provided by the following prescription: the coordinates $z_n\,(t)$ are the N eigenvalues of the solution $U\,(t)$ of (4.152) determined by the following initial data:

$$U_{nm}(0) = \delta_{nm}\, z_n(0), \tag{4.154a}$$

$$\dot{U}_{nm}(0) = -\delta_{nm}\,\left[z_n^2(0) - a^2\right] + \left[\dot{z}_n(0) + z_n^2(0) - a^2\right]^{1/2}\left[\dot{z}_m(0) + z_m^2(0) - a^2\right]^{1/2}. \tag{4.154b}$$

Note that the $N \times N$ matrix $U(0)$ is *diagonal*, while the $N \times N$ matrix $\dot{U}(0)$ is the sum of a *diagonal* matrix and a *dyadic* matrix. ⊡

Notation: $\delta_{nm} \equiv \delta_{n,m}$ is the Kronecker delta symbol, $\delta_{nm} = 1$ if $n = m$, $\delta_{nm} = 0$ if $n \neq m$.

To obtain the *entirely isochronous* variant of this N-body problem one starts from the equations of motion

$$\zeta_n'' = 2\,\zeta_n^3 + 2\sum_{m=1, m\neq n}^{N} \frac{\left(\zeta_n' + \zeta_n^2\right)\left(\zeta_m' + \zeta_m^2\right)}{\zeta_n - \zeta_m}, \tag{4.155}$$

which correspond to (4.153) with $a = 0$ and with the merely notational replacement of the dependent variables $z_n(t)$ with the dependent variables $\zeta_n(\tau)$ (and, of course, now the appended primes denote differentiations with respect to τ). One then applies the trick (4.13) with $\lambda = 1$. This yields the *autonomous* equations of motion

$$\ddot{z}_n - 3i\omega\dot{z}_n - 2\omega^2 z_n = 2z_n^3 + 2\sum_{m=1, m\neq n}^{N} \frac{\left(\dot{z}_n - i\omega z_n + z_n^2\right)\left(\dot{z}_m - i\omega z_m + z_m^2\right)}{z_n - z_m}. \tag{4.156}$$

The solution of the initial-value problem is then obviously given by the solution (via *Proposition 4.2.2-16*) of the problem (4.155) and by the trick relations (4.13), that clearly also imply

$$\zeta_n(0) = z_n(0), \quad \zeta'(0) = \dot{z}_n(0) - i\,\omega z_n(0). \tag{4.157}$$

Equivalently, the solution of this model (4.156) is clearly given by the following

Proposition 4.2.2-17. The dependent variables $z_n(t)$ that solve the initial-value problem for the *Newtonian* N-body problem (4.156) are the N eigenvalues of the $N \times N$ matrix $U(t)$ evolving according to the *solvable* matrix evolution equation

$$\ddot{U} - 3\,i\,\omega\,\dot{U} - 2\,\omega^2\,U = 2\,U^3 \tag{4.158}$$

and being moreover characterized by the following initial data:

$$U_{nm}(0) = \delta_{nm}\, z_n(0), \tag{4.159a}$$

$$\dot{U}_{nm}(0) = -\delta_{nm}\,\left[z_n^2(0)\right] + \left[\dot{z}_n(0) - i\,\omega z_n(0) + z_n^2(0)\right]^{1/2} \times \left[\dot{z}_m(0) - i\,\omega z_m(0) + z_m^2(0)\right]^{1/2}. \tag{4.159b}$$

Again, note that the matrix $U(0)$ is *diagonal*, while the matrix $\dot{U}(0)$ is the sum of a *diagonal* matrix and a *dyadic* matrix. Of course the *solvable* character of the matrix evolution equation (4.158) is implied by the relation of this matrix ODE to the *solvable* matrix ODE (4.152) (with $a = 0$) via the matrix version of the trick formula (4.13) with $\lambda = 1$. ⊡

Remark 4.2.2-18. Since all the solutions $U(t)$ of the matrix evolution equation (4.152) are *meromorphic* functions of the independent variable t, *all* the *nonsingular* solutions $U(t)$ of the matrix evolution equation (4.158) are *periodic* with period $T = 2\pi/\omega$,

$$U(t+T) = U(t). \tag{4.160}$$

The *singular* solutions of (4.158) are *exceptional*, corresponding to a set of initial data having *vanishing* measure with respect to the set of *generic* initial data. ⊡

Hence, as immediate consequence of *Proposition 4.2.2-17* and of this *Remark 4.2.2-18* there holds the following

Proposition 4.2.2-19. All the solutions of the N-body problem (4.156) (except those exceptional ones that run into a collision of two or more particles, corresponding to *nongeneric* initial data) are *completely periodic* with a period which is an *integer multiple* of T. [For a more detailed formulation see *Remark 4.2.2-2*]. ⊡

This proposition displays the *entirely isochronous* character of the N-body problem (4.156), indeed it justifies considering it as one more instance of *nonlinear harmonic oscillators.*

Let us now report the alternative formulations of these N-body problems, without any elaboration about their derivation—which should be pretty clear on the basis of previous developments (or see the original literature [79]).

The alternative formulation of the N-body problem (4.153) reads

$$\begin{aligned}&\ddot{c}_m + 2\,(m-1)\,\dot{c}_{m+1} - 2\,c_1\,\dot{c}_m + 2\,(N+1-m)\,a^2\,\dot{c}_{m-1}\\ &\quad + (m+2)\,(m-3)\,c_{m+2} - 2\,(m-1)\,c_1\,c_{m+1} + 2\,\left[m\,(N+2-m)\,a^2\right.\\ &\quad \left.+\dot{c}_1 - c_1^2 + 3\,c_2\right]\,c_m - 2\,(N+1-m)\,a^2\,c_1\,c_{m-1}\\ &\quad + (N+2-m)\,(N+1-m)\,a^4\,c_{m-2} = 0,\\ &\qquad m = 1,\dots,N, \quad c_0 = 1, \quad c_{-1} = c_{N+1} = c_{N+2} = 0.\end{aligned} \tag{4.161}$$

Remark 4.2.2-20. The ODE of this system with $m = 0$ is identically satisfied; the ODE with $m = N+1$ is also satisfied provided one sets $c_{N+3} = 0$, and even the ODE with $m = N+2$ is identically satisfied if one moreover sets $c_{N+4} = 0$. ⊡

Remark 4.2.2-21. A superficial look at this system of ODEs, (4.161), might suggest that it is a *linear* system of evolution equations for the quantities $c_m(t)$; but this is, of course, *not* the case, due to the presence of the quantities $c_1(t)$ and

$c_2(t)$. Indeed, the highly *nonlinear* character of this system is already evident by looking at the $N=2$ case, when it yields the following (*solvable*!) *fourth-order* ODE for $f(t) \equiv c_1(t)$:

$$f'''' f^2 - 2 f''' f' f^2 - 2 f''' f^3 - 2 (f'')^2 f + 2 f'' (f')^2 + 4 f'' f' f^2 - 2 f'' f^4 \\ - 4 (f')^2 f^3 + 4 f' f^5 + 4 a^2 (f'' f^2 - 2 f' f^3) = 0 \tag{4.162}$$

(here merely for typographical convenience differentiations are denoted by appended primes rather than superimposed dots). ⊡

The solution of the system of ODEs (4.161) is given by the following

Proposition 4.2.2-22. The dependent variables $c_m(t)$ that solve the initial-value problem for the system of nonlinear ODEs (4.161) are the N coefficients of the monic polynomial $\psi(z,t)$, of degree N in the variable z, given by the formula

$$\psi(z,t) = z^N + \sum_{m=0}^{N} c_m(t)\, z^{N-m} = \det[z - U(t)], \tag{4.163}$$

where the $N \times N$ matrix $U(t)$ evolves according to the *solvable* matrix evolution equation (4.152) and is moreover characterized by the initial data (4.154), with the initial values $z_n(0)$, $\dot{z}_n(0)$ related to the initial values $c_m(0), \dot{c}_m(0)$ by the formulas

$$\prod_{n=1}^{N} [z - z_n(0)] = \sum_{m=0}^{N} c_m(0)\, z^{N-m}, \quad c_0 = 1, \tag{4.164a}$$

$$-\sum_{n=1}^{N} \dot{z}_n(0) \prod_{m=1, m \neq n}^{N} [z - z_n(0)] = \sum_{m=1}^{N} \dot{c}_m(0)\, z^{N-m}. \; ⊡ \tag{4.164b}$$

To obtain an *entirely isochronous* version of this alternative N-body problem, (4.161) (with $a=0$)—or, equivalently, an alternative version of the *entirely isochronous* N-body problem (4.156) (with $a=0$)—we use the following version of the trick

$$c_m(t) = (-i)^m \exp(m\,i\,t)\, \gamma_m(\tau), \tag{4.165a}$$

$$\tau = i\,[1 - \exp(i\,t)]. \tag{4.165b}$$

Here and below (within this **Example 4.2.2-9**) we set, without significant loss of generality,

$$\omega = 1, \quad \text{implying} \quad T = 2\,\pi. \tag{4.166}$$

The quantities $\gamma_m(\tau)$ are the dependent variables of the model (4.161) with $a=0$, up to the (purely notational) change consisting in calling the independent variable τ (instead of t) and the dependent variables γ_m (instead of c_m), so that these variables satisfy the following system of ODEs:

$$\gamma_m'' + 2\,(m-1)\,\gamma_{m+1}' - 2\,\gamma_1\,\gamma_m' + (m+2)\,(m-3)\,\gamma_{m+2}$$
$$-2\,(m-1)\,\gamma_1\,\gamma_{m+1} + 2\,\left(\gamma_1' - \gamma_1^2 + 3\,\gamma_2\right)\,\gamma_m = 0$$
$$m = 1, \ldots, N, \quad \gamma_0 = 1, \quad \gamma_{-1} = \gamma_{N+1} = \gamma_{N+2} = 0, \tag{4.167}$$

where, of course, appended primes denote differentiations with respect to the independent variable τ (which we allow to be *complex*, see (4.165b)).

Then by applying the trick (4.165) to the system (4.167) the following new system of *nonlinear* ODEs is obtained:

$$\ddot{c}_m + 2\,(m-1)\,i\,\dot{c}_{m+1} - (2\,m+1+2\,c_1)\,i\,\dot{c}_m - (m+2)\,(m-3)\,c_{m+2}$$
$$+\,2\,(m-1)\,(m+1+c_1)\,c_{m+1} + \left[-m\,(m+1) + 2\,i\,\dot{c}_1 - 2\,(m-1)\,c_1\right.$$
$$\left.+2\,c_1^2 - 6\,c_2\right]\,c_m = 0, \quad m = 1, \ldots, N, \quad c_0 = 1, \quad c_{-1} = c_{N+1} = c_{N+2} = 0. \tag{4.168}$$

Remark 4.2.2-23. The prefactor $(-i)^m$ in (4.165a) is of course unessential, it has been introduced merely to give a marginally nicer look to this system (4.168). With this version, (4.165), of the trick the relation among the particle coordinates satisfying the equations of motion of the *entirely isochronous* N-body problem (4.156) and the quantities $c_m(t)$ satisfying this system of ODEs (4.168) now reads

$$\psi(z,t) = \prod\left[z - z_n(t)\right] = \sum_{m=0}^{N} (i)^m\, c_m(t)\, z^{N-m}, \quad c_0 = 1. \tag{4.169}$$

Note that we introduced here the *monic* polynomial $\psi(z,t)$ having as its N zeros the N dependent variables $z_n(t)$ satisfying (4.156) and as its N coefficients $c_m(t)$ the N dependent variables $c_m(t)$ satisfying (4.168). ⊡

This model, (4.168), is obviously just as *solvable* as the previous one, (4.161), indeed the solution of its initial-value problem can be obtained from the solution of the corresponding problem for (4.161) via the formulas (4.165) that clearly imply the following relations among the initial data of the two models:

$$c_m(0) = (-i)^m\, \gamma_m(0), \tag{4.170a}$$
$$\dot{c}_m(0) - m\,i\,c_m(0) = (-i)^m\, \gamma_m'(0). \tag{4.170b}$$

Equivalently, the solution of this model (4.168) is clearly given by the following

Proposition 4.2.2-24. The dependent variables $c_m(t)$ that solve the initial-value problem for the system of *nonlinear* ODEs (4.168) are the N coefficients of the polynomial $\psi(z,t)$, see (4.169), which is itself given by the formula

$$\psi(z,t) = \det\left[z - U(t)\right], \tag{4.171}$$

where the $N \times N$ matrix $U(t)$ evolves now according to the *solvable* matrix evolution equation (4.158) and is moreover characterized by the initial data (4.159) with the initial values $z_n(0)$, $\dot{z}_n(0)$ related to the initial values $c_m(0), \dot{c}_m(0)$ by the following formulas implied by (4.169),

$$\prod_{n=1}^{N} \left[z - z_n(0)\right] = \sum_{m=0}^{N} (i)^m \; c_m(0)\, z^{N-m}, \quad c_0 = 1, \tag{4.172a}$$

$$-\sum_{n=1}^{N} \left[\dot{z}_n(0) - i\, z_n(0)\right] \prod_{m=1, m\neq n}^{N} \left[z - z_n(0)\right] = \sum_{m=1}^{N} (i)^m \; \dot{c}_m(0)\, z^{N-m}. \; \boxdot \tag{4.172b}$$

As an immediate consequence of this *Proposition 4.2.2-24* and of *Remark 4.2.2-18* there holds the following

Proposition 4.2.2-25. All the *nonsingular* solutions of the system of ODEs (4.168) are *completely periodic* with period $2\,\pi$,

$$c_m(t + 2\,\pi) = c_m(t), \tag{4.173}$$

while the *singular* solutions are *exceptional*, corresponding to a set of initial data having *vanishing* measure with respect to the set of *generic* initial data. $\boxdot$

This proposition displays the *entirely isochronous* character of the N-body problem (4.168), indeed it justifies considering it one more instance of *nonlinear harmonic oscillators*.

Example 4.2.2-10 In this example we report, quite tersely, very recent results [69] [70] yielding new classes of *solvable* many-body problems of *goldfish* type—as well as *entirely isochronous* many-body models obtained from these—which are characterized by a novel feature: their *solvable*, respectively, their *entirely isochronous*, character only holds for a subset of solutions identified by the requirement that the initial data satisfy certain restrictions (generally then conserved throughout the motion). For the first two classes of these models [69] the presentation provided herein is rather terse: in particular, the origin of the *isochronous* behavior is not fully explained. For the third and last class [70] a more complete treatment is provided (also in view of a subsequent use of these results in Chapter 6); it might also help to understand the *isochronous* behavior in the two first cases, since the mechanism underlying its emergence is quite analogous in all three cases.

The first class of these models [69] is characterized by the following (6-parameter: see below the 6 *a priori arbitrary* constants A_j) *Newtonian* equations of motion:

$$\ddot{z}_n = A_1 \dot{z}_n + A_2 + A_3 z_n - \frac{2A_4}{z_n}$$
$$+ \sum_{m=1,\, m \neq n}^{N} \left\{ (z_n - z_m)^{-1} \left[2\dot{z}_n \dot{z}_m + A_5 (\dot{z}_n + \dot{z}_m) z_n \right.\right.$$
$$\left.\left. + 2 \left(A_4 - A_2 z_n + A_6 z_n^2 \right) \right] \right\}. \tag{4.174}$$

Here the *novel* element—not included in the previously known class of *solvable* many-body models of *goldfish* type, see for instance Section 2.3.3 of Reference [37], or the terse treatment provided at the beginning of this Section 4.2.2, see (4.53)—is that associated with the constant A_4. But the *solvable* character of this many-body model only prevails provided the coordinates $z_n(t)$ satisfy the constraint

$$\sum_{n=1}^{N} \frac{1}{z_n(t)} = 0. \tag{4.175a}$$

It can be easily seen that this constraint is compatible with the evolution equations (4.174); hence it is automatically satisfied, in the context of the *initial-value* problem, provided the initial data for the N-body problem (4.174) satisfy the following two conditions:

$$\sum_{n=1}^{N} \frac{1}{z_n(0)} = 0, \quad \sum_{n=1}^{N} \frac{\dot{z}_n(0)}{[z_n(0)]^2} = 0. \tag{4.175b}$$

And it is moreover known [69] that this many-body model, (4.174), is *entirely isochronous* provided the six constants A_j it features satisfy the following four restrictions:

$$A_1 = (k_1 - N k_2)\, i\omega, \tag{4.176a}$$
$$A_3 = \frac{1}{2} (k_2 + k_3) \left[\left(N - \frac{1}{2} \right) (k_2 - k_3) - k_1 \right] \omega^2, \tag{4.176b}$$
$$A_5 = k_2 i\omega, \tag{4.176c}$$
$$A_6 = \frac{k_3^2 - k_2^2}{4} \omega^2, \tag{4.176d}$$

where ω is a *positive* constant, $\omega > 0$ and the three numbers k_1, k_2, k_3 are *integers*, unrestricted (positive, negative or vanishing) except for the requirements

$$k_1 \neq -m\, k_3 \quad \text{for} \quad m = 1, \ldots, N. \tag{4.176e}$$

and

$$k_1 + k_3 \neq 0; \quad 2k_1 + m(k_3 - k) \neq 0 \quad \text{for} \quad m = 1, \ldots, N. \tag{4.176f}$$

Note that the two constants A_2 and A_4 remain completely arbitrary. Also note that, if the two integers k_1 and k_2 vanish—entailing that A_1 and A_5 vanish,

$A_1 = A_5 = 0$, and $A_3 = -(2N-1)\,A_6$, the only remaining restriction on A_6 being that it be a *positive* real number, $A_6 > 0$—then the equations of motion (4.174) become *real* (of course, provided *real* values are also assigned to the two *a priori arbitrary* constants A_2 and A_4).

And there also is a second, different class of *entirely isochronous* systems, that is obtained from (a subclass of) the *solvable* system (4.174) with (4.175a) via the standard "trick" procedure (see for instance [69]), as we now explain. To this end we take as starting point the following special case of the model (4.174):

$$\zeta_n'' = -\frac{2\,A_4}{\zeta_n} + 2 \sum_{m=1,\, m\neq n}^{N} \frac{\zeta_n'\,\zeta_m' + A_4}{\zeta_n - \zeta_m}, \tag{4.177}$$

which corresponds to (4.174) with *all* the constants A_j set to zero except A_4 (and moreover the purely notational replacement—convenient for what shall immediately follow—of the dependent variables $z_n(t)$ with the dependent variables $\zeta_n(\tau)$, and of the independent variable t with the independent variable τ, appended primes denoting of course differentiations with respect to this new independent variable τ). We then apply the usual trick with $\lambda = -1$, namely we set

$$z_n(t) = \exp(-i\,\omega\,t)\;\zeta_n(\tau)\,, \quad \tau = \frac{\exp(i\,\omega\,t) - 1}{i\,\omega}, \tag{4.178}$$

with ω a *positive* constant, and we thereby obtain the new N-body model characterized by the (*autonomous*) *Newtonian* equations of motion

$$\ddot{z}_n = -i\omega\dot{z}_n - \frac{2A_4}{z_n} + 2 \sum_{m=1,m\neq n}^{N} \left\{(z_n - z_m)^{-1}[(\dot{z}_n + i\omega z_n)(\dot{z}_m + i\omega z_m) + A_4]\right\}. \tag{4.179}$$

Due to the way this model has been obtained it is clear that it describes an *entirely isochronous* N-body problem—provided its initial data are constrained by the conditions (4.175b), guaranteeing the validity of (4.175a) (both for this model and for the model (4.177)). It is also easy to verify that this model is *not* a special case of the *entirely isochronous* N-body problem identified above, i.e. (4.174) with (4.176) (and of course with (4.175a)).

The second class of these models [69] is characterized by the following (7-parameter: see below the 7 *a priori arbitrary* constants B_j) *Newtonian* equations of motion:

$$\ddot{z}_n = B_1\,\dot{z}_n - (N-1)\;B_2\,z_n - 2\,(N-1)\;B_3\,z_n^2 + B_4\,\frac{\dot{z}_n}{z_n}$$
$$+ \sum_{m=1,\, m\neq n}^{N} \left\{(z_n - z_m)^{-1}\;[2\,\dot{z}_n\,\dot{z}_m + (\dot{z}_n + \dot{z}_m)\;(B_4 + B_5\,z_n)\right.$$
$$\left. + B_6\,(\dot{z}_n\,z_m + \dot{z}_m\,z_n)\;z_n + 2\,\left(B_7\,z_n + B_2\,z_n^2 + B_3\,z_n^3\right)]\right\}. \tag{4.180}$$

Here the *novel* element—not included in the previously known class of *solvable* many-body models of *goldfish* type, see for instance Section 2.3.3 of Reference [37] or the terse treatment provided at the beginning of this Section 4.2.2, see (4.53)—is now that associated with the constant B_4. But again the *solvable* character of this many-body model only prevails provided the coordinates $z_n(t)$ satisfy a constraint, which in this case reads

$$\sum_{n=1}^{N} \frac{\dot{z}_n(t)}{z_n(t)} = 0. \tag{4.181a}$$

And it is again easily seen that this constraint is compatible with the evolution equations (4.180); hence it is automatically satisfied, in the context of the *initial-value* problem, provided the initial data for the N-body problem (4.180) satisfy the following single condition:

$$\sum_{n=1}^{N} \frac{\dot{z}_n(0)}{z_n(0)} = 0. \tag{4.181b}$$

A subclass of this many-body models that is *entirely isochronous* is characterized by the following restrictions [69]:

$$B_1 = (k_1 - N k_2)\, i\omega, \tag{4.182a}$$

$$B_2 = \frac{k_3^2 - k_2^2}{4}\omega^2, \tag{4.182b}$$

$$B_3 = 0, \tag{4.182c}$$

$$B_5 = k_2 i\omega, \tag{4.182d}$$

$$B_6 = 0, \tag{4.182e}$$

with the two constants B_4 and B_7 *arbitrary* and the three *integers* k_1, k_2 and k_3 also *arbitrary* except for the following restriction,

$$k_2 + k_3 = 0 \quad \text{or} \quad 2k_1 = N(k_2 - k_3), \tag{4.182f}$$

as well as the inequalities (4.176e) and (4.176f). Of course the property of *isochrony* only holds in the *solvable* case, which is characterized by the restriction (4.181b) on the initial data.

And, as in the previous case, there also is a second, different class of *entirely isochronous* systems, that obtains, as above, via the standard "trick" procedure (see (4.178)), from the subclass of the *solvable* system (4.180) with all constants B_j vanishing except B_4. It reads

$$\ddot{z}_n = -i\omega\dot{z}_n + B_4\frac{\dot{z}_n + i\omega z_n}{z_n} + \sum_{m=1,\, m\neq n}^{N} \frac{2(\dot{z}_n + i\omega z_n)(\dot{z}_m + i\omega z_m) + B_4[\dot{z}_n + \dot{z}_m + i\omega(z_n + z_m)]}{z_n - z_m}, \tag{4.183a}$$

with B_4 an *arbitrary* constant and with the constraint (4.181) replaced now by the condition

$$\sum_{n=1}^{N} \frac{\dot{z}_n(t)}{z_n(t)} = -Ni\omega. \tag{4.183b}$$

Finally, the third class of these models [70] is characterized by the following (4-parameter) *Newtonian* equations of motion:

$$\ddot{z}_n = -a_1 \dot{z}_n + a_2 z_n \frac{z_n^2 - 5}{z_n^2 - 1} - 2a_3 \frac{z_n^2 + 1}{z_n^2 - 1} - 2a_4 z_n$$
$$+ 2 \sum_{m=1, m \neq n}^{N} \frac{\dot{z}_n \dot{z}_m + a_2 + a_3 z_n + a_4 \left(z_n^2 - 1\right)}{z_n - z_m}, \quad n = 1, \ldots, N, \tag{4.184a}$$

where the four "coupling constants" a_j are *a priori arbitrary*. The *solvable* character of this N-body problem hinges upon the following four restrictions on its *initial* data:

$$\sum_{n=1}^{N} \frac{1}{z_n(0) \pm 1} = 0, \quad \sum_{n=1}^{N} \frac{\dot{z}_n(0)}{\left[z_n(0) \pm 1\right]^2} = 0, \tag{4.184b}$$

which are then sufficient to guarantee that, throughout the time evolution,

$$\sum_{n=1}^{N} \frac{1}{z_n(t) \pm 1} = 0, \tag{4.184c}$$

implying that for this model it is justified to imagine that only the evolution of $N-2$ particles is determined by the *Newtonian* equations of motion (4.184a), while the evolution of the remaining two is determined by these conditions, see (4.184c).

Then the evolution of the N "particle coordinates" $z_n(t)$—generally taking place in the *complex* z-plane—coincides with the evolution of the N zeros of the following monic polynomial of degree N in the variable z:

$$\psi(z,t) = \pi_N(z) + \sum_{m=1}^{N-3} \left[c_m(t)\, \pi_{N-m}(z)\right] + c_N(t), \tag{4.185a}$$

$$\pi_m(z) = z^m - \varepsilon_m \frac{m}{2} z^2 - \varepsilon_{m+1} m z, \quad m = 0, 1, \ldots, N, \tag{4.185b}$$

$$\varepsilon_m = 1 \text{ if } m \text{ is } \textit{even}, \quad \varepsilon_m = 0 \text{ if } m \text{ is } \textit{odd}, \tag{4.185c}$$

whose $N-2$ coefficients $c_m(t)$ evolve as follows:

$$c_m(t) = \sum_{\ell=1;\ \ell\neq N-1,N-2}^{N} \left\{\gamma^{(\ell,+)} u_m^{(\ell,+)} \exp\left[\lambda^{(\ell,+)} t\right] + \gamma^{(\ell,-)} u_m^{(\ell,-)} \exp\left[\lambda^{(\ell,-)} t\right]\right\},$$

$$m = 1, \ldots, N-3 \text{ and } m = N, \tag{4.186a}$$

$$\lambda^{(\ell,\pm)} = \frac{-a_1 \pm \Delta_\ell}{2}, \quad \Delta_\ell^2 = a_1^2 + 4\ell\left[a_2 + (2N-\ell-3)\, a_4\right],$$

$$\ell = 1, \ldots, N-3, N. \tag{4.186b}$$

Note that the coupling constant a_3 does not appear explicitly in these formulas, but of course all four coupling constants a_j do play a role in determining the quantities $u_m^{(\ell,\pm)}$ appearing in the right-hand side of (4.186a).

These results entail that, for generic (possibly *complex*) values of the "coupling constants" a_j, the asymptotic behavior at large time of the solutions of the N-body problem (6.10) can be inferred from the treatment provided in appendix G of [37] (entitled "Asymptotic behavior of the zeros of a polynomial whose coefficients diverge exponentially"). In the special case when Δ_ℓ vanishes for some relevant value of ℓ, the asymptotic behavior at large time of the polynomial $\psi(z,t)$ (see (4.185a) with (4.186)) generally also contains a term linear in t—which might become dominant as $t \to \infty$ if all the other terms vanish exponentially in this limit. While clearly, if a_1 and (for the relevant values of ℓ) Δ_ℓ are *imaginary*, then *all* motions of the N-body problem (6.10) remain *confined* for all time. And finally if

$$a_1 = ik_1\omega, \quad a_2 = k_4\left[(2N-3)\, k_4 - k_1\right]\omega^2, \quad a_4 = k_4^2\omega^2, \tag{4.187a}$$

yielding

$$\lambda^{(\ell,+)} = i\ell k_4\omega, \quad \lambda^{(\ell,-)} = -i\,(\ell k_4 + k_1)\,\omega, \quad \ell = 1, \ldots, N-3, N, \tag{4.187b}$$

with ω an arbitrary *positive* constant and k_1, k_4 two *arbitrary integers* respecting the inequalities

$$k_4 \neq 0; \quad k + \ell\, k_4 \neq 0 \text{ and } k_1 \neq 2\,\ell\, k_4, \quad \ell = 1, \ldots, N-3, N \tag{4.187c}$$

(to avoid the vanishing of $\lambda^{(\ell,+)}$ or $\lambda^{(\ell,-)}$ or the equality $\lambda^{(\ell,+)} = \lambda^{(\ell,-)}$), then clearly the N-body problem (6.10) is *entirely isochronous*: its *generic* (*complex*!) solutions—in its *entire* phase space, except possibly for a subregion of vanishing measure characterized by solutions that become *singular* due to the collision of two or more particles—are *completely periodic,*

$$z_n\left(t + \tilde{T}\right) = z_n(t), \quad n = 1, 2, \ldots, N, \tag{4.187d}$$

with a common period $\tilde{T}$ (possibly not the primitive period in all sectors of phase space) which is a (generally small [100]) *integer* multiple of a basic period, itself a *rational* multiple (depending in an obvious manner from the *integers* k_1 and k_4) of the standard period $T = 2\pi/\omega$ associated with the circular frequency ω.

Clearly this discussion entails that the conditions (4.187a) are *sufficient* to guarantee that the N-body problem (4.184), with N arbitrary ($N \geq 3$), is *entirely isochronous*; for $N > 3$ (but not for $N = 3$, see below) they are also *necessary*.

It is remarkable that the *qualitative* behavior of this N-body problem, (4.184), turns out to be quite independent of the value of the coupling constant a_3.

For an additional discussion of the behavior of this many-body problem, (4.184), when it features the phenomenology of *asymptotic isochrony*, see Chapter 6.

Example 4.2.2-11 The *entirely isochronous* N-body problems of *goldfish* type reported in this example were arrived at [58] [59] via a somewhat different route than those described in the previous cases. Here the point of departure are certain *integrable* models of *goldfish* type taken from the literature, which are then ω-modified via the trick and thereby transformed into *entirely isochronous* models. We only report here the *Newtonian* equations of motion of (some of) these *entirely isochronous* N-body problems; a qualitative difference from those discussed above is the fact that these models belong to the class of Toda-type many-body problems featuring only "nearest-neighbor" interactions (with the neighborhood defined in terms of the particle labels; see below). Readers interested in more details are advised to look at the original literature [58] [59], which mainly focussed on the identification of the equilibrium configurations for these models, on their behavior near equilibria, and thereby on the derivation of *Diophantine* relations (for these, see Appendix C).

The simplest of the *entirely isochronous* N-body problems considered in [58] is characterized by the following *Newtonian* equations of motion:

$$\begin{aligned} &\ddot{z}_n - (2\,\lambda + 1)\; i\,\omega\, \dot{z}_n - \lambda\,(\lambda + 1)\;\omega^2\, z_n \\ &\quad = -\left(\dot{z}_n - i\,\lambda\,\omega\, z_n\right)\left(\frac{\dot{z}_{n-1} - i\,\lambda\,\omega\, z_{n-1}}{z_n - z_{n-1}} + \frac{\dot{z}_{n+1} - i\,\lambda\,\omega\, z_{n+1}}{z_n - z_{n+1}}\right), \end{aligned} \tag{4.188}$$

where λ is a *rational*, but otherwise *arbitrary*, constant, and the rest of the notation is self-explanatory. In these equations, as usual, the index n runs from 1 to N; therefore, some additional prescriptions must be given to assign the "extremal" coordinates z_0 and z_{N+1}. The two prescriptions that guarantee the *entirely isochronous* character of this model are the "periodic" one,

$$z_0\,(t) = z_N\,(t)\,, \quad z_{N+1}\,(t) = z_1\,(t)\,, \tag{4.189a}$$

and the *free ends* one, which can be formally implemented via the assignment

$$\dot{z}_0\,(t) = z_0\,(t) = 0, \quad \dot{z}_{N+1}\,(t) = z_{N+1}\,(t) = 0. \tag{4.189b}$$

Indeed in both cases the corresponding unmodified models—which obtain from these by setting $\omega = 0$—are not only *integrable* but actually *solvable* (see for instance, [45]); and from the analyticity properties of these *solvable* models the *entirely isochronous* character of the N-body problems (4.188) with (4.189a) or (4.189b) is apparent.

Two quite analogous (yet different) *entirely isochronous* models also discussed in [58] are characterized by the equations of motion

$$\ddot{z}_n - i\,\omega\,\dot{z}_n - \lambda\,\omega^2\,z_n = \frac{\dot{z}_n^{\,2}}{z_n} - (\dot{z}_n - i\lambda\omega z_n)\left(\frac{\dot{z}_{n-1} - i\lambda\omega z_{n-1}}{z_n - z_{n-1}} + \frac{z_n\,(\dot{z}_{n+1} - i\lambda\omega z_{n+1})}{z_{n+1}\,(z_n - z_{n+1})}\right), \quad (4.190)$$

respectively

$$\ddot{z}_n - (2\,\lambda + 1)\,i\,\omega\,\dot{z}_n - \lambda\,(\lambda + 1)\,\omega^2\,z_n = (\dot{z}_n - i\,\lambda\,\omega\,z_n)^2\left(\frac{1}{z_n - z_{n-1}} + \frac{1}{z_n - z_{n+1}}\right), \quad (4.191)$$

again to be supplemented by either one of the conditions (4.189). The investigation of these models in the neighborhood of their *genuine* (i.e., with $z_n \neq z_{n\pm1}$) equilibrium configurations—which only exist in the *free ends* case—yields *Diophantine* findings (see Appendix C).

In [59] the unmodified model taken as starting point is (a special case of) an *integrable* system of Toda-type discovered (in a slightly different version: see below) by Alexey Shabat and Ravil Yamilov [128] [129] [140]. Its equations of motion read as follows:

$$\zeta_n'' = \left(\zeta_n'^{\,2} + c\,\zeta_n^{\,k}\right)\left(\frac{1}{\zeta_n - \zeta_{n+1}} + \frac{1}{\zeta_n - \zeta_{n-1}}\right) - \frac{k\,c}{2}\,\zeta_n^{\,k-1},$$
$$\zeta_n \equiv \zeta_n(\tau), \quad \zeta_n' \equiv \zeta_n'(\tau) \equiv \frac{d\zeta_n(\tau)}{d\tau}, \quad (4.192)$$

The constant c could be rescaled away, but we prefer to keep it in evidence; as for the index k, it is required to take one of the following five *integer* values: $k = 0, 1, 2, 3$ or 4. But it is easily seen that via the change of dependent variables $\zeta_n \longmapsto 1/\zeta_n$ this model (4.192) goes into a completely analogous one up to the corresponding change $k \longmapsto 4 - k$. Hence, it is sufficient to restrict attention to the three models with $k = 0,\ 1,\ 2$. Moreover—as can be easily verified—the trick is only applicable to this model (in the sense of yielding an *autonomous* ω-modified system) for $k \neq 2$. Hence, in the following we focus on the two cases with $k = 0$ and $k = 1$.

The original treatment of Shabat and Yamilov [128] [129] [140] considered the system (4.192) in the case in which the index n ranges over *all* integers, and demonstrated the existence in such case of an *infinite* number of symmetries, implying *complete integrability.* We focus instead, as usual in this book, on the N-body model in which the index n ranges from 1 to N, and we must, therefore, supplement the system (4.192) with *boundary conditions* specifying the values to be attributed to the coordinates $\zeta_0(t)$ respectively $\zeta_{N+1}(t)$ which appear in the right-hand sides of (4.192) for $n = 1$ respectively for $n = N$. In [59] two different prescriptions were considered: the *periodic* one, analogous to (4.189a)—for which the N-body problem (4.192) is indeed known to be still *integrable* [103] [102]—and the *free ends* one, which entails that the equations of motion (4.192) are only valid as written for $1 < n < N$, while for $n = 1$ respectively for $n = N$ the terms in their right-hand side containing ζ_0 respectively ζ_{N+1} in the denominator must be omitted, an assignment which can as well be formally implemented (see (4.192)) by the following assignment:

$$\zeta_0(\tau) = \zeta_{N+1}(\tau) = \infty. \tag{4.193}$$

Through the trick (4.13) with $\lambda = 2/(k-2)$ the following ω-modified (*autonomous* and *entirely isochronous*) versions of the system (4.192) are obtained: for $k = 0$,

$$\ddot{z}_n + i\,\omega\,\dot{z}_n = \left[(\dot{z}_n + i\,\omega\,z_n)^2 + c\right]\left(\frac{1}{z_n - z_{n+1}} + \frac{1}{z_n - z_{n-1}}\right), \tag{4.194}$$

and for $k = 1$,

$$\ddot{z}_n + 3\,i\,\omega\dot{z}_n - 2\omega^2 z_n = \left[(\dot{z}_n + 2\,i\,\omega\,z_n)^2 + c\,z_n\right]\left(\frac{1}{z_n - z_{n+1}} + \frac{1}{z_n - z_{n-1}}\right) - \frac{c}{2}. \tag{4.195}$$

The index n runs again from 1 to N, and these two systems of ODEs, (4.194) and (4.195), must be again complemented by prescriptions specifying the values taken by the coordinates $z_0(t)$ respectively $z_{N+1}(t)$, which appear in the right-hand sides of these ODEs for $n = 1$ respectively $n = N$. The canonical assignments are again those mentioned above: the *periodic* one,

$$z_0(t) = z_N(t), \quad z_{N+1}(t) = z_1(t), \tag{4.196}$$

and the *free ends* one, which again entails that the equations of motion (4.194) and (4.195) are only valid as written for $1 < n < N$, while for $n = 1$ respectively $n = N$ the terms in the right-hand side of the equations of motion containing z_0 respectively z_N in the denominator must be omitted, so that these systems read as follows: for $k = 0$,

$$\ddot{z}_1 + i\,\omega\,\dot{z}_1 = \frac{(\dot{z}_1 + i\,\omega\,z_1)^2 + c}{z_1 - z_2}, \tag{4.197a}$$

$$\ddot{z}_n + i\,\omega\,\dot{z}_n = \left[(\dot{z}_n + i\,\omega\,z_n)^2 + c\right]\left(\frac{1}{z_n - z_{n+1}} + \frac{1}{z_n - z_{n-1}}\right),$$
$$n = 2, \ldots, N-1, \tag{4.197b}$$

$$\ddot{z}_N + i\,\omega\,\dot{z}_N = \frac{(\dot{z}_N + i\,\omega\,z_N)^2 + c}{z_N - z_{N-1}}, \tag{4.197c}$$

and for $k=1$,

$$\ddot{z}_1 + 3\,i\,\omega\,\dot{z}_1 - 2\,\omega^2\,z_1 = \frac{\left(\dot{z}_1 + 2\,i\,\omega\,z_1\right)^2 + c\,z_1}{z_1 - z_2} - \frac{c}{2}, \tag{4.198a}$$

$$\ddot{z}_n + 3i\omega\dot{z}_n - 2\omega^2 z_n = \left[\left(\dot{z}_n + 2i\omega z_n\right)^2 + c z_n\right]\left(\frac{1}{z_n - z_{n+1}} + \frac{1}{z_n - z_{n-1}}\right) - \frac{c}{2},$$
$$n = 2, \dots, N-1, \tag{4.198b}$$

$$\ddot{z}_N + 3\,i\,\omega\,\dot{z}_N - 2\,\omega^2\,z_N = \frac{\left(\dot{z}_N + 2\,i\,\omega\,z_N\right)^2 + c\,z_N}{z_N - z_{N-1}} - \frac{c}{2}. \tag{4.198c}$$

In [59] the equilibrium configurations of these models are discussed: it turns out that such *genuine* configurations (i.e., with $z_n \neq z_{n\pm1}$) only exist in the *free ends* cases (see (4.197) and (4.198)) and, in the $k=1$ case (4.198), only if, moreover, N is *odd*. The behavior of these models in the neighborhood of these equilibrium configurations confirms the *entirely isochronous* character of the *generic* solution of both these models in the *free ends* case, and it leads to *Diophantine* relations (see Appendix C).

4.2.3 Nonlinear oscillators

Typical systems of *first-order* ODEs characterizing *nonlinear oscillators* read as follows:

$$\dot{x}_n - i\,k_n\,\omega\,x_n = f_n(\underline{x}, \underline{y}), \quad n = 1, \dots, N,$$
$$\dot{y}_m + i\,j_m\,\omega\,y_m = g_m(\underline{x}, \underline{y}), \quad m = 1, \dots, M. \tag{4.199}$$

Here the N-vector $\underline{x}$, respectively the M-vector $\underline{y}$, have as components the $N+M$ *complex* dependent variables $x_n \equiv x_n(t)$, $y_m \equiv y_m(t)$; the $N+M$ parameters k_n, j_m are all *nonnegative integers* (or they could be *nonnegative rational* numbers); and the $N+M$ *complex* functions f_n, g_m are restricted by the following conditions (which are *sufficient* to guarantee the *isochrony* of this dynamical system):

$$f_n(\underline{x}, \underline{y}) \text{ and } g_m(\underline{x}, \underline{y}) \text{ are } \textit{holomorphic} \text{ at } \underline{x} = 0,\ \underline{y} = 0; \tag{4.200a}$$

$$\lim_{\varepsilon \to 0}\left[\varepsilon^{-1}\,\underline{f}(\varepsilon\,\underline{x}, \varepsilon\,\underline{y})\right] = \underline{0}, \quad \lim_{\varepsilon \to 0}\left[\varepsilon^{-1}\,\underline{g}(\varepsilon\,\underline{x}, \varepsilon\,\underline{y})\right] = \underline{0}; \tag{4.200b}$$

$$\underline{f}(\underline{x}, \underline{y}) \text{ and } \underline{g}(\underline{x}, \underline{y}) \text{ are } \textit{polynomial} \text{ in the } y_m; \tag{4.200c}$$

$$\lim_{\varepsilon \to 0}\left[\varepsilon^{-1-k_n}\,f_n(\varepsilon^{\underline{k}}\,\underline{x}, \varepsilon^{-\underline{j}}\,\underline{y})\right] = \text{nondivergent}, \quad n = 1, \dots, N; \tag{4.200d}$$

$$\lim_{\varepsilon \to 0}\left[\varepsilon^{-1+j_m}\,g_m(\varepsilon^{\underline{k}}\,\underline{x}, \varepsilon^{-\underline{j}}\,\underline{y})\right] = \text{nondivergent}\ , \quad m = 1, \dots, M. \tag{4.200e}$$

In the conditions (4.200d) and (4.200e) the notation $\varepsilon^{\underline{k}}\,\underline{x}$ indicates, of course, the N-vector of components $\varepsilon^{k_n}\,x_n$, and likewise $\varepsilon^{-\underline{j}}\,\underline{y}$ is the M-vector of components $\varepsilon^{-j_m}\,y_m$.

Note that this dynamical system, (4.199), includes the *Hamiltonian* case characterized by the restrictions

$$N = M, \quad k_n = j_n, \quad f_n(\underline{x}, \underline{y}) = \frac{\partial V(\underline{x}, \underline{y})}{\partial y_n}, \quad g_n(\underline{x}, \underline{y}) = -\frac{\partial V(\underline{x}, \underline{y})}{\partial x_n}, \tag{4.201}$$

which imply that the equations of motion (4.199) are just the *Hamiltonian* equations entailed by the Hamiltonian function

$$H(\underline{y}, \underline{x}) = i\,\omega \sum_{n=1}^{N} k_n\, x_n\, y_n + V(\underline{x}, \underline{y}), \tag{4.202}$$

where x_n are the N canonical coordinates and y_n are the corresponding canonical momenta. *Isochronicity* is now guaranteed by the following conditions on the function $V(\underline{x}, \underline{y})$:

$$V(\underline{x}, \underline{y}) \text{ is } \textit{holomorphic} \text{ at } \underline{x} = 0, \ \underline{y} = 0; \tag{4.203a}$$

$$\lim_{\varepsilon \to 0} \left[\varepsilon^{-2}\, V(\varepsilon\, \underline{x}, \varepsilon\, \underline{y})\right] = \underline{0}; \tag{4.203b}$$

$$V(\underline{x}, \underline{y}) \text{ is } \textit{polynomial} \text{ in the } y_n; \tag{4.203c}$$

$$\lim_{\varepsilon \to 0} \left[\varepsilon^{-1}\, V(\varepsilon^{\underline{k}}\, \underline{x}, \varepsilon^{-\underline{p}}\, \underline{y})\right] = \text{nondivergent}\ . \tag{4.203d}$$

These findings have been proved in the standard manner, based on the trick; for details see [64].

4.2.4 *Two Hamiltonian systems*

Most of the unmodified systems of ODEs described above are *Hamiltonian*, namely they can be obtained in the standard manner from a given Hamiltonian function (as explicitly shown in some cases); and in some—but certainly not in all—cases this property remains true for the ω-modified *isochronous* systems obtained via the application of (some appropriate version of) the trick. In Chapter 5 we present a new kind of tricks directly applicable to a *Hamiltonian* system and capable to generate an ω-modified version of its *Hamiltonian* which generates *isochronous* equations of motion. In this Section 4.2.4 we exhibit—without any proofs and with minimal discussion (for these, including the exhibition of specific examples, see [46])—two Hamiltonian models which yield *isochronous* equations of motion.

The *first* is defined by the *Hamiltonian*

$$H(\underline{p}, \underline{z}) = \omega \sum_{n=1}^{N} \left[\cosh(p_n)\, \left(a_n^2 + z_n^2\right)^{1/2}\right] + V(\underline{z}), \tag{4.204a}$$

$$V(\underline{z}) = \sum_{k=1}^{K} V^{(-2\,k)}(\underline{z}), \tag{4.204b}$$

$$V^{(-2\,k)}(c\,\underline{z}) = c^{-2\,k}\, V^{(-2\,k)}(\underline{z}), \tag{4.204c}$$

where the N (possibly *complex*) constants a_n^2 are *arbitrary* and the dependence of the functions $V^{(-2k)}(\underline{z})$ on the (N components of the) N-vector $\underline{z}$ is required to be *analytic* and to satisfy the scaling property (4.204c), but is otherwise as well *arbitrary*. The corresponding *Newtonian* equations of motion read as follows:

$$\ddot{z}_n + \omega^2 z_n = -\left[\dot{z}_n^2 + \omega^2 \left(a_n^2 + z_n^2\right)\right]^{1/2} \frac{\partial V(\underline{z})}{\partial z_n}, \quad n = 1, \ldots, N. \qquad (4.205)$$

Remark 4.2.4-1. We assume here that $\omega > 0$; indeed for $\omega = 0$ these equations of motion become

$$\ddot{z}_n = -\dot{z}_n \frac{\partial V(\underline{z})}{\partial z_n}, \quad n = 1, \ldots, N, \qquad (4.206)$$

which are compatible, but more general, than the trivial equations of motion $\dot{z}_n = 0$ entailed by the *Hamiltonian* (4.204a) with $\omega = 0$. This is due to the fact that, to derive the *Newtonian* equations of motion (4.205) from the *Hamiltonian* equations of motion entailed by the *Hamiltonian* (4.204), one must perform an operation (dividing by ω) which is not permissible when $\omega = 0$. ⊡

Although these *Newtonian* equations of motion (4.205) become *real* if the constants a_n^2 and the function $V(\underline{z})$ are *real*, the property of *isochrony* holds in any case—as usual—only in the *complex* domain. To prove it one uses the version of the trick reading

$$z_n(t) = \exp(-i\omega t)\, \zeta_n(\tau), \quad \tau = \frac{\exp(2i\omega t) - 1}{2i\omega}, \quad n = 1, \ldots, N. \qquad (4.207)$$

The evolution equations satisfied by the functions $\zeta_n(\tau)$ are then easily obtained from (4.205) with (4.204c):

$$\zeta_n'' = -\left[\omega^2 a_n^2 - 2i\omega \zeta_n' \zeta_n + (1 + 2i\omega\tau)\left(\zeta_n'\right)^2\right]^{1/2} \cdot$$
$$\cdot \sum_{k=1}^{K} (1 + 2i\omega\tau)^{k-1} \frac{\partial V^{(-2k)}(\underline{\zeta})}{\partial z_n}, \quad n = 1, \ldots, N, \qquad (4.208)$$

and using the fact that their right-hand side is *holomorphic* in τ in the neighborhood of $\tau = 0$ the property of *isochrony* can be proven [46].

The *second* of these models is defined by the *Hamiltonian*

$$H(\underline{p}, \underline{z}) = \sum_{n=1}^{N} [\varphi(p_n)\, z_n] + W(\underline{z}) \qquad (4.209)$$

where the function $\varphi(p)$ is required to satisfy the ODE

$$\varphi(p) = [\varphi'(p) + i(1+\mu)\,\omega]^{(1+\mu)/2}\, [\varphi'(p) - i(1-\mu)\,\omega]^{(1-\mu)/2} \qquad (4.210)$$

with μ a *real rational* number different from minus unity, $\mu = p/q \neq -1$, and satisfying either one (or both) of the following two inequalities:

$$\mu > -3 \quad \text{or} \quad \mu < -(1+2K)\,, \tag{4.211}$$

where K is an *arbitrary nonnegative* integer that, together with μ, characterizes the function $W(\underline{z})$ via the formula

$$W(\underline{z}) = \sum_{k=0}^{K} W^{(a_k)}(\underline{z}) \tag{4.212a}$$

with

$$a_k = -\frac{2k}{1+\mu} \tag{4.212b}$$

and with $W^{(a)}(\underline{z})$ a function the dependence of which on the (N components of the) N-vector $\underline{z}$ is *analytic* and only required to satisfy the scaling property

$$W^{(a)}(c\,\underline{z}) = c^a\, W^{(a)}(\underline{z})\,. \tag{4.212c}$$

The corresponding *Newtonian* equations of motion read as follows:

$$\begin{aligned}\ddot{z}_n + 2\,i\,\mu\,\omega\,\dot{z}_n + \left(1-\mu^2\right)\omega^2\,z_n &= -\left[\dot{z}_n + i\,(1+\mu)\,\omega\,z_n\right]^{(1-\mu)/2}\cdot\\ \cdot\left[\dot{z}_n - i\,(1-\mu)\,\omega\,z_n\right]^{(1+\mu)/2}&\frac{\partial W(\underline{z})}{\partial z_n}, \quad n=1,\ldots,N.\end{aligned} \tag{4.213}$$

To provide a neat example of these Newtonian equations of motion we note that, with the assignment $\mu = 1$, the ODE (4.210) can be solved to yield

$$\varphi(p) = 2\,i\,\omega + \exp(p), \tag{4.214a}$$

where, without significant loss of generality, we set to unity the *a priori* arbitrary constant multiplying the exponential in the right-hand side. Hence the corresponding *Hamiltonian* (4.209) can in this case be explicitly exhibited:

$$H(\underline{p},\underline{z}) = \sum_{n=1}^{N}\left[2\,i\,\omega + \exp(p_n)\right] z_n + W(\underline{z}). \tag{4.214b}$$

The corresponding *Newtonian* equations of motion read of course (see (4.213))

$$\ddot{z}_n + 2\,i\,\omega\,\dot{z}_n = -\dot{z}_n\,\frac{\partial W(\underline{z})}{\partial z_n}, \quad n=1,\ldots,N, \tag{4.214c}$$

where the function $W(\underline{z})$ must satisfy now the scaling property (4.212) with

$$a_k = -k, \quad k=1,\ldots,K. \tag{4.214d}$$

An assignment of $W(\underline{z})$ that satisfies this condition and that moreover entails that the right-hand side of these *Newtonian* equations of motion (4.214c) only feature two-body forces reads

$$W(\underline{z}) = \frac{1}{2} \sum_{m,n=1,m\neq n}^{N} \sum_{k=1}^{K} \frac{g_{nm}^{(k)} \left(z_n - z_m\right)^{-k}}{k}. \tag{4.215a}$$

Indeed the corresponding *Newtonian* equations of motion (4.214c) then read

$$\ddot{z}_n + 2\, i\, \omega\, \dot{z}_n = \dot{z}_n \sum_{m=1,m\neq n}^{N} \sum_{k=2}^{K} \frac{g_{nm}^{(k)}}{\left(z_n - z_m\right)^k}, \quad n = 1, \ldots, N. \tag{4.215b}$$

Here the "coupling constants" $g_{nm}^{(k)}$ are of course *arbitrary* (possibly *complex*), except for the obvious (see (4.215a)) symmetry property $g_{nm}^{(k)} = (-)^k\, g_{mn}^{(k)}$. Note that these *Newtonian* equations of motion, (4.215b), are *translation-invariant*; and ponder on their similarity with, as well as their difference from, the (also *Hamiltonian*, and also *isochronous*) Newtonian equations of motion (4.36b).

Actually in [46] it is shown that the following *Newtonian* equations of motion (more general than (4.213)),

$$\ddot{z}_n + 2\, i\, \mu_n\, \omega\, \dot{z}_n + \left(1 - \mu_n^2\right) \omega^2\, z_n = - \left[\dot{z}_n + i \left(1 + \mu_n\right) \omega\, z_n\right]^{(1-\mu_n)/2} \cdot$$
$$\cdot \left[\dot{z}_n - i \left(1 - \mu_n\right) \omega\, z_n\right]^{(1-\mu_n)/2} \frac{\partial\, W(\underline{z})}{\partial\, z_n}, \quad n = 1, \ldots, N, \tag{4.216}$$

where the N constants μ_n are all *rational* numbers—that clearly reduce to (4.213) if $\mu_n = \mu$—are as well *Hamiltonian*, and that they are *isochronous* provided the function $W(\underline{z})$ satisfies the scaling property

$$W(\underline{z}) = \sum_{k=1}^{K} W^{(a_{1k},\ldots,a_{Nk})}(\underline{z}), \tag{4.217a}$$

$$W^{(a_{1k},\ldots,a_{Nk})}(c_1 z_1, \ldots, c_N z_N) = \left[\prod_{n=1}^{N} \left(c_{nk}\right)^{a_{nk}}\right]^{\frac{1}{N}} W^{(a_{1k},\ldots,a_{Nk})}(\underline{z}), \quad k = 1, \ldots, K, \tag{4.217b}$$

with

$$\frac{1}{N} \sum_{n=1}^{N} a_{nk} \left(1 + \mu_n\right) = -2\, k, \quad n = 1, \ldots, N, \quad k = 1, \ldots, K. \tag{4.217c}$$

[These scaling properties reduce, of course, to (4.212) if $\mu_n = \mu$ and $a_{nk} = a_k$]. However, in this more general case to guarantee *isochrony* an additional condition is required: there should exist N numbers A_n such that the following NK

inequalities hold:

$$-A_n + \frac{1}{N} \sum_{m=1}^{N} A_m \, a_{mk} \geq k + \frac{1}{2} (1 - \mu_n), \quad n = 1, \ldots, N, \quad k = 1, \ldots, K. \tag{4.218}$$

These last conditions, however, are not too restrictive. For instance, by setting

$$A_n = -\frac{1}{2} b \, (1 + \mu_n) + c, \quad n = 1, \ldots, N \tag{4.219a}$$

with b and c two *a priori arbitrary* constants, they become (via (4.217c))

$$b\left[k + \frac{1}{2}(1 + \mu_n)\right] + c(1 - \bar{a}_k) \geq k + \frac{1}{2} \, (1 - \mu_n) \,, \quad n = 1, \ldots, N, \quad k = 1, \ldots, K, \tag{4.219b}$$

where we introduced the convenient definition

$$\bar{a}_k = \frac{1}{N} \sum_{n=1}^{N} a_{nk}. \tag{4.219c}$$

And it is clear that these conditions, (4.219b), can be satisfied (at least) in any one of the following cases (and we indicate in each case in square brackets via which assignment of the two arbitrary constants b, c):

$$\text{case (i):} \quad \mu_n > -3, \;\; n = 1, \ldots, N, \quad \left[b \geq \max\left(\frac{2\,k + 1 - \mu_n}{2\,k + 1 + \mu_n}\right), \;\; c = 0\right], \tag{4.220a}$$

$$\text{case (ii):} \; \mu_n < -\,(1 + 2\,K)\,, \;\; n = 1, \ldots, N, \quad \left[b \leq \min\left(\frac{2\,k + 1 - \mu_n}{2\,k + 1 + \mu_n}\right), \;\; c = 0\right], \tag{4.220b}$$

$$\text{case (iii):} \quad \bar{a}_k < 1, \;\; k = 1, \ldots, K, \quad \left[b = 0, \quad c \geq \max\left(\frac{k + \frac{1}{2}\,(1 - \mu_n)}{1 - \bar{a}_k}\right)\right], \tag{4.220c}$$

$$\text{case (iv):} \quad \bar{a}_k > 1, \;\; k = 1, \ldots, K, \quad \left[b = 0, \; c \leq \min\left(\frac{k + \frac{1}{2}\,(1 - \mu_n)}{1 - \bar{a}_k}\right)\right]. \tag{4.220d}$$

The first two of these four cases correspond to the condition (4.211) when $\mu_n = \mu$; and the last two of these four cases imply that these conditions can always be satisfied if $K = 1$, unless $\bar{a}_k = 1$, in which case the sufficient conditions for *isochrony* are provided by the two cases (i) and (ii) (namely, the rational numbers μ_n should be either *all* smaller, or *all* larger, than -3).

4.3 Two-dimensional systems

In the introductory Chapter 1 it was indicated how a *one-dimensional complex* model can be reformulated as a *two-dimensional real* model (see (1.7)). And in the introductory part of this Chapter 4 the ambiguity was outlined of the distinction among *one-dimensional complex* models and *two-dimensional real* models. This connection has been treated at considerable length in the literature, see in particular Chapter 4 ("Solvable and/or integrable many-body problems in the plane, obtained by complexification") of the book [37], as well as the literature quoted there (see in particular Section 4.N of that book, and Section 4.N below). Hence in this Section 4.3 we do not add anything more to what was written above and limit our presentation, without any commentary, to exhibit the *two-dimensional real* versions of some of the *one-dimensional complex* models treated above. Consistently with the focus of this book we only exhibit ω-modified *isochronous* models—whose *nonisochronous* version can in any case be easily obtained by just setting the (otherwise *positive*) constant ω to *zero*.

In the following *real* two-vectors are denoted by superimposed arrows and their Cartesian and circular coordinates are denoted in the standard manner, say

$$\vec{r}_n \equiv (x_n,\, y_n) \equiv (r_n,\, \theta_n)\,; \;\; x_n = r_n \cos(\theta_n)\,, \;\; y_n = r_n \sin(\theta_n)\,; \;\;\; r_n^2 = x_n^2 + y_n^2\,, \tag{4.221a}$$

with the standard connection among the *complex one-dimensional* and the *real two-dimensional* notations, say

$$z_n = x_n + i\, y_n = r_n \exp(i\,\theta_n)\,. \tag{4.221b}$$

Sometimes it is convenient to imagine the plane in which the motions take place to be immersed in *three-dimensional* space, so that

$$\vec{r}_n \equiv (x_n,\, y_n,\, 0)\,, \quad \hat{k} \wedge \vec{r}_n = (-y_n,\, x_n,\, 0) \quad \text{with} \quad \hat{k} \equiv (0,\, 0,\, 1)\,, \tag{4.221c}$$

$$\vec{r}_n \cdot \vec{r}_m = x_n\, x_m + y_n\, y_m = r_n\, r_m \cos(\theta_n - \theta_m)\,, \tag{4.221d}$$

$$\hat{k} \wedge \vec{r}_n \cdot \vec{r}_m = x_n\, y_m - y_n\, x_m = -r_n\, r_m \sin(\theta_n - \theta_m)\,. \tag{4.221e}$$

Occasionally the abbreviated notation

$$\vec{r}_{nm} \equiv \vec{r}_n - \vec{r}_m, \quad r_{nm}^2 = r_n^2 + r_m^2 - 2\,\vec{r}_n \cdot \vec{r}_m \tag{4.221f}$$

will be used.

In some cases the *real two-dimensional* version was already reported, see for instance (1.7) and (4.37).

To write the *real two-dimensional* version of (4.43a) it is preferable to use circular coordinates:

$$\ddot{r}_n - \dot{\theta}_n^2\, r_n + 3\,\omega\,\dot{\theta}_n\, r_n - 2\,\omega^2\, r_n$$
$$= \sum_{j,k,m=1}^{N} A_{njkm}\, r_j\, r_k\, r_m \cos(\theta_j + \theta_k + \theta_m - \theta_n - 2\,\gamma_{njkm}), \tag{4.222a}$$

$$r_n\,\ddot{\theta}_n + \left(2\,\dot{\theta}_n - 3\,\omega\right)\,\dot{r}_n$$
$$= \sum_{j,k,m=1}^{N} A_{njkm}\, r_j\, r_k\, r_m\, \sin(\theta_j + \theta_k + \theta_m - \theta_n - 2\,\gamma_{njkm}). \qquad (4.222b)$$

This system of equations of motion have been obtained from (4.43a) via the assignment $a_{njkm} = A_{njkm}\,\exp\left(-2\,i\,\gamma_{njkm}\right)$; they are *rotation-invariant* in the sense of being invariant under the transformation $\theta_n \mapsto \theta_n + \theta$, $\gamma_{njkm} \mapsto \gamma_{njkm} + \theta$, with θ an *arbitrary* constant (to justify this transformation one might wish to consider the N^4 quantities γ_{njkm} as additional dependent variables satisfying the trivial equation of motion $\dot{\gamma}_{njkm} = 0$).

Similarly, the *real two-dimensional* version of (4.45a) reads

$$\ddot{r}_n - \dot{\theta}_n^{\,2}\, r_n + 5\,\omega\,\dot{\theta}_n\, r_n - 6\,\omega^2\, r_n = \sum_{j,k=1}^{N} C_{njk}\, r_j\, r_k\, \cos(\theta_j + \theta_k - \theta_n - \gamma_{njk}), \qquad (4.223a)$$

$$r_n\,\ddot{\theta}_n + \left(6\,\dot{\theta}_n - 5\,\omega\right)\,\dot{r}_n = \sum_{j,k=1}^{N} C_{njk}\, r_j\, r_k\, \sin(\theta_j + \theta_k - \theta_n - \gamma_{njk}). \qquad (4.223b)$$

This system of equations of motion have been obtained from (4.45a) via the assignment $c_{njk} = C_{njk}\,\exp\left(-i\,\gamma_{njk}\right)$; again, they are *rotation-invariant* in the sense of being invariant under the transformation $\theta_n \mapsto \theta_n + \theta$, $\gamma_{njk} \mapsto \gamma_{njk} + \theta$, with θ an *arbitrary* constant (and again, to justify this transformation one might wish to consider the N^3 quantities γ_{njk} as additional dependent variables satisfying the trivial equation of motion $\dot{\gamma}_{njk} = 0$).

The two-dimensional versions of (4.48) and (4.82) are too well known (see for instance [37]) to report them here.

The two-dimensional version of the *entirely isochronous* many-body problem (4.98) reads

$$\ddot{\vec{r}}_n - 3\,\omega\,\hat{k} \wedge \dot{\vec{r}}_n - 2\,\omega^2\,\vec{r}_n = 2\left[\dot{\vec{r}}_n\,(\vec{a}\cdot\vec{r}_n) + \vec{r}_n\left(\vec{a}\cdot\dot{\vec{r}}_n\right)\right.$$
$$\left. - \vec{a}\left(\dot{\vec{r}}_n\cdot\vec{r}_n\right)\right] - 2\,\omega\,\hat{k}\wedge\left[2\,\vec{r}_n\,(\vec{a}\cdot\vec{r}_n) - \vec{a}\,r_n^2\right]$$
$$+ 2\sum_{m=1, m\neq n}^{N} r_{nm}^{-2}\left\{\dot{\vec{r}}_n\left(\dot{\vec{r}}_m\cdot\vec{r}_{nm}\right) + \dot{\vec{r}}_m\left(\dot{\vec{r}}_n\cdot\vec{r}_{nm}\right)\right.$$
$$- \vec{r}_{nm}\left(\dot{\vec{r}}_n\cdot\dot{\vec{r}}_m\right) - \omega\,\hat{k}\wedge\left[\dot{\vec{r}}_n\,(\vec{r}_m\cdot\vec{r}_{nm}) + \dot{\vec{r}}_m\,(\vec{r}_n\cdot\vec{r}_{nm})\right.$$

$$-\vec{r}_n\left(\left[\dot{\vec{r}}_n+\dot{\vec{r}}_m\right]\cdot\vec{r}_m\right)+\vec{r}_m\left(\left[\dot{\vec{r}}_n+\dot{\vec{r}}_m\right]\cdot\vec{r}_m\right)\Big]$$
$$+\omega^2\left[\vec{r}_n r_m^2-\vec{r}_m r_n^2\right]\Big\}. \tag{4.224}$$

To obtain this equation the relation among the (*complex*) constant a in (4.98) and the (*real*) vector $\vec{a}$ appearing here is

$$\vec{a}\equiv(\mathrm{Re}\,a,\;-\mathrm{Im}\,a,\;0)\,; \tag{4.225}$$

note the minus sign in this formula! Thanks to this assignment, this system of Newtonian equations of motion in the plane is *covariant*. It is not *rotation-invariant* because the *constant* two-vector $\vec{a}$ identifies a preferred direction in the plane; this can be remedied by considering $\vec{a}$ as an auxiliary variable satisfying the trivial equation of motion $\dot{\vec{a}}=0$.

The *two-dimensional* version of the *entirely isochronous* system of *nonlinear harmonic oscillators* (4.119) can be obtained in an analogous manner, and can also be cast in *covariant* form via (4.225); we leave its explicit derivation as a simple task for the diligent reader. Instead, we display the *two-dimensional* version of the *entirely isochronous* system of *nonlinear harmonic oscillators* (4.124):

$$\ddot{\vec{c}}_m-(2\,m+1)\,\omega\,\hat{k}\wedge\dot{\vec{c}}_m-2\,m\,\dot{\vec{c}}_{m+1}-\omega^2\,m\,(m+1)\,\vec{c}_m$$
$$+\,2\,m\,(m+1)\,\omega\,\hat{k}\wedge\vec{c}_{m+1}+(m+2)\,(m-1)\,\vec{c}_{m+2}$$
$$=-2\left\{\vec{c}_1\left[\vec{b}\cdot\left(\dot{\vec{c}}_m-m\,\vec{c}_{m+1}\right)\right]+\vec{c}_2\left(\vec{b}\cdot\vec{c}_m\right)\right.$$
$$\left(\dot{\vec{c}}_m-m\,\vec{c}_{m+1}\right)\left(\vec{b}\cdot\vec{c}_1\right)+\vec{c}_m\left(\vec{b}\cdot\vec{c}_2\right)$$
$$\left.-\vec{b}\left[\vec{c}_1\cdot\left(\dot{\vec{c}}_m-m\,\vec{c}_{m+1}\right)+\vec{c}_2\cdot\vec{c}_m\right]\right\}$$
$$+\,2\,m\,\omega\,\hat{k}\wedge\left[\vec{c}_1\left(\vec{b}\cdot\vec{c}_m\right)+\vec{c}_m\left(\vec{b}\cdot\vec{c}_1\right)-\vec{b}\left(\vec{c}_1\cdot\vec{c}_m\right)\right],$$
$$m=1,\ldots,N, \tag{4.226a}$$

with the boundary conditions

$$\vec{c}_{N+1}=\vec{c}_{N+2}=0. \tag{4.226b}$$

To obtain this *covariant* system of *two-dimensional Newtonian* equations of motion we first set in (4.124) $c_m(t)=b\,\gamma_m\,(t)$, and we then made the transition from *one complex* dimension to *two real* dimensions by introducing the vectors $\vec{c}_m\equiv(\mathrm{Re}\,\gamma_m,\,\mathrm{Im}\,\gamma_m,\,0)$ and $\vec{b}\equiv(\mathrm{Re}\,b,\,-\mathrm{Im}\,b,\,0)$ (note the minus sign in the last definition!). The *covariant* character of these Newtonian equations of motion entails that this N-body problem—describing an assembly of *nonlinear harmonic oscillators* (see (4.124c)) moving in the *horizontal* plane—is *rotation-invariant*, provided one treats the vector $\vec{b}$ as an auxiliary variable satisfying the trivial (non-evolution) equation $\dot{\vec{b}} = 0$.

We trust these examples are sufficient to indicate how to perform also in other case—including all those treated above—the transition from *one-dimensional complex* models to (equivalent) *two-dimensional real* ones—and, in particular, how this can often be done so as to obtain models characterized by *covariant* equations of motion, interpretable as *rotation-invariant* N-body problems in the plane.

4.4 Three-dimensional systems

A strategy to manufacture many-body problems "amenable to exact treatments" in *three-dimensional* space is described in Chapter 5 ("Many-body systems in ordinary (three-dimensional) space: solvable, integrable, linearizable problems") of [37], where several examples are also provided. The main idea of this strategy is to start from *matrix* evolution equations amenable to exact treatment and then to parametrize the matrices in terms of *three-vectors* (and possibly also of *scalars*) in a manner which is compatible with the evolution equation under consideration and indeed transforms it into a set of *covariant* equations of motion of *Newtonian* type for *three-vector* dependent variables (possibly also involving *scalar* dependent variables). The models thereby obtained are therefore generally characterized by *Newtonian* equations of motion having a *covariant* look in *three-dimensional* space, allowing them to be interpreted as *rotation-invariant* many-body problems; in some cases these models are moreover *translation-invariant.* Some of these many-body problems are *isochronous*, but no *systematic* attempt was made in [37] to identify models having this property.

In this Section 4.4 we take advantage of results obtained in [37] in order to exhibit several representative examples of *isochronous* many-body problems in *three-dimensional* space; often the fact that these models are amenable to exact treatments entails that their *isochronous* versions feature the property of *isochrony* in their *entire* natural phase space (perhaps up to a *lower-dimensional* set of *nongeneric* solutions)—thereby qualifying as *entirely isochronous* models, or instances of *nonlinear harmonic oscillators* in *ordinary* (*three-dimensional*) space. We also present some instances of many-body problems in *three-dimensional* space which only feature the property of *isochrony* in an (open, *fully dimensional*) region of their natural phase space: of course, an immense class of such systems can be manufactured via the trick, the few presented below are merely some representative examples. All the examples reported below are obtained via applications of the standard trick—applications which we deem by now too obvious to require any elaboration below (but see Section 4.N for some bibliographic indications).

Clearly, the main purpose of the following terse presentation is to wet the appetite of readers interested in *isochronous* many-body problems in *ordinary* (*three-dimensional*) space, possibly in view of physical applications—on the understanding that they may also derive on their own other models, of specific interest to them, by taking advantage of techniques analogous to those employed herein (generally relying on the trick, see below).

Notation: in this Section 4.4 superimposed arrows denote *three-vectors*, say $\vec{r} \equiv (x,\, y,\, z)$, and $r^2 \equiv x^2 + y^2 + z^2$.

Example 4.4-1 A *solvable, entirely isochronous* one-body problem is characterized by the *rotation-invariant Newtonian* equation of motion

$$\ddot{\vec{r}} - (2\,\lambda + 1)\, i\, \omega\, \dot{\vec{r}} - \lambda\, (\lambda + 1)\, \omega^2\, \vec{r}$$
$$= \frac{2\left(\dot{\vec{r}} - i\,\lambda\,\omega\,\vec{r}\right)\left[\left(\dot{\vec{r}} - i\,\lambda\,\omega\,\vec{r}\right)\cdot\vec{r}\right] - \vec{r}\left[\left(\dot{\vec{r}} - i\,\lambda\,\omega\,\vec{r}\right)\cdot\left(\dot{\vec{r}} - i\,\lambda\,\omega\,\vec{r}\right)\right]}{r^2}, \tag{4.227}$$

where λ is a *rational* number, $\lambda = p\,/\,q$ with p and q two *coprime* integers and $q > 0$. Here of course the dependent *three-vector* variable $\vec{r}\,(t)$ is necessarily *complex*.

Indeed, as can be verified by direct calculation, the solution of its initial-value problem reads

$$\vec{r}\,(t) = \exp\left(i\lambda\omega t\right)\exp\left(v\tau\right)\left\{\vec{r}\,(0)\cos\left(v\tau\right) + \left[u\vec{r}\,(0) - \dot{\vec{r}}\,(0) + i\lambda\omega\vec{r}\,(0)\right]\frac{\sin\left(v\tau\right)}{v}\right\}, \tag{4.228a}$$

$$u = \frac{\left[\dot{\vec{r}}\,(0) - i\lambda\omega\vec{r}\,(0)\right]\cdot\vec{r}\,(0)}{r^2\,(0)}, \tag{4.228b}$$

$$\vec{v} = \frac{\vec{r}\,(0)\wedge\dot{\vec{r}}\,(0)}{r^2\,(0)}, \tag{4.228c}$$

$$\tau = \frac{\exp\left(i\,\omega\,t\right) - 1}{i\,\omega}. \tag{4.228d}$$

This neat formula shows clearly that $\vec{r}\,(t)$ is *periodic* with period $\tilde{T} = q\,T$.

Example 4.4-2 A *solvable, entirely isochronous* 3-body problem is characterized by the *rotation-invariant Newtonian* equations of motion

$$\ddot{\vec{r}}_n - (2\,\lambda + 1)\, i\, \omega\, \dot{\vec{r}}_n - \lambda\, (\lambda + 1)\, \omega^2\, \vec{r}_n$$
$$= \frac{a}{\vec{r}_1\cdot\vec{r}_2\wedge\vec{r}_3}\sum_{m=1,2,3\,\mathrm{mod}\,(3)}\left(\dot{\vec{r}}_m - i\,\lambda\,\omega\,\vec{r}_m\right)\left[\left(\dot{\vec{r}}_n - i\,\lambda\,\omega\,\vec{r}_n\right)\cdot\vec{r}_{m+1}\wedge\vec{r}_{m+2}\right],$$
$$n = 1,\, 2,\, 3 \tag{4.229}$$

(same notation as in **Example 4.4-1**).

Example 4.4-3 A *solvable, entirely isochronous* 4-body problem is characterized by the *rotation-invariant Newtonian* equations of motion

$$\ddot{\vec{r}}_n - (2\,\lambda + 1)\,i\,\omega\,\dot{\vec{r}}_n - \lambda\,(\lambda+1)\,\omega^2\,\vec{r}_n$$
$$= \frac{a}{D} \sum_{m=1,2,3,4 \mod(4)} \left\{ (-)^m \left(\dot{\vec{r}}_m - i\,\lambda\,\omega\,\vec{r}_m \right) \cdot \right.$$
$$\left. \cdot \left[\left(\dot{\vec{r}}_n - i\,\lambda\,\omega\,\vec{r}_n \right) \cdot (\vec{r}_{m+1} - \vec{r}_{m+2}) \wedge (\vec{r}_{m+2} - \vec{r}_{m+3}) \right] \right\},$$
$$n = 1,\, 2,\, 3,\, 4, \tag{4.230a}$$

where

$$D = (\vec{r}_2 - \vec{r}_1) \cdot (\vec{r}_3 - \vec{r}_1) \wedge (\vec{r}_4 - \vec{r}_1)\,, \tag{4.230b}$$

and the rest of the notation is as in **Example 4.4-1**. Note that for $\lambda = 0$ this 4-body problem is also *translation-invariant.*

Example 4.4-4 An *integrable, entirely isochronous* scalar-vector N-body problem is characterized by the *rotation-invariant Newtonian* equations of motion

$$\ddot{\vec{r}}_n - i\,\omega\,\dot{\vec{r}}_n - 2\,\omega^2\,\vec{r}_n = c \left[\left(\dot{\vec{r}}_n - i\,\omega\,\vec{r}_n \right) (\rho_{n+1} - 2\,\rho_n + \rho_{n-1}) \right.$$
$$\left. + (\dot{\rho}_n - i\,\omega\,\rho_n)\,(\vec{r}_{n+1} - 2\,\vec{r}_n + \vec{r}_{n-1}) + \left(\dot{\vec{r}}_n - i\,\omega\,\vec{r}_n \right) \wedge (\vec{r}_{n+1} - \vec{r}_{n-1}) \right], \tag{4.231a}$$

$$\ddot{\rho}_n - i\,\omega\,\dot{\rho}_n - 2\,\omega^2\,\rho_n$$
$$= c \left[\dot{\rho}\,(\rho_{n+1} - 2\,\rho_n + \rho_{n-1}) - \left(\dot{\vec{r}}_n - i\,\omega\,\vec{r}_n \right) \cdot (\vec{r}_{n+1} - 2\,\vec{r}_n + \vec{r}_{n-1}) \right]. \tag{4.232a}$$

Note the nearest-neighbor character of these interactions. Of course these equations of motion must be supplemented by border conditions specifying the values of the dependent variables when their indices take the values $n = 0$ and $n = N+1$. Periodic conditions,

$$\vec{r}_0 = \vec{r}_N,\ \rho_0 = \rho_N, \quad \vec{r}_{N+1} = \vec{r}_1,\ \rho_{N+1} = \rho_1,$$

as well as *free ends* conditions,

$$\vec{r}_0 = \vec{r}_{N+1} = 0,\ \rho_0 = \rho_{N+1} = 0,$$

certainly maintain the *integrable* character of this model and thereby guarantee the *complete periodicity* of the *generic* solutions of these ω-modified models.

Example 4.4-5 A *solvable, entirely isochronous* scalar-vector N^2-body problem is characterized by the *rotation-invariant* Newtonian equations of motion

$$\ddot{\vec{r}}_{nm} - i\,\omega\,\dot{\vec{r}}_{nm} - 2\,\omega^2\,\vec{r}_{nm}$$
$$= \sum_{m_1,m_2=1}^{N} g_{m_1 m_2}\Big[\Big(\dot{\vec{r}}_{nm_1} - i\,\omega\,\vec{r}_{nm_1}\Big)\wedge\vec{r}_{m_2 m} + 2\,\vec{r}_{nm_1}\wedge\Big(\dot{\vec{r}}_{m_2 m} - i\,\omega\,\vec{r}_{m_2 m}\Big)\Big]$$
$$+ \sum_{m_1,m_2,m_3,m_4=1}^{N} g_{m_1 m_2}\,g_{m_3 m_4}\,\big\{\vec{r}_{nm_1}\,\big[(\vec{r}_{m_2 m_3}\cdot\vec{r}_{m_4 m}) - \rho_{m_2 m_3}\,\rho_{m_4 m}\big]$$
$$+\vec{r}_{m_2 m_3}\big[(\vec{r}_{nm_1}\cdot\vec{r}_{m_4 m}) - \rho_{nm_1}\,\rho_{m_4 m}\big] + \vec{r}_{m_4 m}\big[(\vec{r}_{nm_1}\cdot\vec{r}_{m_2 m_3}) - \rho_{nm_1}\,\rho_{m_2 m_3}\big]\big\}\,, \tag{4.233a}$$

$$\ddot{\rho}_{nm} - i\,\omega\,\dot{\rho}_{nm} - 2\,\omega^2\,\rho_{nm}$$
$$= \sum_{m_1,m_2=1}^{N} g_{m_1 m_2}\,\Big[\Big(\dot{\vec{r}}_{nm_1} - i\,\omega\,\vec{r}_{nm_1}\Big)\cdot\vec{r}_{m_2 m} - \big(\dot{\rho}_{nm_1} - i\,\omega\,\rho_{nm_1}\big)\,\rho_{m_2 m}$$
$$+2\,\vec{r}_{nm_1}\cdot\Big(\dot{\vec{r}}_{m_2 m} - i\,\omega\,\vec{r}_{m_2 m}\Big) - 2\,\rho_{nm_1}\,\big(\dot{\rho}_{m_2 m} - i\,\omega\,\rho_{m_2 m}\big)\Big]$$
$$+ \sum_{m_1,m_2,m_3,m_4=1}^{N} g_{m_1 m_2}\,g_{m_3 m_4}\,\big\{\rho_{nm_1}\,\big[(\vec{r}_{m_2 m_3}\cdot\vec{r}_{m_4 m}) - \rho_{m_2 m_3}\,\rho_{m_4 m}\big]$$
$$+\rho_{m_2 m_3}\,(\vec{r}_{nm_1}\cdot\vec{r}_{m_4 m}) + \rho_{m_4 m}\,(\vec{r}_{nm_1}\cdot\vec{r}_{m_2 m_3})$$
$$-(\vec{r}_{nm_1}\wedge\vec{r}_{m_2 m_3})\cdot\vec{r}_{m_4 m}\big\}\,. \tag{4.233b}$$

Note that this system features N^2, essentially *arbitrary*, (*complex*) coupling constants g_{nm} (the only restriction is that the $N \times N$ matrix with matrix elements g_{nm} be invertible). The *solvable* character of this model entails that *all* its solutions are *completely periodic* with period $T = 2\,\pi\,/\,\omega$.

Example 4.4-6 This next-to-last example is characterized by the N *Newtonian* equations of motion

$$\ddot{\vec{r}}_n + i\,\Omega\,\dot{\vec{r}}_n + 2\,\Omega^2\,\vec{r}_n = \sum_{m=1,m\neq n}^{N} \frac{M_m\,\vec{r}_{mn}}{r_{mn}^3}\,. \tag{4.234}$$

Here, as always in this Section 4.4, superimposed arrows denote 3-vectors, and we use the short-hand notation $\vec{r}_{mn} \equiv \vec{r}_m - \vec{r}_n$. Actually the property of *isochrony* of this ω-modified N-body problem would hold no less if the dependent variables were S-vectors, with S an arbitrary positive integer; the interest of focussing on ordinary, *three-dimensional* space is because these equations of motion become those of the classical gravitational N-body problem when Ω vanishes. This system, (4.234), is indeed obtained from that classical problem by applying the standard trick with $\lambda = -2\,/\,3$ (and then setting $\Omega = \omega\,/\,3$). The fact that this system—whose dependent variables $\vec{r}_n(t)$ are of course *complex*—is *isochronous* is (perhaps) remarkable; it is indeed easy to show [40] that there exists an open, *fully dimensional* region in the phase space of this N-body problem (4.234) where *all* solutions are *completely periodic* with period $\tilde{T} = 2\pi/\Omega$. [The alert reader will note that this same model was already treated in Section 4.1.2, see (4.41)].

Example 4.4-7 The last example we report is characterized by the following N^2 scalar-vector *Newtonian* equations of motion with *quadratic* nonlinearity:

$$\ddot{\vec{r}}_{nm} - 5\,i\,\omega\,\dot{\vec{r}}_{nm} - 6\,\omega^2\,\vec{r}_{nm}$$
$$= i \sum_{m_1,m_2,m_3,m_4=1}^{N} \eta_{nmm_1m_2m_3m_4} \left[\rho_{m_1m_2}\vec{r}_{m_3m_4} + \rho_{m_3m_4}\,\vec{r}_{m_1m_2} + \vec{r}_{m_1m_2} \wedge \vec{r}_{m_3m_4}\right], \tag{4.235a}$$

$$\ddot{\rho}_{nm} - 5\,i\,\omega\,\dot{\rho}_{nm} - 6\,\omega^2\,\rho_{nm}$$
$$= i \sum_{m_1,m_2,m_3,m_4=1}^{N} \eta_{nmm_1m_2m_3m_4} \left[\rho_{m_1m_2}\,\rho_{m_3m_4} - (\vec{r}_{m_1m_2} \cdot \vec{r}_{m_3m_4})\right]. \tag{4.235b}$$

Here the N^6 constants $\eta_{nmm_1m_2m_3m_4}$ are *arbitrary* (possibly *complex*), and the rest of the notation is, we trust, self-explanatory. The *isochronous* character of this many-body problem is demonstrated at the end of the following Section 4.5.

4.5 Multi-dimensional systems

We begin this Section 4.5 by essentially repeating the proof—already provided in Section 2.2—of the *entirely isochronous* character of the many-body models in multi-dimensional space (1.12) and (1.13)—namely, the fact that the *generic* solution of these systems is *completely periodic* with period T, see (1.2) with (1.3). It is thus seen that these findings are special cases of a more general result. We then show that much more general many-body systems in multi-dimensional space are also *isochronous*, although for these systems the property of *complete periodicity* holds only in a part of their natural phase space (yet in an open, *fully dimensional*, region, as required for them to qualify as *isochronous*).

Let us recall to begin with a *lemma* due to V. I. Inozemtsev (already reported above as *Lemma 2.2-1*).

Lemma 4.5-1 The *general* solution of the matrix evolution equation

$$U'' = c\,U^3 \tag{4.236}$$

—with $U \equiv U(\tau)$ a *square matrix* of *arbitrary* rank, and c an *arbitrary* scalar constant—is a *meromorphic* function of its independent variable τ. ⊡

We then apply the trick (with $\lambda = 1$) to this matrix ODE by setting

$$W(t) = \exp(i\,\omega\,t)\; U(\tau), \tag{4.237a}$$

$$\tau = \frac{\exp(i\,\omega\,t) - 1}{i\,\omega}, \tag{4.237b}$$

entailing that $W(t)$ satisfies the matrix ODE

$$\ddot{W} - 3\,i\,\omega\,\dot{W} - 2\,\omega^2\,W = c\,W^3; \tag{4.238}$$

while the *meromorphic* character of the *general* solution $U(\tau)$ of the matrix ODE (4.236) clearly implies *complete periodicity* of the *generic* solution of this *isochronous* ODE (4.238),

$$W(t+T) = W(t), \quad T = \frac{2\,\pi}{\omega}. \tag{4.239}$$

It is as well clear that the solutions of this *isochronous* ODE (4.238) that are *not* periodic because they run into a singularity as functions of the *real* independent variable t ("time") are *exceptional*—corresponding to a *lower dimensional* set of initial data—as entailed by the related phenomenon (see (4.237a)) that a pole of the corresponding solution $U(\tau)$ fall *exactly*, in the *complex* τ-plane, on the circle C, of radius $1/\omega$ and centered at $\tau_C = i/\omega$, on which the variable τ travels counterclockwise round and round as the *real* variable t evolves from $t = 0$ towards $t = \infty$ (see (4.237b)).

The two *nonlinear harmonic oscillators* models (1.12) and (1.13) are then merely special instances of this general class of matrix evolution equations (of *arbitrary* rank!) (4.238), obtained by appropriate reparameterizations (see for instance Section 5.6.5 of [37]) of the matrix W in terms of the NMS components of the NM S-vectors $\underline{z}_{nm}$.

Clearly other *entirely isochronous* many-body problems, or equivalently *nonlinear harmonic oscillators* models, can be obtained via this same approach by using other parameterizations of the matrix W.

A much larger class of *isochronous* many-body problems in multi-dimensional space are clearly contained as special cases (with polynomial right-hand sides) within the **Examples 4.2.1-5** and **4.2.1-6** treated above. A convenient way to obtain some of them (featuring *covariant* S–vector *Newtonian* equations of motion) is to start from the *matrix* generalization of the findings reported in these two previous examples, which can be formulated as follows (their proofs are obvious generalizations of the appropriate subcases of a *matrix* generalization of *Lemma 4.1.1-1*, resulting from the replacement of the N-vector $\underline{z}$ with the matrix W).

The generalization of (4.48) states that an *isochronous* matrix ODE reads as follows:

$$\ddot{W} - 3\,i\,\omega\,\dot{W} - 2\,\omega^2\,W = \sum_{k=1}^{K} F^{(2+k)}(W), \tag{4.240a}$$

where the K matrix-valued functions $F^{(2+k)}(W)$—in addition to depend *analytically* on *all* the matrix elements of W—are required to satisfy the scaling property

$$F^{(2+k)}(c\,W) = c^{2+k}\,F^{(2+k)}(W), \quad k = 1, \ldots, K, \tag{4.240b}$$

where c is a *scalar* parameter. The *open* domain of initial data that yield *isochronous* evolutions of $W(t)$ (with period T, see (3.8b)) is characterized by matrices $W(0)$ the components of which are *quite small* (in modulus), while the components of the matrices $\dot{W}(0)$ are as well *quite small* (in modulus) and thereby also consistent with the (approximate) condition

$$\dot{W}(0) - i\,\omega\,W(0) \approx 0. \tag{4.241}$$

An interesting special case is again that in which the functions $F^{(2+k)}(W)$ are *polynomial* (of course *homogeneous*, of degree $2+k$) in the matrix W; and again particularly interesting is the even more special case with $K=1$, that describes an assembly of oscillators interacting via (*one-body velocity-dependent*) *linear* forces and (*three-body velocity-independent*) *cubic* forces, reading as follows:

$$\ddot{W} - 3\,i\,\omega\,\dot{W} - 2\,\omega^2\,W = \sum_{\ell=1}^{L} A^{(\ell)}\,W\,B^{(\ell)}\,W\,C^{(\ell)}\,W\,D^{(\ell)}, \tag{4.242}$$

where the constant matrices $A^{(\ell)}$, $B^{(\ell)}$, $C^{(\ell)}$, $D^{(\ell)}$ are *arbitrary* and the *positive integer* L is *arbitrary* as well (but an excessively large L entails no additional generality: see below).

Let us moreover assume that all these matrices, W, $A^{(\ell)}$, $B^{(\ell)}$, $C^{(\ell)}$, $D^{(\ell)}$, have a *block structure*, indicating with the same symbol, but with (upper, bracketed) indices n, m, the matrices located at the position n, m within that block structure; and let us also assume that all the constant matrices $A^{(\ell)(nm)}$, $B^{(\ell)(nm)}$, $C^{(\ell)(nm)}$, $D^{(\ell)(nm)}$ are just multiples of the *unit* matrix,

$$A^{(\ell)(nm)} = a^{(\ell)(nm)}\,\mathbf{1}, \;\; B^{(\ell)(nm)} = b^{(\ell)(nm)}\,\mathbf{1},$$
$$C^{(\ell)(nm)} = c^{(\ell)(nm)}\,\mathbf{1}, \;\; D^{(\ell)(nm)} = d^{(\ell)(nm)}\,\mathbf{1}. \tag{4.243}$$

In this manner the matrix equation (4.242) becomes

$$\ddot{W}^{(nm)} - 3\,i\,\omega\,\dot{W}^{(nm)} - 2\,\omega^2\,W^{(nm)} = \sum_{m_1,m_2,m_3,m_4,m_5,m_6=1}^{N} \eta_{nmm_1m_2m_3m_4m_5m_6}\,W^{(m_1m_2)}\,W^{(m_3m_4)}\,W^{(m_5m_6)}, \tag{4.244a}$$

$$\eta_{nmm_1m_2m_3m_4m_5m_6} = \sum_{\ell=1}^{L} a^{(\ell)(nm_1)}\,b^{(\ell)(m_2m_3)}\,c^{(\ell)(m_4m_5)}\,d^{(\ell)(m_6m)}, \tag{4.244b}$$

where N is the block-rank of the matrix W, the N^2 square matrices $W^{(nm)}$ have the same *arbitrary* rank (not yet assigned), and the N^8 *scalar* constants

$\eta_{nmm_1m_2m_3m_4m_5m_6}$ can also be *arbitrarily* assigned, thanks to the arbitrariness of the $4\,L\,N^2$ constants $a^{(\ell)(nm)}$, $b^{(\ell)(nm)}$, $c^{(\ell)(nm)}$, $d^{(\ell)(nm)}$ (since L is also *a priori arbitrary*). Through an appropriate parameterization of these matrices $W^{(nm)}$ in terms of S-vectors one can then transform these matrix ODEs into *covariant Newtonian* equations of motions in S-dimensional space.

Likewise, the generalization of (4.44) states that an *isochronous* matrix ODE reads as follows:

$$\ddot{W} - 5\,i\,\omega\,\dot{W} - 6\,\omega^2\,W = \sum_{k=1}^{K} F^{(\frac{3+k}{2})}(W), \tag{4.245a}$$

where the matrix-valued functions $F^{(\frac{3+k}{2})}(W)$—in addition to depend *analytically* on *all* the matrix elements of W—are required to satisfy the scaling property

$$F^{(\frac{3+k}{2})}(c\,W) = c^{\frac{3+k}{2}}\,F^{(\frac{3+k}{2})}(W), \quad k = 1, \ldots, K, \tag{4.245b}$$

where c is a *scalar* parameter. The *open* domain of initial data that yield *isochronous* evolutions of $W\,(t)$ (with period $T = 2\,\pi\,/\,\omega$) is again characterized by matrices $W(0)$ the components of which are *quite small* (in modulus), while the components of the matrices $\dot{W}(0)$ are as well *quite small* (in modulus) and thereby also consistent with the (approximate) condition

$$\dot{W}(0) - 2\,i\,\omega\,W(0) \approx 0. \tag{4.246}$$

An interesting special case is again that in which the functions $F^{(\frac{3+k}{2})}(W)$ are *polynomial* (of course, *homogeneous*, of degree $(3+k)\,/\,2)$ in the matrix W; and again particularly interesting is the even more special case with $K = 1$, that describes an assembly of oscillators interacting via (*one-body velocity-dependent*) *linear* forces and (*two-body velocity-independent*) *quadratic* forces, for instance, reading as follows:

$$\ddot{W} - 5\,i\,\omega\,\dot{W} - 6\,\omega^2\,W = \sum_{\ell=1}^{L} A^{(\ell)}\,W\,B^{(\ell)}\,W\,C^{(\ell)}, \tag{4.247}$$

where the constant matrices $A^{(\ell)}$, $B^{(\ell)}$, $C^{(\ell)}$ are *arbitrary* and the *positive integer* L is *arbitrary* as well.

Let us moreover assume again that all these matrices, W, $A^{(\ell)}$, $B^{(\ell)}$, $C^{(\ell)}$, have a *block structure*, indicating with the same symbol, but with (upper, bracketed) indices n, m, the matrices located at the position n, m within that block structure; and let us also assume that all the constant matrices $A^{(\ell)(nm)}$, $B^{(\ell)(nm)}$, $C^{(\ell)(nm)}$ are just multiples of the *unit* matrix

$$A^{(\ell)(nm)} = a^{(\ell)(nm)}\,\mathbf{1}, \; B^{(\ell)(nm)} = b^{(\ell)(nm)}\,\mathbf{1}, \; C^{(\ell)(nm)} = c^{(\ell)(nm)}\,\mathbf{1}. \tag{4.248}$$

In this manner the matrix equation (4.247) becomes

$$\ddot{W}^{(nm)} - 5\,i\,\omega\,\dot{W}^{(nm)} - 6\,\omega^{2}\,W^{(nm)}$$

$$= \sum_{m_1,m_2,m_3,m_4=1}^{N} \eta_{nmm_1m_2m_3m_4}\,W^{(m_1m_2)}\,W^{(m_3m_4)}, \tag{4.249a}$$

$$\eta_{nmm_1m_2m_3m_4} = \sum_{\ell=1}^{L} a^{(\ell)(nm_1)}\,b^{(\ell)(m_2m_3)}\,c^{(\ell)(m_4m)}, \tag{4.249b}$$

where N is the block-rank of the matrix W, the N^2 square matrices $W^{(nm)}$ have the same *arbitrary* rank (not yet assigned), and the N^6 *scalar* constants $\eta_{nmm_1m_2m_3m_4}$ can also be *arbitrarily* assigned, thanks to the arbitrariness of the $3\,L\,N^2$ constants $a^{(\ell)(nm)}$, $b^{(\ell)(nm)}$, $c^{(\ell)(nm)}$ (since L is also *a priori arbitrary*). By an appropriate parameterization of these matrices $W^{(nm)}$ in terms of S-vectors one can then transform these matrix ODEs into *covariant* Newtonian equations of motions in S-dimensional space. For instance, assuming the square matrices $W^{(nm)}$ to have rank 2 and using for them the simple parameterization

$$W^{(nm)} = \rho_{nm} + i\,\vec{r}_{nm}\cdot\vec{\sigma}, \tag{4.250}$$

where superimposed vectors identify 3-vectors and the 3 components of the 3-vector $\vec{\sigma}$ are the three Pauli matrices, one easily obtains the result reported above as **Example 4.4-7**.

4.N Notes to Chapter 4

The presentation of Section 4.1 follows quite closely [50]. Note the analogy of the proof presented in that Section 4.1 to that presented at the end of Section 2.2 (as already indicated in Section 2.N, the standard result on the region of *holomorphy* of solutions of analytic ODEs can be found, for instance, in Section 13.21 of [105]).

For **Example 4.1.2-1** see also [37], [38], [86] and [42]; these results were partly reported already in Chapter 1 and in Section 2.2.

For **Example 4.1.2-2** see also [37], [61], [66], [45]. Results of this kind were reported already in Chapter 1 and in Section 2.2, and note that the class of evolution ODEs (4.33) includes the ω-modified version of (3.29), see (3.71). See also Section 4.2.2.

The *isochronous* character of the ω-modified gravitational N-body problem mentioned in the context of **Example 4.1.2-3** (see (4.41)) was pointed out in [40]; this remarkable system is also presented as **Example 4.4-6**.

Some results mentioned in the context of **Example 4.1.2-5** were already reported in Section 2.2 (in particular those suggesting the notion of *nonlinear harmonic oscillators* [75]); see also Section 4.4.

Also the results of **Example 4.1.2-6** are considered again in Section 4.4.

For the standard "Poincaré–Dulac" techniques mentioned at the end of Section 2.2 see for instance [90] [135] [94] [97], as well as the discussion of this issue in the last section of [64].

In connection with the treatment of Section 4.1 let us remark that, while in the ω-modified dynamical systems considered there the *linear* part of the evolution is the *same* for every component of the evolving N-vector $\underline{z}(t)$, see (4.9), it is actually possible indeed easy, in the context of this approach, to dispense with such a restriction; as done in [64] in an analogous, but more general, context than that considered herein. The N-vector equation of motion (4.9) gets then replaced by the (more general) N equations of motion

$$\ddot{z}_n - i\,(2\,\lambda_n + 1)\,\omega\,\dot{z}_n - \lambda_n\,(\lambda_n + 1)\,\omega^2\,z_n = \sum_{k=1}^{K} F_n^{(a_{nk},\,b_{nk})}(z_m,\,\dot{z}_m - i\,\lambda_m\,\omega\,z_m), \tag{4.251}$$

where the N numbers λ_n are required to be *rational* but are otherwise arbitrary and the $N\,K$ functions $F_n^{(a_{nk},\,b_{nk})}(\underline{z},\,\underline{\dot{z}})$ must satisfy scaling properties that generalize (4.8). We refer for such an extension to the dissertation by Mauro Mariani [112], which also includes the analogous treatment of dynamical systems characterized by *first-order*, rather than *second-order*, ODEs; and see also the *isochronous* systems treated at the end of Section 4.2.4, where it is moreover shown how this approach can be used even when the scaling (4.8) does *not* apply.

A standard reference for the material identified by the title of Section 4.2.1 is [37]; see also Section 1.N.

The presentation of Section 4.2.2 is mainly based on [21]; the original goldfish model (4.46) was introduced and shown to be solvable in [29]; the name "goldfish" was introduced in a contribution [36] to a conference celebrating the 60th birthday of V. E. Zakharov and has been subsequently extensively employed, see for instance [66] [45] [65] [47] [48] [20] [21] [79]; Zakharov's quote is from [143]. The papers in which the recent results reported in Section 4.2.2 were obtained are [65] (published with much delay due to mishandling by the journal it was firstly submitted to) [47] [48] [76] [20] [21] [79] and also [58] [59]. With respect to the developments reported in Section 4.2.2 it seems appropriate to mention that the general idea of investigating how the *eigenvalues* of a matrix change when the matrix changes in some continuous manner (for instance some, or several, of its matrix elements are linear functions of an evolving parameter) is quite natural, hence it has been investigated in several applicative contexts, see for instance [122] [141] [142] [116]. Papers where this investigation has been made in order to investigate *integrable* dynamical systems, and which therefore should be considered as precursors of the method described in this Section 4.2.2, include primarily papers by M. A. Olshanetsky and A. M. Perelomov [118] [119] [120] [123] [121] (see also Section 2.1.3.2, entitled "The technique of solution of Olshanetsky and Perelomov (OP)", of [37]) and, more recently, by J. Arnlind and J. Hoppe [5]; see also [7] [125] [117]. Papers where this approach was used but the auxiliary variables rather than being eliminated were provided with some "physical" interpretation (as some kind of spin variables) are primarily [98] and [138] [139]; see also [8] [9] [109] [6].

The *second ansatz* (4.83) was guessed [65] on the basis on an analogous guess made originally in [14] in connection with the search and discovery of the second Lax matrix for certain *integrable* N-body problems belonging to the Ruijsenaars–Schneider class (see for instance [37]).

For more details on the material in **Example 4.2.2-2** see [65].

For more details on the material in **Example 4.2.2-3** respectively **Example 4.2.2-4** see [47] respectively [48]. For the proof of *Proposition 4.2.2-4* see Appendix B.

For more details on the material in **Example 4.2.2-5** see [99].

For more details on the material in **Example 4.2.2-6** see [76]. For the proof of *Proposition 4.2.2-8* see Appendix B.

For more details on the material in **Examples 4.2.2-7, 4.2.2-8** respectively **4.2.2-9** see [20], [21] respectively [79]. In these references, *Propositions* are reported (and proven) analogous to *Propositions 4.2.2-1* and *4.2.2-5* (see **Examples 4.2.2-4** and **4.2.2-6**). Note that in this monograph—as prototypes of analogous treatments—we only report (in Appendix B) the proofs of the two *Propositions 4.2.2-1* and *4.2.2-5.*

For more details on the material in **Example 4.2.2-10** see [69] [70] (the requirements (4.176f)—needed to guarantee that the integer k in all time-dependences of type $\exp(ik\omega t)$ *not* vanish, $k \neq 0$—had been inadvertently omitted in [69]).

For more details on the material in **Example 4.2.2-11** see [58] and [59]: in particular (4.188), (4.190) respectively (4.191) correspond to eqs. (1.4), (2.19) respectively (2.25) (with $c = -1$) of [58]. For the original models whose ω-modified versions are discussed in this example see [26] [127] [45] and [128] [140] [129] [103] [102] (or see [37]). Models featuring only "nearest neighbor" interactions such as those considered in this example are often called "Toda-type" models, because the first *nonlinear integrable* many-body problem of this type was invented by M. Toda, see for instance [132].

The results summarized in Section 4.2.3 are based on [64].

For more details on the results summarized in Section 4.2.4 see [46].

For the treatment of Section 4.3 see [34] and Chapter 4 of [37] (including the review of relevant literature in Section 4.N of that book).

The result of **Example 4.4-1** follows by applying the standard trick to the simpler version of the first example of Section 5.1 of [37] that obtains by setting in it $a = b = 0$ and $c = 1$ (see the eqs. (5.1-1) and (5.1-4) of [37]). The alert reader will use the findings reported in Section 5.6.1 of [37] to obtain and discuss the analogous result in the more general case in which c is an *arbitrary* (possibly even *complex*) constant. For another version of this model which is *entirely isochronous* even in the *real* case see Section 5.6.1 of [37].

The results of **Examples 4.4-2** and **4.4-3** follow in an analogous manner from the examples given as eqs. (5.6.1-5) and (5.6.1-5) in [37]. For previous treatments of quite analogous systems see [77] [78].

The result of **Example 4.4-4** is obtained by applying the standard trick with $\lambda = 1$ to the example characterized by the equations of motion (5.6.3-9) of [37].

The result of **Example 4.4-5** is obtained by applying the standard trick with $\lambda = 1$ to the example characterized by the equations of motion (5.6.3-15) of [37] with the $3\,N^2$ constants a_{nm}, b_{nm}, c_{nm} set to *zero* and $d_{nm} = g_{nm}$ (merely a notational change).

For more details on the finding reported as **Example 4.4-6** see [40]; and see also **Example 4.1.2-3**.

Lemma 4.5-1 is a consequence of the explicit—if quite involved—general solution of (4.236) in terms of elliptic sigma functions obtained by V. I. Inozemtsev [106].

For a detailed presentation of the idea to manufacture treatable many-body problems characterized by *covariant* Newtonian equations of motion in multi-dimensional space by starting from treatable matrix ODEs see Chapter 5 of [37]. This approach was exploited previously in [15] [16] [17] and subsequently in [108] [18] [19]; for a convenient reparameterization of matrices in terms of vectors yielding *covariant* equations see also [57]. Standard formulas involving Pauli matrices, useful in connection with the standard parameterization (4.250), can be found in [37] (see, in particular, Appendix H of that book).

5

ISOCHRONOUS HAMILTONIAN SYSTEMS ARE NOT RARE

In the preceding part of this monograph quite a few *isochronous* systems have been manufactured, whose equations of motion are of *Newtonian* type, allowing to interpret these models as N-body problems. These models are generally ω-modified systems obtained by applying the trick to unmodified systems, themselves generally interpretable as N-body problems but not being *isochronous*. Some of these unmodified N-body problems are *Hamiltonian*, namely their *Newtonian* equations of motion can be obtained in the standard manner in a *Hamiltonian* framework. In some cases the *isochronous* ω-modified N-body problems obtained in this manner are as well *Hamiltonian*, but often this does not appear to be the case. For instance, as shown in Chapter 1, the *Newtonian* equations of motion of the unmodified (generalized) *goldfish* model,

$$\ddot{z}_n = 2 \sum_{n=1,\, m \neq n}^{N} \frac{a_{nm}\, \dot{z}_n\, \dot{z}_m}{z_n - z_m}, \tag{5.1}$$

are obtained in the standard manner from the (not normal) *Hamiltonian*

$$H\left(\underline{p}, \underline{z}\right) = \sum_{n=1}^{N} \left[\exp\left(c\, p_n\right) \prod_{m=1, m \neq n}^{N} \left(z_n - z_m\right)^{-a_{nm}} \right], \tag{5.2}$$

while the *Newtonian* equations of motion of the corresponding ω-modified N-body problem,

$$\begin{aligned} &\ddot{z}_n - (2\,\lambda + 1)\, i\, \omega\, \dot{z}_n - \lambda\, (\lambda + 1)\, \omega^2\, z_n \\ &\quad = 2 \sum_{n=1,\, m \neq n}^{N} \frac{a_{nm}\, \left(\dot{z}_n - i\,\lambda\,\omega\, z_n\right)\left(\dot{z}_m - i\,\lambda\,\omega\, z_m\right)}{z_n - z_m}, \end{aligned} \tag{5.3}$$

which as we know is *isochronous* provided λ is a *rational* number, seem obtainable from a Hamiltonian *only* if $\lambda = 0$ or if $\lambda = -1$ (see (1.5a), with ω replaced by $-\omega$ if $\lambda = -1$).

In this Chapter 5 we show how it is possible to ω-modify a *Hamiltonian* $H\left(\underline{p}, \underline{q}\right)$ so that the ω-modified *Hamiltonian* $\tilde{H}\left(\underline{p}, \underline{q}; \omega\right)$ thereby obtained is *isochronous*, namely its *Hamiltonian* equations of motion feature an open, *fully dimensional*, region in their phase space where *all* their solutions are *completely periodic* with the standard period

$$T = \frac{2\,\pi}{\omega}. \tag{5.4}$$

This is achieved via a new kind of trick—nontrivially related to the previous one—that, as shown in the following Section 5.1, is applicable to a *quite large* class of *Hamiltonian* systems—justifying the title of this chapter. But for definiteness we mainly focus our presentation on the *Hamiltonian* describing a quite general (nonrelativistic) N-body problem—with *arbitrary* interactions, except for the *sole* restriction that they be *translation invariant*, which as we show is *sufficient* for the applicability of this approach.

In the subsequent Section 5.2 we introduce the notion of *partially isochronous* Hamiltonian systems, for which the *complete periodicity* property with a *fixed* period prevails in a region of phase space having a bit *less than full dimensionality* in phase space—for instance, it holds in a phase space region of codimension one or two. Such systems are perhaps interesting because of the variety of tricks that may be used to produce their ω-modified structure, allowing for a larger applicative potentiality, see below.

In Section 5.3 we tersely outline the treatment of more general Hamiltonians.

Examples of *isochronous*, and *partially isochronous*, N-body problems are presented in Section 5.4.

Finally a novel, quite effective, trick to generate *entirely isochronous* (autonomous) *Hamiltonians* is presented in Section 5.5 and its subsections. For historical reasons this last development is presented at the end of this Chapter—it emerged after a first version of this monograph had been completed—but let us re-emphasize that it appears particularly powerful, especially in its version applicable to the nonrelativistic many-body problem with *translation-invariant*, but otherwise *arbitrary*, interactions.

5.1 Another trick

In this section, to illustrate our main finding, we focus on the dynamical system describing the (quite general) nonrelativistic N-body problem, as characterized by the *Hamiltonian*

$$H(\underline{p},\,\underline{q}) = \frac{1}{2}\sum_{n=1}^{N} p_n^2 + V\left(\underline{q}\right) + C, \tag{5.5}$$

where the *arbitrary* constant C is introduced for convenience (see below) and the potential $V\left(\underline{q}\right) \equiv V\left(q_1,\, q_2, ..., q_N\right)$ is assumed to be *translation invariant*:

$$V\left(\underline{q} + a\right) \equiv V\left(q_1 + a,\, q_2 + a, ..., q_N + a\right) = V\left(\underline{q}\right), \tag{5.6a}$$

with a an arbitrary constant, entailing (for infinitesimal a)

$$\sum_{n=1}^{N} \frac{\partial V(\underline{q})}{\partial q_n} = 0. \tag{5.6b}$$

This is the standard *Hamiltonian* for the nonrelativistic N-body problem, entailing the standard *Hamiltonian* equations of motion

$$\dot{q}_n = p_n, \quad \dot{p}_n = -\frac{\partial V(\underline{q})}{\partial q_n}, \tag{5.7}$$

as well as the corresponding *Newtonian* equations of motions

$$\ddot{q}_n = -\frac{\partial V(\underline{q})}{\partial q_n}, \tag{5.8}$$

the total momentum conservation formula

$$P(\underline{p}) = \sum_{n=1}^{N} p_n, \tag{5.9a}$$

$$\dot{P} = 0, \tag{5.9b}$$

and the "center of mass" evolution

$$Q(\underline{q}) = \frac{1}{N} \sum_{n=1}^{N} q_n, \tag{5.10a}$$

$$Q(t) = Q(0) + \frac{P}{N} t. \tag{5.10b}$$

Notational reminder: here and hereafter N is a positive integer ($N \geq 2$), the (*real*) independent variable t has the significance of *physical time*, superimposed dots indicate differentiations with respect to this variable, indices such as n, m run from 1 to N unless otherwise indicated, underlined symbols denote N-vectors, for instance $\underline{q} \equiv (q_1, ..., q_N)$, and we use (here and below) the shorthand notation according to which, say, $Q(t) \equiv Q\left[\underline{q}(t)\right]$. Note that, for simplicity, we assume here all the particles to have the same mass, and the motion to occur in one-dimensional space; but these restrictions can be easily dispensed with, see below.

Let us emphasize that, while in the standard "physical interpretation" of this model the "particle coordinates" $q_n(t)$ are *real*, in the following—as done heretofore in our treatment of *isochronous* systems—we shall generally work with *complex* dependent variables, so that the time evolutions we consider take place in the *complex* plane. Of course by introducing the *real* and *imaginary* parts of these *complex* variables, say $q_n(t) = x_n(t) + i\, y_n(t)$, and likewise the *real* and *imaginary* parts of all other quantities (such as canonical momenta and coupling constants), one can always reformulate the problem under consideration in terms of *real* variables only. Thereby what was originally a *one-dimensional* model gets transformed into a model describing evolutions in a plane; likewise, a model originally describing evolutions in S-*dimensional* space gets thereby transformed into a model describing evolutions in $(2\,S)$-*dimensional* space. Whether such models—describing evolutions in an ambient space with *double* the number of dimensions than the original models—can be formulated in *covariant* form is a question that can only be treated case-by-case. On the other hand, it is generally possible to reformulate a *Hamiltonian* problem from a *complex* to a *real* environment. The formula providing, from the *complex Hamiltonian* $H\,(p_n,\,q_n)$ written in terms of N *complex* canonical momenta p_n and N *complex* canonical coordinates q_n, a *real Hamiltonian* $h\,(\hat{p}_n,\,\check{p}_n;\,\hat{q}_n,\,\check{q}_n)$ featuring $2\,N$ *real* canonical momenta $\hat{p}_n,\,\check{p}_n$ and $2\,N$ *real* canonical coordinates $\hat{q}_n,\,\check{q}_n$, reads

$$\mathrm{Re}\,[H\,(\hat{p}_n - i\,\check{p}_n,\,\hat{q}_n + i\,\check{q}_n)] = h\,(\hat{p}_n,\,\check{p}_n;\,\hat{q}_n,\,\check{q}_n)\,; \tag{5.11a}$$

note the minus sign in the left-hand side. It is indeed well known (see for instance [37]) that, provided the original *Hamiltonian* H is an *analytic* (but not necessarily a singularity-free) function of the canonical momenta and coordinates p_n and q_n—as we always assume—the *complex* Hamiltonian equations of motion entailed by $H\,(p_n,\,q_n)$ are *completely equivalent* to the *real* Hamiltonian equations of motion entailed by $h\,(\hat{p}_n,\,\check{p}_n;\,\hat{q}_n,\,\check{q}_n)$—via the standard identification of the *real* canonical coordinates $\hat{q}_n$ respectively $\check{q}_n$ as the *real* respectively *imaginary* parts of the *complex* coordinates q_n, and of the corresponding *real* canonical momenta $\hat{p}_n$ respectively $\check{p}_n$ as the *real* parts respectively the *opposite* of the *imaginary* parts of the *complex* momenta p_n:

$$p_n = \hat{p}_n - i\,\check{p}_n, \quad q_n = \hat{q}_n + i\,\check{q}_n. \tag{5.11b}$$

Before introducing the *new* approach let us tersely review the application of the *previous* trick in the context of the N-body problem characterized by the *Hamiltonian* (5.5). We recall that the trick consists of the following change of dependent and independent variables:

$$\tilde{q}_n(t) = \exp(i\,\lambda\,\omega\,t)\,q_n(\tau), \tag{5.12a}$$

$$\tau = \frac{\exp(i\,\omega\,t) - 1}{i\,\omega}, \tag{5.12b}$$

where λ is a *rational* constant to be chosen appropriately (see below). Note that this change of variables entails the following relations among the initial data for the "old" dependent variables q_n and the "new" dependent variables $\tilde{q}_n$:

$$q_n(0) = \tilde{q}_n(0), \quad q_n'(0) = \dot{\tilde{q}}(0) - i\,\lambda\,\omega\,\tilde{q}_n(0). \tag{5.13}$$

Notation: here and hereafter primes denote differentiations with respect to the argument of the function they are appended to, and of course $q_n'(0)$ denotes the value of $q_n'(\tau) \equiv d\,q_n(\tau)\,/\,d\,\tau$ at $\tau = 0$.

The insertion of this change of variables, (5.12), in the (*autonomous*) dynamical system characterized by the *Newtonian* equations of motion (5.8) (with t formally replaced by τ), yields the *autonomous* dynamical system

$$\ddot{\tilde{q}}_n - (2\,\lambda + 1)\; i\,\omega\,\dot{\tilde{q}}_n - \lambda\;(\lambda + 1)\;\omega^2\,\tilde{q}_n = -\frac{\partial\,V\,(\underline{\tilde{q}})}{\partial\,\tilde{q}_n}, \tag{5.14a}$$

provided the potential $V(\underline{q})$ satisfies the scaling property

$$V\,(c\,\underline{q}) = c^{\gamma}\,V\,(\underline{q}) \tag{5.14b}$$

(where c is an arbitrary constant and γ is the exponent characterizing the scaling property) and correspondingly the parameter λ is assigned as follows:

$$\lambda = \frac{2}{\gamma - 2}. \tag{5.14c}$$

This new ω-modified system of (*autonomous*) ODEs (5.14), determining the time evolution of the new dependent variables $\tilde{q}_n \equiv \tilde{q}_n(t)$, is the ω-dependent dynamical system and, as we by now well know, it is generally *isochronous* provided the parameter λ is *real* and *rational*. Note that λ is indeed *real* and *rational* provided γ, see (5.14b), is itself *real* and *rational* (and different from 2, $\gamma \neq 2$; see (5.14c)). Also note that the dynamical variables $\tilde{q}_n(t)$ are now necessarily (unless $\lambda = -1/2$, see (5.14a)) evolving in the *complex* $\tilde{q}$-plane; this shall be the case as well for all the evolutions considered hereafter. Of course, in order that derivatives with respect to *complex* variables (such as τ, see (5.12b)) make good sense, one must deal (as we *always* do) with *analytic* functions.

Let us tersely review the argument showing that the time evolution entailed by the equations of motion (5.14) (with λ *real* and *rational*) is indeed *isochronous*. (i) The change of dependent variable (5.12b) entails that, as the (*real*) "physical time" variable t evolves from $t = 0$ onwards, the (*complex*) time-like variable τ goes round and round, in the *complex* τ-plane, on the circle C centered at i/ω and having radius $1/\omega$, traveling a full circle (from $\tau = 0$ to $\tau = 0$) in every time interval T, see (5.4). (ii) Hence, if the functions $q_n(\tau)$ of the *complex* variable τ are *holomorphic* in the (closed) disk D enclosed by the circle C, the corresponding variables $\tilde{q}_n(t)$ (see (5.12) and (5.4)) are λ-*periodic* with period T,

$$\tilde{q}_n(t + T) = \exp(2\,i\,\pi\,\lambda)\,\tilde{q}_n(t), \tag{5.15}$$

and this entails that they are indeed *periodic* with period T if λ is an *integer*, or with a period which is an *integer multiple* of T if λ is a (*real* and) *rational* number. (iii) On the other hand, the functions $q_n(\tau)$ are just the analytic continuation for *complex time* of the solutions of the *Newtonian* equations of motion (5.8)—corresponding to the formal replacement of the *real* variable t by the *complex* variable τ, or equivalently to the replacement of the equations of motion (5.8) with the analogous equations

$$q_n'' = -\frac{\partial V\,(q)}{\partial\, q_n}. \tag{5.16}$$

(iv) It is then clear that, provided the initial data (see (5.13)) for these equations of motion are assigned so as to avoid that their right-hand sides be *singular* (at $\tau = 0$), their solutions $q_n(\tau)$ are certainly *holomorphic* functions of the *complex* variable τ for $|\tau| < \rho$ for some $\rho > 0$, namely (at least) inside a circle of *positive* radius ρ centered at the origin ($\tau = 0$) in the *complex* τ-plane—as implied by the standard theorem guaranteeing the existence, uniqueness and analyticity of the solutions of analytic (systems of) ODEs. The (minimum) value of ρ depends on the *initial* values of the right-hand sides of the equations (of motion in the *complex* time-like variable τ) (5.16), on the distances in the (*complex*) q-plane of the *initial* assignments $q_n(0)$ of the dependent variables from the values of these variables that cause the right-hand sides of the ODEs (5.16) to become *singular*, and on the (moduli of) the *initial* assignments $q_n'(0)$ of the derivatives of the dependent variables (see (5.13)). But it stands to reason (and the diligent reader by now well knows) that there generally exists a set of initial data $\tilde{q}_n(0)$ and $\dot{\tilde{q}}_n(0)$ having *full dimensionality* in the (phase) space of these dependent variables such that (via (5.13)) the corresponding initial data $q_n\,(0)$, $q_n'\,(0)$ entail that

$$\rho > \frac{2}{\omega}. \tag{5.17}$$

(v) This implies that the time evolution of the dependent variables $\tilde{q}_n(t)$ resulting from such initial data is *completely periodic*, since this inequality, (5.17), implies that the circle of radius ρ centered at the origin in the *complex* τ-plane *encloses* the circular disk D, implying the *holomorphy* of the functions $q_n(\tau)$ in this disk, hence the *λ-periodicity of $\tilde{q}_n\,(t)$*, see (5.15).

This concludes our terse review—in the specific context of the standard N-body problem, originally characterized by the *Newtonian* equations of motion (5.8)—of the argument associating to a given dynamical system an ω-modified system—in this case, characterized by the *Newtonian* equations of motions (5.14)—having the remarkable property to be *isochronous*. As indicated by the argument reviewed above, the main condition required to obtain such an *isochronous* system is validity of the scaling property (5.14b) with γ *real* and *rational* ($\gamma \neq 2$). And let us re-emphasize that, as implied by the above argument and as already noted in this monograph, in order that the ω-modified system of *Newtonian* equations (5.14) be *isochronous* it is *not* required that the original unmodified system from which it is obtained (namely, the system of *Newtonian* equations of motion (5.8) corresponding to (5.14) with $\omega = 0$, see (5.16)) be *integrable* (nor even *autonomous*, see for instance Section 4.1)—but let us also recall that, if that system (5.8) is indeed *integrable*, this often implies that *all* its solutions $q_n(\tau)$ are *meromorphic* functions of τ, entailing the *complete λ-periodicity* (5.15) of *all* the *nonsingular* solutions $\tilde{q}_n(t)$ of (5.14a), namely that the *isochrony* region of this ω-modified system (5.14a) coincides with its *entire* phase space (possibly except for a *lower dimensional* set), so that the system is *entirely isochronous.*

But let us conclude this terse review by emphasizing that it is far from obvious—and indeed often *not* true—that the new ω-modified *Newtonian* equations of motion (5.14) be *Hamiltonian*, i.e. obtainable from a *Hamiltonian.*

We now report a *new* technique producing, from a given *Hamiltonian* $H\left(\underline{p},\, \underline{q}\right)$, an ω-modified *Hamiltonian* $\tilde{H}\left(\underline{p},\, \underline{q};\, \omega\right)$ yielding *isochronous* equations of motion. The main requirement on the original *Hamiltonian* $H\left(\underline{p},\, \underline{q}\right)$ for the applicability of this technique is that the dynamics it entails allow the identification of a "collective variable," explicitly defined in terms of the Hamiltonian canonical coordinates and momenta, whose time evolution is essentially identical to *time* itself. If we denote such a variable as $\Theta(\underline{p},\, \underline{q})$, it will be characterized by the formula

$$\left[H(\underline{p},\, \underline{q}),\ \Theta(\underline{p},\, \underline{q})\right] = 1. \tag{5.18}$$

Notation: here and hereafter the *Poisson bracket* $[F,\, G]$ of two functions $F(\underline{p},\, \underline{q})$, $G(\underline{p},\, \underline{q})$ of the canonical coordinates and momenta is defined in the standard manner,

$$\left[F(\underline{p},\, \underline{q}),\ G(\underline{p},\, \underline{q})\right] = \sum_{n=1}^{N} \left[\frac{\partial\, F(\underline{p},\, \underline{q})}{\partial\, p_n}\, \frac{\partial\, G(\underline{p},\, \underline{q})}{\partial\, q_n} - \frac{\partial\, F(\underline{p},\, \underline{q})}{\partial\, q_n}\, \frac{\partial\, G(\underline{p},\, \underline{q})}{\partial\, p_n}\right]. \tag{5.19}$$

The ω-modified Hamiltonian $\tilde{H}(\underline{p},\, \underline{q}; \omega)$ yielding *isochronous* motions is then defined as follows:

$$\tilde{H}(\underline{p},\, \underline{q}; \omega) = u(\underline{p},\, \underline{q}; \omega)\, H(\underline{p},\, \underline{q}), \tag{5.20a}$$

$$u(\underline{p},\, \underline{q}; \omega) = 1 + i\, \omega\, \Theta(\underline{p},\, \underline{q}). \tag{5.20b}$$

Hereafter we only consider time evolutions yielded by this ω-modified Hamiltonian $\tilde{H}(\underline{p},\, \underline{q}; \omega)$: therefore, the time evolution of functions of the *Hamiltonian* variables p_n and q_n—such as $H(\underline{p},\, \underline{q})$, $P(\underline{p})$, $Q(\underline{q})$, $u(\underline{p},\, \underline{q}; \omega)$, see (5.5), (5.9a), (5.10a), (5.20b)—obtains via the time-dependence of these *Hamiltonian* variables $p_n \equiv p_n(t)$, $q_n \equiv q_n(t)$ implied by the *Hamiltonian* equations (see below) associated with the *Hamiltonian* $\tilde{H}(\underline{p},\, \underline{q}; \omega)$; hence the time evolution of these collective variables is given by the standard formula characterizing the *Hamiltonian* evolution of any function $F(t) = F\left[\underline{p}(t),\, \underline{q}(t)\right]$ of the *Hamiltonian* variables,

$$\dot{F} = \left[\tilde{H},\ F\right]. \tag{5.21}$$

In particular—as implied by this formula via (5.20) and (5.18)—the quantities $H(t)$ respectively $u(t; \omega)$ now evolve as follows:

$$\dot{H} = -i\, \omega\, H \tag{5.22a}$$

entailing

$$H(t) = H(0)\, \exp\left(-i\,\omega\, t\right), \tag{5.22b}$$

respectively

$$\dot{u} = i\,\omega\, u \tag{5.23a}$$

entailing

$$u(t;\omega) = u(0;\omega)\, \exp\left(i\,\omega\, t\right). \tag{5.23b}$$

Note the consistency, via (5.20a), of the time evolutions (5.22b) and (5.23b) with the obvious fact that the Hamiltonian $\tilde{H}$ is a constant of motion.

Clearly the class of *Hamiltonians* $\tilde{H}(\underline{p},\, \underline{q};\omega)$, see (5.20), is quite vast. But hereafter we mainly focus, for definiteness, on the (unmodified) N-body *Hamiltonian* (5.5), taking advantage of the fact that in this case, as a consequence of its *Galilean invariance*, see (5.5) and (5.6), a collective variable Θ satisfying (5.18) is provided by the following explicit formula:

$$\Theta(\underline{p},\, \underline{q}) = \frac{N\, Q\left(\underline{q}\right)}{P\left(\underline{p}\right)}, \tag{5.24}$$

see (5.9a) and (5.10a). Of course to this quantity one could add an arbitrary function of P, and also of other quantities, if any, which Poisson-commute with the (unmodified) *Hamiltonian* $H(\underline{p},\, \underline{q})$ and are therefore constants of motion for the evolution determined by this *Hamiltonian*; but for simplicity we refrain from doing so in the following.

We now demonstrate, for this specific case, the *isochronous* character of the ω-modified Hamiltonian $\tilde{H}$, as given by (5.20) with (5.5) and (5.24), (5.9a), (5.10a). But before proceeding with this proof let us emphasize that to prove the *isochronous* character of this ω-modified *Hamiltonian* (5.20) *no* additional property of the original N-body Hamiltonian H, see (5.5), is required besides *Galilean invariance*, in particular, *no* scaling property such as that entailed by the condition (5.14b)—which was instead essential to allow the transformation via the trick (5.12) from the (*autonomous*) *Newtonian* equations of motion (5.8) to the (as well *autonomous*) ω-modified *Newtonian* equations of motion (5.14). On the other hand, the *Newtonian* equations of motion entailed by the ω-modified Hamiltonian $\tilde{H}$, see (5.20), while having the appealing properties to be themselves (of course!) *Hamiltonian* and *isochronous*, seem somewhat less susceptible of a "physical interpretation" than the equations, see (5.14), yielded by the standard approach based on the application of the trick (5.12); indeed these novel *Newtonian* equations of motion cannot be generally written in quite explicit form, although the corresponding *Hamiltonian* equations can be explicitly exhibited and are relatively neat, see below.

The time evolution associated with the Hamiltonian $\tilde{H}(\underline{p}, \underline{q}; \omega)$ (see (5.20) with (5.5)) is clearly given by the following *Hamiltonian* equations of motion:

$$\dot{q}_n = \left[q_n, \tilde{H}\right] = \frac{\partial \tilde{H}}{\partial p_n} = u\, p_n + \frac{(1-u)\, H}{P}, \tag{5.25a}$$

$$\dot{p}_n = \left[p_n, \tilde{H}\right] = -\frac{\partial \tilde{H}}{\partial q_n} = -u\, \frac{\partial V\left(\underline{q}\right)}{\partial q_n} - \frac{i\,\omega\, H}{P}. \tag{5.25b}$$

From the latter Hamiltonian equations, (5.25b), we get (via (5.9a) and (5.6b))

$$\dot{P} = \left[P, \tilde{H}\right] = -\frac{N\, i\, \omega\, H}{P}, \tag{5.26a}$$

entailing, via (5.22b),

$$P(t) = P(0) \left[\frac{1 - \alpha \exp\left(-i\, \omega\, t\right)}{1-\alpha}\right]^{1/2}. \tag{5.26b}$$

Here and below we use the short-hand notation

$$\alpha = \frac{2\, N\, H(0)}{2\, N\, H(0) - P^2(0)}. \tag{5.27}$$

Likewise, from the Hamiltonian equations of motion (5.25a) and (5.10a) we get

$$\dot{Q} = \frac{u\, P}{N} + \frac{(1-u)\, H}{P}, \tag{5.28a}$$

hence, via (5.20b), (5.24) and (5.23b),

$$Q(t) = \left[\frac{Q(0) \exp\left(i\, \omega\, t\right)}{P(0)} + \frac{\exp\left(i\, \omega\, t\right) - 1}{N\, i\, \omega}\right] P(t), \tag{5.28b}$$

as well as (from (5.28a) and (5.20b), (5.24))

$$H = -\frac{P^2 \left(\dot{Q} - i\, \omega\, Q - \frac{P}{N}\right)}{N\, i\, \omega\, Q}. \tag{5.29}$$

Finally, we recall that the time evolution of $H(t)$ and $u(t; \omega)$ is provided by the explicit expressions (5.22) and (5.23), of course now with

$$H(0) = \frac{1}{2} \sum_{n=1}^{N} p_n^2(0) + V\left[\underline{q}(0)\right] + C, \tag{5.30}$$

$$u(0; \omega) = 1 + N\, i\, \omega\, \frac{Q(0)}{P(0)} = 1 + i\, \omega\, \frac{\sum_{n=1}^{N} q_n(0)}{\sum_{n=1}^{N} p_n(0)}. \tag{5.31}$$

It clearly follows from its explicit expression (5.26b) that $P(t)$ has (*primitive*) period T, see (5.4), if $|\alpha| < 1$ and instead period $2T$ if $|\alpha| > 1$:

$$P(t+T) = P(t) \quad \text{if } |\alpha| < 1, \tag{5.32a}$$
$$P(t+2T) = P(t) \quad \text{if } |\alpha| > 1. \tag{5.32b}$$

In the special case when the modulus of α is just unity, $|\alpha| = 1$, the total momentum $P(t)$, see (5.26b), vanishes at some finite (*real*) value $t = t_s$, $P(t_s) = 0$, and clearly at this time t_s the equations of motion (5.25a) and (5.25b) become *singular* due to the blow-up of their right-hand sides. This is not surprising, in view of the appearance of P in the *denominator* in the definition of our Hamiltonian $\tilde{H}(\underline{p}, \underline{q}; \omega)$, see (5.20) with (5.24).

Of course, see (5.28b), the periodicity properties of the total momentum $P(t)$ are shared by the center-of-mass coordinate $Q(t)$,

$$Q(t+T) = Q(t) \quad \text{if } |\alpha| < 1, \tag{5.33a}$$
$$Q(t+2T) = Q(t) \quad \text{if } |\alpha| > 1. \tag{5.33b}$$

Hence we conclude that the motion of the center of mass of our system is *isochronous*, the primitive *isochrony period* being T or $2T$ depending whether the initial data entail that the modulus of the quantity α, see (5.27), is smaller or larger than unity (but recall that $H(t)$ and $u(t;\omega)$ are always *periodic* with period T, see (5.22) and (5.23)).

Let us now turn to the *relative* motion. To this end let us set

$$p_{n,m} = p_n - p_m, \qquad q_{n,m} = q_n - q_m. \tag{5.34}$$

One notes by a routine computation (see (5.24)) that

$$[\Theta,\, p_{n,m}] = 0, \qquad [\Theta,\, q_{n,m}] = 0, \tag{5.35}$$

from which it immediately follows that

$$\dot{q}_{n,m} = \left[\tilde{H},\, q_{n,m}\right] = [H,\, q_{n,m}]\; u = u(0)\; [H,\, q_{n,m}] \exp(i\,\omega\, t), \tag{5.36a}$$
$$\dot{p}_{n,m} = \left[\tilde{H},\, p_{n,m}\right] = [H,\, p_{n,m}]\; u = u(0)\; [H,\, p_{n,m}] \exp(i\,\omega\, t). \tag{5.36b}$$

Note that here we have more ODEs than *independent* unknowns. However, these ODEs all follow from the equations of motion (5.25) hence they are certainly *consistent*: of course only $N-1$ of the ODEs (5.36a) are *independent*, and likewise only $N-1$ of the ODEs (5.36b).

To analyze the time evolution of these dependent variables $p_{n,m}(t)$ and $q_{n,m}(t)$ it is now convenient to perform the following change of (independent and dependent) variables (namely, essentially again the old trick, see (5.12), but now in the simpler version with $\lambda = 0$):

$$q_{n,m}(t) = \xi_{n,m}(\tau), \quad p_{n,m}(t) = \pi_{n,m}(\tau), \tag{5.37a}$$

where the new (*complex*) time-like variable τ is again defined as above (see (5.12b)):

$$\tau = \frac{\exp(i\,\omega\,t) - 1}{i\,\omega}. \tag{5.37b}$$

Note that this definition, (5.37), entails that the initial data for the new dependent variables $\xi_{n,m}$ and $\pi_{n,m}$ coincide with the initial data for the original problem:

$$\xi_{n,m}(0) = q_{n,m}(0), \quad \pi_{n,m}(0) = p_{n,m}(0). \tag{5.38}$$

The time evolution of the dependent variables $\xi_{n,m}(\tau)$ and $\pi_{n,m}(\tau)$ is now given, from (5.36) via (5.37), by the following (*autonomous*) equations of motion:

$$\xi'_{n,m} = u(0;\omega)\,\pi_{n,m}, \tag{5.39a}$$

$$\pi'_{n,m} = -u(0;\omega)\,\frac{\partial V\,(\underline{\xi})}{\partial\,\xi_{n,m}}, \tag{5.39b}$$

where, of course, appended primes indicate differentiations with respect to the new (*complex*) independent variable τ. (We hope the attentive reader will pardon the abuse entailed by our use of the notation $\frac{\partial V(\underline{\xi})}{\partial \xi_{n,m}}$ in the right-hand side of the last equation, and will understand the significance of this notation, which is, of course, permissible thanks to the *translation-invariant* character of the potential $V(\underline{q})$, see (5.6), entailing that $V(\underline{q})$ is only a function of the difference of the particle coordinates, see (5.34) and (5.37a)). And now, by the standard argument associated with the trick—as tersely reviewed above, see the discussion following (5.16), the adaptation of which to the present circumstances is too obvious to require a detailed treatment—one easily concludes that the (*nonautonomous*) dynamical system characterized by the equations of motion (5.36) is *isochronous*, there being an open set of initial data $q_{n,m}(0)$, $p_{n,m}(0)$, having *full dimensionality* in the space of these data, such that the resulting motions are *completely periodic* with period T,

$$q_{n,m}(t+T) = q_{n,m}(t), \quad p_{n,m}(t+T) = p_{n,m}(t). \tag{5.40}$$

To formulate our final result we need coordinates respectively momenta describing the motion *relative* to the center of mass $Q\,(t)$ respectively the total momentum $P\,(t)$. Finding *canonical* coordinates that do this is notoriously awkward, but fortunately not necessary here. Let us therefore introduce the *non-canonical* coordinates (*not* to be confused with the coordinates introduced above when reviewing the "old" trick, see (5.12))

$$\tilde{p}_n = p_n - P, \quad \tilde{q}_n = q_n - Q, \tag{5.41}$$

describing the *relative* motion. It then follows from (5.40) that $\tilde{p}_n$ and $\tilde{q}_n$ are all *isochronous*, as they are clearly expressible in terms of the $q_{n,m}$ and the $p_{n,m}$:

$$\tilde{q}_n(t+T) = \tilde{q}_n(t), \quad \tilde{p}_n(t+T) = \tilde{p}_n(t). \tag{5.42}$$

And from this formula, via (5.41) and (5.32), (5.33), we arrive at our fundamental conclusion, namely that the time evolution entailed by the ω-modified Hamiltonian (5.20) possesses an *open*, *fully dimensional*, set of initial data $\underline{p}(0)$, $\underline{q}(0)$ such that the solutions of the corresponding initial-value problem are *isochronous,*

$$q_n(t+T) = q_n(t), \quad p_n(t+T) = p_n(t) \quad \text{if } |\alpha| < 1, \tag{5.43a}$$

$$q_n(t+2T) = q_n(t), \quad p_n(t+2T) = p_n(t) \quad \text{if } |\alpha| > 1, \tag{5.43b}$$

with the constant α defined in terms of the initial data by (5.27).

This completes the proof of the *isochronous* character of the *Hamiltonian* (5.20) with (5.5) and (5.24), (5.10a), (5.9a).

Let us also exhibit the *Newtonian* equations of motion associated with this *isochronous* dynamical system, that are of course obtained by time-differentiating the first, (5.25a), of the two *Hamiltonian* equations, by then using the second, (5.25b), as well as (5.24), (5.22b) and (5.26a), to get rid of all time-differentiated terms in the right-hand side of the resulting equations, and by finally using again (5.25a) to get rid (to the extent possible, see below) of p_n so as to get *Newtonian* equations of motion that feature in their left-hand sides the "accelerations" $\ddot{q}_n$ and in their right-hand sides the corresponding "forces" expressed in terms of the positions q_m and the velocities $\dot{q}_m$ of the "particles" (moving in the *complex* q-plane). We thus obtain the following (of course *autonomous*) equations of motion:

$$\ddot{q}_n = i\,\omega\,\dot{q}_n - \left(1 + N\,i\,\omega\,\frac{Q}{P}\right)^2 \frac{\partial V\,(\underline{q})}{\partial\,q_n} + F, \tag{5.44a}$$

$$F = -\frac{2\,i\,\omega\,H}{P}\left(1 + \frac{N^2\,i\,\omega\,Q\,H}{2\,P^3}\right) = \ddot{Q} - i\,\omega\,\dot{Q}. \tag{5.44b}$$

Note that the "collective force" F acts equally on *all* the coordinates q_n, and it is moreover easily seen from the results reported above that the time-dependence of this force is *periodic* (with period T or $2T$, see the second version of (5.44b) and (5.33)).

But these equations of motion, (5.44), do not quite have yet *Newtonian* form, because the right-hand sides of (5.44a), namely the "forces", are not yet expressed just in terms of the coordinates q_m and the velocities $\dot{q}_m$: a dependence on the canonical momenta p_m still lingers, albeit only via the collective coordinates $P \equiv P\left(\underline{p}\right)$ and $F \equiv F\left(\underline{p}, \underline{q}\right)$, see (5.9a) and (5.44b) with (5.5). Actually, the second of this collective coordinates can be rather neatly expressed in terms of the first (via (5.29) and (5.44)):

$$F = \frac{P^2 - \left[N\left(\dot{Q} - i\,\omega\,Q\right) - 2\,P\right]^2}{N^2\,Q}. \tag{5.45}$$

But in order to express P in terms of the coordinates q_n and of the velocities $\dot{q}_n$—the first of which actually only enter via $Q\left(\underline{q}\right)$ and $V\left(\underline{q}\right)$, see (5.10a) and (5.5), while the second enter via $\dot{Q} = Q\left(\dot{\underline{q}}\right)$ and the collective coordinate K,

$$K \equiv K\left(\dot{\underline{q}}\right) = \frac{1}{2}\sum_{n=1}^{N}\left(\dot{q}_n\right)^2, \tag{5.46}$$

see below—one should solve the following algebraic equation (of *fifth* degree in P), obtained from (5.5) via (5.25a) with (5.20b) and (5.24) (to obtain p_n in terms of $\dot{q}_n$, Q, P, H) and then via (5.29) (to express H in terms of P, Q and $\dot{Q}$):

$$P^2\left\{-\frac{N\,S^2}{2} + N\,i\,\omega\,Q\,s\,(s-1)\,S + K\right\} = (N\,\omega\,s\,Q)^2\,(V + C), \tag{5.47a}$$

$$S = \dot{Q} - i\,\omega\,s\,Q = \dot{Q} - i\,\omega\,Q - \frac{P}{N}, \tag{5.47b}$$

$$s = 1 + \frac{P}{N\,i\,\omega\,Q}. \tag{5.47c}$$

However if the initial data entail that

$$H(0) \equiv H\left(\underline{p}(0), \underline{q}(0)\right) = 0, \tag{5.48a}$$

yielding via (5.22b)

$$H(t) = 0, \tag{5.48b}$$

the *Newtonian* equations of motion take instead the much simpler, completely explicit, form

$$\ddot{q}_n - i\,\omega\,\dot{q}_n = -\left[\frac{\sum_{n=1}^{N}\dot{q}_n}{\sum_{n=1}^{N}\left(\dot{q}_n - i\,\omega\,q_n\right)}\right]^2 \frac{\partial V\left(\underline{q}\right)}{\partial q_n}, \tag{5.49}$$

as entailed by (5.44a) (with $F = 0$, see (5.44b) with (5.48)) and (5.28a) (with (5.48), (5.20b) and (5.24) yielding $P = N\,(\dot{Q} - i\,\omega\,Q)$); while in the special examples in which the original Hamiltonian H reduces to the free Hamiltonian (i.e., $V + C = 0$, see below), the equation (5.47) for P becomes a *cubic*.

Remark 5.1-1. In view of the significant simplification that obtains (here, and see also below) when the initial data entail validity of the condition (5.48a), let us note that the requirement that this condition hold restricts the initial data in the (*complex*) phase space to a hypersurface of (*complex*) codimension one. But this hypersurface can be shifted at will by varying the parameter C in the definition (5.5) of the original Hamiltonian H. One way to justify doing so is to consider C itself as an additional dynamical variable rather than a given constant; then the absence in the Hamiltonian $\tilde{H}(\underline{p},\, \underline{q}; \omega)$ of the corresponding canonical momentum entails that this quantity C is indeed *time independent*, namely, that C maintains for all time its initial value; which can then be assigned, together with the assignment of all the other initial data, so that (5.48a) hold. Note however that, while varying C does not affect the dynamics generated by the original Hamiltonian H, see (5.5), it does modify the dynamics yielded by the ω-modified Hamiltonian $\tilde{H}$, see (5.20). ⊡

Remark 5.1-2. Note that if $H\,(0)$ vanishes (see (5.48a)), then the constant α, see (5.27), also vanishes, $\alpha = 0$, satisfying thereby the condition $|\alpha| < 1$. Hence, whenever the simplifying condition (5.48a) holds, the basic *primitive* period of *isochrony* is T rather than $2\,T$: see (5.32a), (5.33a), (5.43a) and (5.4). ⊡

Let us end this Section 5.1 by mentioning that, although for simplicity we limited consideration above to the *one-dimensional* N-body problem with *equal-mass* particles, lifting both these restrictions is a trivial task [80]: of course, in the *multi-dimensional* case there is considerable more freedom in the definition of the quantity Θ, thanks to the possibility to use in the formula (5.24) any component of the *multi-dimensional* quantities $\vec{P}$ and $\vec{Q}$—or, not to spoil *rotation invariance*, one could replace, in the definition (5.24), P with $\vec{P} \cdot \vec{Q} / \sqrt{\vec{Q} \cdot \vec{Q}}$ and Q with $\sqrt{\vec{Q} \cdot \vec{Q}}$.

5.2 Partially isochronous Hamiltonian systems

In this section we discuss—again in the context of the N-body problem characterized by the Hamiltonian (5.5)—variants of the technique described in the preceding Section 5.1, yielding modified Hamiltonians with enough *completely periodic* solutions to justify calling them *partially isochronous* (see the definition of this term given above, near the end of the introductory part to this Chapter 5).

Let us take as starting point of our treatment the modified Hamiltonian

$$\breve{H}\,(\underline{p},\, \underline{q}) = U\,(P,\, Q)\; H\,(\underline{p},\, \underline{q})\,. \tag{5.50}$$

Notation: largely the same as above, see in particular (5.5), (5.9a) and (5.10a). The novelty is that, for the time being, we do not commit ourselves to a specific form of the "modifying multiplicative coefficient" U, although we do assume that it depends on the canonical coordinates and momenta *only* via the collective coordinates $Q \equiv Q(\underline{q})$ and $P \equiv P(\underline{p})$; of course for the special choice $U(P, Q) = u(P, Q; \omega)$ (see (5.20) with (5.24), (5.9a) and (5.10a)) the treatment of this Section 5.2 reduces to that given in Section 5.1. And let us again emphasize, as we did in Section 5.1, that hereafter we consider the time evolution entailed by this *Hamiltonian* $\breve{H}$, see (5.50), both for the Hamiltonian variables $p_n(t)$ and $q_n(t)$ and for any collective coordinate—such as P, Q, H, U—the time evolution of which obtains via the time dependence of the Hamiltonian variables.

The following evolution equations are then yielded by this *modified Hamiltonian* (5.50):

$$\dot{H} = -\frac{P\,U_Q\,H}{N}, \tag{5.51a}$$

$$\dot{U} = \frac{P\,U_Q\,U}{N}, \tag{5.51b}$$

$$\dot{P} = -U_Q\,H, \tag{5.51c}$$

$$\dot{Q} = \frac{U\,P}{N} + U_P\,H, \tag{5.51d}$$

$$\dot{q}_n = U\,p_n + U_P\,H, \tag{5.51e}$$

$$\dot{p}_n = -U\,\frac{\partial V(\underline{q})}{\partial q_n} - \frac{U_Q\,H}{N}. \tag{5.51f}$$

Note that, here and hereafter, we use the short-hand notation U_Q and U_P to denote the (partial) derivatives of the (yet to be assigned) function U with respect to its two arguments:

$$U_Q \equiv \frac{\partial U}{\partial Q}, \quad U_P \equiv \frac{\partial U}{\partial P}. \tag{5.52}$$

From (5.51c) and (5.51a) we get

$$\dot{P}\,P = \dot{H}, \tag{5.53a}$$

entailing

$$P = \left(2\,H - B^2\right)^{1/2}, \tag{5.53b}$$

$$B^2 = 2\,H(0) - P^2(0). \tag{5.53c}$$

To proceed further one must make a specific choice for the function $U(P, Q)$. We consider below various possibilities.

5.2.1 *A simple variant*

The first "β–modified" Hamiltonian we now consider is characterized by the following variant of the N-body Hamiltonian (5.5),

$$\breve{H}(\underline{p}, \underline{q}; \beta) = \left[1 + i\,\beta\,Q(\underline{q})\right]\,H(\underline{p}, \underline{q}), \tag{5.54}$$

corresponding to (5.50) with

$$U(P, Q) \equiv U(Q; \beta) = 1 + i\,\beta\,Q \tag{5.55a}$$

implying

$$U_Q = i\,\beta, \quad U_P = 0. \tag{5.55b}$$

We now show that there exists also in this case a set of initial data $q_n(0)$ and $p_n(0)$—having a bit *less* than full dimensionality in phase space, being restricted by the requirement

$$H(0) \equiv H\left[\underline{p}(0), \underline{q}(0)\right] = 0 \tag{5.56}$$

(and see in this connection *Remark 5.1-1*)—yielding motions *all* of which are *completely periodic* with the period

$$\breve{T} = \frac{2\,\pi}{\Omega}, \quad \Omega = \frac{\beta\,P(0)}{N} \equiv \frac{\beta\,P\left[\underline{p}(0)\right]}{N}. \tag{5.57}$$

In this formula $P\left(\underline{p}\right)$ is of course defined by (5.9a). Note that this formula entails that the period $\breve{T}$ does depend on the initial data, albeit only via the collective variable $P\left(\underline{p}\right)$, and of course that we must restrict the initial data so that this period $\breve{T}$, hence as well the corresponding circular frequency Ω, be *real*. This entails an *additional* restriction of the initial data $p_n(0)$ to a hypersurface of *real* codimension one. And one can repeat in this respect a completely analogous discussion to that given in the *Remark 5.1-1*, up to replacing the role played in that context by the parameter C with the role played in the present context by the parameter β. Note moreover that in this manner—adjusting at one's convenience the value of β (as it were *a posteriori*, after the initial data $p_n(0)$ have been assigned)—one can impose that $\breve{T}$ have a given, preassigned value, thereby restoring the property of *isochrony*—albeit for a set of initial data having a bit *less* than full dimensionality in phase space, being restricted to lie on a manifold of *complex* codimension *two* in phase space—i.e., the intersection of *two* hypersurfaces each of which of *complex* codimension *one*, that characterized by (5.56) and that characterized by the requirement that the period $\breve{T}$, or equivalently the circular frequency Ω, see (5.57), have a *preassigned real* value.

To prove this result we note that the Hamiltonian $\breve{H}(\underline{p}, \underline{q}; \beta)$, see (5.54), entails the following time evolution equations (see (5.51e), (5.51f), (5.51d), (5.51c), (5.51a) with (5.55)):

$$\dot{q}_n = (1 + i\,\beta\,Q)\,p_n, \tag{5.58a}$$

$$\dot{p}_n = -\,(1 + i\,\beta\,Q)\,\frac{\partial\,V\left(\underline{q}\right)}{\partial\,q_n} - i\,\beta\,H, \tag{5.58b}$$

$$\dot{Q} = \frac{(1 + i\,\beta\,Q)\;P}{N}, \tag{5.58c}$$

$$\dot{P} = -i\,\beta\,H, \tag{5.58d}$$

$$\dot{H} = -\frac{i\,\beta\,H\,P}{N}. \tag{5.58e}$$

The last three equations can be easily integrated for general initial data, but we restrict attention here only to the case characterized by the condition (5.56), in which case the outcome is particularly simple:

$$H(t) = H(0) = 0, \tag{5.59a}$$

$$P(t) = P(0) \equiv P\left[\underline{p}(0)\right], \tag{5.59b}$$

$$1 + i\,\beta\,Q(t) = \left[1 + i\,\beta\,Q(0)\right]\,\exp\left(i\,\Omega\,t\right). \tag{5.59c}$$

Insertion of the first and last of these formulas in (5.58a) and (5.58b) yields the equations that determine the time evolution of the dependent variables $q_n(t)$ and $p_n(t)$:

$$\dot{q}_n = \left[1 + i\,\beta\,Q(0)\right]\,\exp\left(i\,\Omega\,t\right)\,p_n, \tag{5.60a}$$

$$\dot{p}_n = -\left[1 + i\,\beta\,Q(0)\right]\,\exp\left(i\,\Omega\,t\right)\,\frac{\partial V\left(\underline{q}\right)}{\partial q_n}. \tag{5.60b}$$

And from these evolution equations, via the, by now usual, argument based on the trick, one easily concludes that the assertions made at the beginning of this Section 5.2.1 are valid.

Let us end this Section 5.2.1 by pointing out that the *Hamiltonian* $\breve{H}(\underline{p},\,\underline{q};\beta)$, see (5.54), is, perhaps, more susceptible to a "physical interpretation" than the *Hamiltonian* $\tilde{H}(\underline{p},\,\underline{q};\omega)$, see (5.20), inasmuch as $\breve{H}(\underline{p},\,\underline{q};\beta)$, in contrast to $\tilde{H}(\underline{p},\,\underline{q};\omega)$, does *not* feature momentum variables in the denominator. This observation is underscored by the possibility to write in this case quite explicitly the *Newtonian* equations of motion that correspond to the *Hamiltonian* equations (5.58a) and (5.58b):

$$\ddot{q}_n = \frac{i\,\beta\,\dot{Q}\,\dot{q}_n}{1 + i\,\beta\,Q} - \left(1 + i\,\beta\,Q\right)^2\,\frac{\partial V\left(\underline{q}\right)}{\partial q_n} - \frac{i\,\beta\,\left(1 + i\,\beta\,Q\right)\,H}{N}, \tag{5.61a}$$

where, of course, $Q \equiv Q(\underline{q})$ (see (5.10a)) and $H \equiv H\left(\underline{p},\,\underline{q}\right)$ (see (5.5)) with

$$p_n = \frac{\dot{q}_n}{1 + i\,\beta\,Q} \tag{5.61b}$$

(see (5.58a)). And note moreover that the last term in the right-hand side of the *Newtonian* equations of motion (5.61a) is altogether missing when (5.56) holds, see (5.59a).

5.2.2 *A more general variant*

A more general variant—that includes as a special case those treated above—is associated with the observation that—quite generally—the evolution equation (5.51a) with (5.56) entails the vanishing of H for all time, see (5.59a), and that—again quite generally, see (5.51c)—the vanishing of H entails that P is time-independent, see (5.59b). This we assume now, hence we consider again a modified *Hamiltonian* defined by (5.50) (with (5.5)), and initial data restricted so that (5.56) hold. But now we note that (5.59c) does not generally follow from (5.59a) and (5.59b); all that one can assert is that, if (5.59a) and (5.59b) hold, then (5.51b) and (5.51d) take the following simpler form:

$$\dot{U} = -\frac{P(0)\,U_Q\,U}{N}, \quad \dot{Q} = \frac{P(0)\,U}{N}. \tag{5.62}$$

To proceed further, an assumption must again be made on the assignment of $U(P,\,Q)$, although the time independence of P entails that we can now manage with an *ansatz* that allows a certain generality as regards the dependence on this collective variable. For instance let us set

$$U = 1 + \Phi(P)\,Q + \Psi(P)\,Q^2, \tag{5.63a}$$

entailing

$$U_Q = \Phi(P) + 2\,\Psi(P)\,Q \tag{5.63b}$$

(of course the results of the previous Section 5.2.1 are reproduced for $\Phi(P) = i\,\beta$, $\Psi(P) = 0$, although this might involve a nontrivial limiting process; and those of Section 5.1—with (5.48)—are reproduced for $\Phi(P) = N\,i\,\beta\,/\,P$, $\Psi(P) = 0$). Then one can easily obtain the following results:

$$U(t) = A\left\{\sin\left[\frac{\Omega\,(t - t_0)}{2}\right]\right\}^{-2}, \tag{5.64a}$$

$$A = 1 - \frac{\Phi^2(P)}{4\,\Psi(P)}, \tag{5.64b}$$

$$\Omega = \frac{P}{N}\,\left[-\Phi^2(P) + 4\,\Psi(P)\right]^{1/2}, \tag{5.64c}$$

$$\sin(t_0) = \left[\frac{A}{1 + \Phi(P)\,Q(0) + \Psi(P)\,Q^2(0)}\right]^{1/2}. \tag{5.64d}$$

We then note that, via (5.59a) and (5.64), the *Hamiltonian* equations (5.51e) and (5.51f) now read

$$\dot{q}_n = A\left\{\sin\left[\frac{\Omega\,(t - t_0)}{2}\right]\right\}^{-2} p_n, \tag{5.65a}$$

$$\dot{p}_n = -A\left\{\sin\left[\frac{\Omega\,(t - t_0)}{2}\right]\right\}^{-2} \frac{\partial\,V\,(\underline{q})}{\partial\,q_n}, \tag{5.65b}$$

or equivalently

$$\dot{q}_n = \dot{\tau}\,p, \quad \dot{p}_n = -\dot{\tau}\,\frac{\partial\,V\,(\underline{q})}{\partial\,q_n}, \tag{5.66}$$

provided we set

$$\tau(t) = -\left(\frac{2\,A}{\Omega}\right)\cot\left[\frac{\Omega\,(t-t_0)}{2}\right]. \tag{5.67}$$

It is now clear from (5.66)—via a generalized version of the trick, the details of which can be easily filled in by the interested reader, being based on the change of independent variable

$$q_n(t) = \breve{q}_n(\tau), \quad p_n(t) = \breve{p}_n(\tau), \tag{5.68}$$

with the new (*complex*) variable τ related to the (*real*) time variable t by (5.67)—that this modified *Hamiltonian* system is *partially isochronous* (with period $2\,\pi\,/\,\Omega$, see (5.64c)), the initial data having to be restricted, in order to entail *complete periodicity* of the corresponding solutions, by the requirement that (5.56), hence as well (5.59a), hold (a restriction of *complex* codimension *one*; but see *Remark 5.1-1*) and moreover so that Ω be *real* (a restriction of *real* codimension *one*) or have a *preassigned* fixed (of course *real*) value (a restriction of *complex* codimension *one*). This last restriction could, however, be lifted for special choices of the functions $\Phi(P)$ and $\Psi(P)$ that guarantee that Ω have a *fixed real* value independent of P, as is for instance the case if

$$-\Phi^{2}(P) + 4\,\Psi(P) = \left(\frac{N\,\omega}{P}\right)^{2}, \tag{5.69}$$

yielding $\Omega = \omega$, see (5.64c). Note that this restriction, (5.69), is compatible with $\Phi(P)$ and $\Psi(P)$ being both *real* functions (that yield *real* values whenever their argument, P, is *real*), provided $\Psi(P)$ does not vanish identically (as it instead did in the treatment of Section 5.1, which indeed involved the *imaginary* function $\Phi(P) = i\,\omega\,/\,P$). In addition—in order to avoid that the equations of motion run into a *singularity* as the (*real*) time t evolves—the initial data should, of course, *exclude* that the quantity t_0 be *real* (see (5.64d) and (5.65), (5.67)).

5.3 More general Hamiltonians

In this section we outline how the approach introduced above can be applied to *more general Hamiltonians* than that describing the nonrelativistic N-body problem, see (5.5). This discussion allows a more explicit display of the essential aspects of this technique to manufacture ω-modified *Hamiltonians* yielding an *isochronous* dynamics.

Let us therefore consider again an ω-modified *Hamiltonian* defined by (5.20) with the collective variable $\Theta(\underline{p}, \underline{q})$ satisfying (5.18), but now without making any assumption on the *original* Hamiltonian H and therefore without being able to provide an *explicit* form for the dependence of Θ on the Hamiltonian variables. It is nevertheless clear that the time evolution of H and of u is always given by

(5.22) and (5.23), and therefore that the (originally *autonomous*) *Hamiltonian* equations can now be rewritten in the following (*nonautonomous*) version:

$$\dot{q}_n = i\,\omega\,H(0)\,\exp(-i\,\omega\,t)\,\frac{\partial\,\Theta\,(\underline{p},\underline{q})}{\partial\,p_n} + u(0)\,\exp(i\,\omega\,t)\,\frac{\partial\,H\,(\underline{p},\underline{q})}{\partial\,p_n}, \tag{5.70a}$$

$$\dot{p}_n = -i\,\omega\,H(0)\,\exp(-i\,\omega\,t)\,\frac{\partial\,\Theta\,(\underline{p},\underline{q})}{\partial\,q_n} - u(0)\,\exp(i\,\omega\,t)\,\frac{\partial\,H\,(\underline{p},\underline{q})}{\partial\,q_n}. \tag{5.70b}$$

Now the first observation is that, if one restricts attention to initial data such that $H(0)$ vanishes, see (5.48a), one gets again an *isochronous* behavior, since these *Hamiltonian* equations of motion, (5.70), take then the simpler form

$$\dot{q}_n = u(0)\,\exp(i\,\omega\,t)\,\frac{\partial\,H\,(\underline{p},\underline{q})}{\partial\,p_n}, \tag{5.71a}$$

$$\dot{p}_n = -u(0)\,\exp(i\,\omega\,t)\,\frac{\partial\,H\,(\underline{p},\underline{q})}{\partial\,q_n}, \tag{5.71b}$$

to which the by now familiar trick can be applied, entailing *isochrony*. This is, however, a case of *partial isochrony* (see above, near the end of the introductory part to this Chapter 5, for this terminology), because the initial data must be restricted to satisfy the condition (5.48a).

Let us emphasize that this observation—possibly associated with the general possibility to add a constant C to the original *Hamiltonian* (and see in this respect *Remark 5.1-1* to assess how this possibility can facilitate the implementation of the restriction (5.48a))—entails that the class of *Hamiltonians* to which our technique can be successfully applied is indeed *quite large*, at least as regards manufacturing ω-modified *Hamiltonians* which are *partially isochronous*.

In the more general case when $H(0)$ does *not* vanish, we conjecture that the motion will be *isochronous* in quite general circumstances, but we do not try to prove this here. We rather restrict attention to systems such that there exists a collective function Θ satisfying (5.18) that depends on the Hamiltonian coordinates *only* via the two collective coordinates P and Q, see (5.9a) and (5.10a):

$$\Theta\,(\underline{p},\underline{q}) = \Theta\,[P(\underline{p}),Q(\underline{q})] \tag{5.72}$$

(this, of course, limits substantially the class of Hamiltonians to which this approach is applicable). Then one easily sees that there hold the relations (5.35) (with (5.34)) entailing the equations of motion (5.36), hence proceeding as in Section 5.1—i.e., introducing the *Hamiltonian* $\tilde{H}\,(\underline{p},\,\underline{q};\,\omega)$ via (5.20)—one concludes that the *relative* motions are *isochronous*.

To proceed further it is necessary to gain some knowledge on the time evolution of the global quantities $Q(t)$ and $P(t)$. This requires additional information on the explicit form of the function $\Theta\,(P,Q)$, and moreover on the specific form of the original *Hamiltonian*, because the time evolution of these quantities, $Q(t)$ and $P(t)$, is determined by the following evolution equations (implied by (5.70)):

$$\dot{Q} = i\omega\, H(0)\, \exp(-i\,\omega\, t)\, \Theta_P(P, Q) + \frac{u(0)\, \exp(i\,\omega\, t)}{N} \sum_{n=1}^{N} \frac{\partial\, H\,(\underline{p}, \underline{q})}{\partial\, p_n}, \tag{5.73a}$$

$$\dot{P} = -\frac{i\,\omega\, H(0)\, \exp(-i\,\omega\, t)\, \Theta_Q(P, Q)}{N} - u(0)\, \exp(i\,\omega\, t) \sum_{n=1}^{N} \frac{\partial\, H\,(\underline{p}, \underline{q})}{\partial\, q_n}. \tag{5.73b}$$

5.4 Examples

In this section we present tersely some examples of the application of the approach described in the preceding sections of this Chapter 5.

Example 5.4-1 The very simplest case obtains for the choice

$$V(\underline{q}) = 0 \tag{5.74a}$$

in (5.5), i.e. for the original *Hamiltonian*

$$H(\underline{p},\, \underline{q}) = \frac{1}{2} \sum_{n=1}^{N} p_n^2 + C, \tag{5.74b}$$

yielding via (5.20) with (5.24) the ω-modified *Hamiltonian*

$$\tilde{H}(\underline{p},\, \underline{q}; \omega) = \left[1 + N\, i\, \omega\, \frac{Q\,(\underline{q})}{P\,(\underline{p})}\right] \left[\frac{1}{2} \sum_{n=1}^{N} p_n^2 + C\right]. \tag{5.74c}$$

Here and hereafter the collective coordinates Q and P are of course defined by (5.10a) and (5.9a).

The *Hamiltonian* respectively the (almost but not quite: see above) *Newtonian* equations of motions yielded by this ω-modified *Hamiltonian*, (5.74c), read

$$\dot{q}_n = \left[1 + N\, i\, \omega\, \frac{Q}{P}\right] p_n - \frac{N\, i\, \omega\, Q\, H}{P^2}, \tag{5.75a}$$

$$\dot{p}_n = -\frac{i\,\omega\, H}{P}, \tag{5.75b}$$

respectively

$$\ddot{q}_n - i\,\omega\, \dot{q}_n = \frac{P^2 - \left[N\left(\dot{Q} - i\,\omega\, Q\right) - 2\, P\right]^2}{N^2\, Q}. \tag{5.76}$$

And it is easily seen that the solution of the corresponding initial-value problem reads

$$p_n(t) = p_n(0) + \frac{P(0)}{N} \left\{ \left[\frac{1 - \alpha \exp(-i\,\omega\,t)}{1 - \alpha} \right]^{1/2} - 1 \right\}, \tag{5.77a}$$

$$q_n(t) = q_n(0) + \frac{N\,Q(0)}{P(0)} \left[\exp(i\,\omega\,t)\, p_n(t) - p_n(0) \right] + p_n(t)\, \frac{\exp(i\,\omega\,t) - 1}{i\,\omega}, \tag{5.77b}$$

with α given by (5.27), and of course (see (5.75a)) with

$$p_n(0) = \left[1 + N\,i\,\omega\, \frac{Q(0)}{P(0)} \right]^{-1} \left[\dot{q}_n(0) + \frac{N\,i\,\omega\,Q(0)\,H(0)}{P^2(0)} \right] \tag{5.77c}$$

in the context of the initial-value problem for the (almost but not quite: see above) *Newtonian* equations of motions (5.76).

The *entirely isochronous* character of this *Hamiltonian* N-body problem is now evident: its phase space is divided into two parts, that characterized by $|\alpha| < 1$ and by solutions *completely periodic* with primitive period T, see (5.4), and that characterized by $|\alpha| > 1$ and by solutions *completely periodic* with period $2\,T$, while on the *separatrix* between these two phase-space regions, characterized by the formula

$$|\alpha| = 1, \quad \alpha = \exp(i\,\vartheta), \quad \mathrm{Im}(\vartheta) = 0, \tag{5.78}$$

the *Hamiltonian* equations of motion become *singular* at the (*real*) time $t_s = \vartheta/\omega \mod(T)$ due to the vanishing at that time of the collective coordinate P, $P(t_s) = 0$, see (5.26b) and (5.75).

Example 5.4-2 As second example let us consider the standard N-body problem (5.5) characterized by the following *two-body homogeneous* potentials:

$$V(\underline{q}) = \frac{1}{2} \sum_{n,m=1,\, n \neq m}^{N} V_{nm}(q_n - q_m), \quad V_{nm}(q) = \frac{g_{nm}\, q^{2\,k}}{2\,k}, \tag{5.79}$$

where the "coupling constants" g_{nm} satisfy the symmetry condition $g_{nm} = g_{mn}$ and k is an arbitrary *integer* (*positive* or *negative*, but *nonvanishing,* and *different from unity,* $k \neq 1$, to exclude the uninteresting case of *linear* Newtonian equations). This choice allows a comparison with the *isochronous* equations of motion of *Newtonian* type (5.14) obtained via the application of the previous ("old") trick (5.12) to the *Newtonian* equations yielded by the original Hamiltonian H, see (5.5) with (5.79)—that in this case read as follows:

$$\ddot{\tilde{q}}_n - \left(\frac{k+1}{k-1} \right) i\,\omega\, \dot{\tilde{q}}_n - \left[\frac{k}{(k-1)^2} \right] \tilde{q}_n = - \sum_{m=1,\, m \neq n}^{N} g_{nm}\, (\tilde{q}_n - \tilde{q}_m)^{2\,k-1}. \tag{5.80}$$

The (*autonomous*) *Newtonian* equations of motion (5.44) yielded by the approach of this Chapter 5 read instead

$$\ddot{q}_n - i\,\omega\,\dot{q}_n = -u^2 \sum_{m=1,\, m\neq n}^{N} g_{nm}\,(q_n - q_m)^{2\,k-1} + F, \tag{5.81}$$

with u respectively F expressed in terms of $Q \equiv Q(\underline{q})$, $P \equiv P(\underline{q})$ and $H \equiv H(\underline{p}, \underline{q})$, see (5.10a), (5.9a) and (5.5) with (5.79), by the formulas (5.20b) with (5.24) respectively (5.44b) (except that these are not quite *Newtonian* equations of motion, as discussed above; the display of such equations, to the extent it is possible, see above, can be left to the diligent reader).

In this case it is possible to provide explicit conditions on the initial data that are *sufficient* (but of course not *necessary*) to guarantee the *isochronous* behavior of the corresponding solution: the task to exhibit them is left to the diligent reader.

Example 5.4-3 As third example we return to the original N-body problem with vanishing potential, see (5.74), but using the variants of Sections 5.2.1 and 5.2.2. We directly report the relevant results, since their derivation is sufficiently straightforward to be left as an exercise for the diligent reader.

The β–modified *Hamiltonian* of Section 5.2.1 reads now as follows:

$$\breve{H}(\underline{p},\, \underline{q};\beta) = \left[1 + \frac{i\,\beta}{N} \sum_{n=1}^{N} q_n\right] \left[\frac{1}{2} \sum_{n=1}^{N} p_n^2 + C\right], \tag{5.82}$$

and, for all initial data satisfying the condition

$$\frac{1}{2} \sum_{n=1}^{N} p_n^2(0) + C = 0, \tag{5.83}$$

it yields the solution

$$p_n(t) = p_n(0), \quad q_n(t) = q_n(0) + \left[1 + \frac{i\,\beta}{N} \sum_{n=1}^{N} q_n(0)\right] p_n(0)\, \frac{\exp{(i\,\Omega\,t)} - 1}{i\,\Omega}, \tag{5.84a}$$

$$\Omega = \frac{\beta}{N} \sum_{n=1}^{N} p_n(0), \tag{5.84b}$$

which is clearly *completely periodic* with period $2\,\pi \,/\, |\Omega|$ provided Ω is *real* (and *nonvanishing*).

The modified *Hamiltonian* of Sections 5.2.2 reads now as follows:

$$\breve{H}(\underline{p},\, \underline{q}) = \left[1 + \Phi(P)\,Q + \Psi(P)\,Q^2\right] \left[\frac{1}{2} \sum_{n=1}^{N} p_n^2 + C\right], \tag{5.85}$$

where of course $P \equiv P(\underline{p})$ and $Q \equiv Q(\underline{q})$, see (5.9a) and (5.10a). For all initial data satisfying the condition (5.83) it yields the solution

$$p_n(t) = p_n(0), \quad q_n(t) = q_n(0) - \frac{2A}{\Omega}\, p_n(0) \left\{ \cot\left[\frac{\Omega(t - t_0)}{2}\right] + \cot\left[\frac{\Omega t_0}{2}\right] \right\}. \tag{5.86}$$

In these formulas we are, of course, using the notation of Section 5.2.2, in particular A, Ω and t_0 are defined by (5.64b), (5.64c) and (5.64d) with $P = P(0)$. This solution is clearly *completely periodic* with period $2\pi / |\Omega|$ provided Ω is *real* (and *nonvanishing*) and t_0 is instead *not real.*

Example 5.4-4 In this example we report tersely some of the results obtained [82] by ω-modifying the one-body model—in two dimensions—characterized by the following Hamiltonian (in self-evident notation):

$$H(\vec{p}, \vec{r}) = \frac{y^2\left(p_x^2 + p_y^2\right)}{2} - C, \quad \vec{p} \equiv (p_x, p_y), \quad \vec{r} \equiv (x, y). \tag{5.87}$$

This Hamiltonian describes free motions in the Poincaré half-plane, i.e. in the half-plane $y > 0$ endowed with the metric

$$ds^2 = \frac{(dx)^2 + (dy)^2}{y^2}, \tag{5.88}$$

namely free motions on a surface of constant negative curvature. It clearly yields the following equations of motion:

$$\dot{x} = [H, x] = \frac{\partial H}{\partial p_x} = y^2 p_x, \tag{5.89a}$$

$$\dot{y} = [H, y] = \frac{\partial H}{\partial p_y} = y^2 p_y, \tag{5.89b}$$

$$\dot{p}_x = [H, p_x] = -\frac{\partial H}{\partial x} = 0, \tag{5.89c}$$

$$\dot{p}_y = [H, p_y] = -\frac{\partial H}{\partial y} = -y\left(p_x^2 + p_y^2\right). \tag{5.89d}$$

Here of course the Poisson-commutator (or Poisson-bracket) of two functions $F(\vec{p}, \vec{r})$, $G(\vec{p}, \vec{r})$ of the canonical variables x, y and the corresponding momenta p_x, p_y is defined in the standard manner:

$$[F(\vec{p}, \vec{r}), G(\vec{p}, \vec{r})] \equiv \frac{\partial F(\vec{p}, \vec{r})}{\partial p_x} \frac{\partial G(\vec{p}, \vec{r})}{\partial x} - \frac{\partial G(\vec{p}, \vec{r})}{\partial p_x} \frac{\partial F(\vec{p}, \vec{r})}{\partial x} + \frac{\partial F(\vec{p}, \vec{r})}{\partial p_y} \frac{\partial G(\vec{p}, \vec{r})}{\partial y} - \frac{\partial G(\vec{p}, \vec{r})}{\partial p_y} \frac{\partial F(\vec{p}, \vec{r})}{\partial y}. \tag{5.90}$$

The constant C (which we clearly assume to be independent of the canonical coordinates and momenta) appearing in the right-hand side of (5.87) has of course no influence on the flow induced by the Hamiltonian $H(\vec{p},\vec{r})$; but it will influence the time evolution induced by the ω-modified Hamiltonian $\tilde{H}(\vec{p},\vec{r};\omega)$, see below.

It can be easily verified that the following quantity,

$$\Theta(\vec{p},\vec{r}) = -\frac{1}{y\left(p_x^2+p_y^2\right)^{1/2}} \operatorname{arctanh}\left[\frac{p_y}{\left(p_x^2+p_y^2\right)^{1/2}}\right], \tag{5.91a}$$

$$\Theta(\vec{p},\vec{r}) = -\frac{1}{\left[2\left(H+C\right)\right]^{1/2}} \operatorname{arctanh}\left\{\frac{y\,p_y}{\left[2\left(H+C\right)\right]^{1/2}}\right\}, \tag{5.91b}$$

has the following Poisson-commutator with $H(\vec{p},\vec{r})$:

$$[H,\Theta] = 1. \tag{5.92}$$

Clearly this Poisson-commutation formula, (5.92), entails for the time evolution (yielded by the Hamiltonian $H(\vec{p},\vec{r})$) of the quantity $\Theta(\vec{p},\vec{r})$ the simple rule

$$\dot{\Theta} = [H,\Theta] = 1, \tag{5.93a}$$

implying

$$\Theta = \Theta_0 + t \tag{5.93b}$$

hence, via (5.91a),

$$\frac{p_y}{\left(p_x^2+p_y^2\right)^{1/2}} = -\tanh\left[y\left(p_x^2+p_y^2\right)^{1/2}\left(\Theta_0+t\right)\right], \tag{5.94a}$$

or equivalently, via (5.87),

$$\frac{p_y}{\left(p_x^2+p_y^2\right)^{1/2}} = -\tanh\left\{\left[2\left(H+C\right)\right]^{1/2}\left(\Theta_0+t\right)\right\}. \tag{5.94b}$$

It is moreover possible [82] to obtain the following formulas displaying the time evolution of the canonical coordinates and momenta, namely the explicit solution of the initial-value problem for the equations of motion (5.89):

$$x(t) = x(0) + \frac{A\left\{\tanh\left[\varphi(t)\right] - \tanh\left[\varphi(0)\right]\right\}}{p_x(0)}, \tag{5.95a}$$

$$y(t) = \frac{A}{p_x(0)\cosh\left[\varphi(t)\right]} = y(0)\,\frac{\cosh\left[\varphi(0)\right]}{\cosh\left[\varphi(t)\right]}, \tag{5.95b}$$

$$p_x(t) = p_x(0), \tag{5.95c}$$

$$p_y(t) = -p_x(0)\sinh\left[\varphi(t)\right] = p_y(0)\,\frac{\sinh\left[\varphi(t)\right]}{\sinh\left[\varphi(0)\right]}, \tag{5.95d}$$

where

$$A = y(0) \left\{[p_x(0)]^2 + [p_y(0)]^2\right\}^{1/2} = \{2\,[H(0) + C]\}^{1/2}, \tag{5.95e}$$

$$\varphi(t) \equiv A\,(\Theta_0 + t)\,. \tag{5.95f}$$

Here, of course, the "initial" value of $H(0)$ is determined by the initial values of the canonical quantities and momenta via (5.87), and the quantity Θ_0 is related to the initial values of the canonical quantities and momenta via, for instance, (5.94), yielding (at $t = 0$)

$$A\,\Theta_0 = -\text{arctanh}\left[\frac{y(0)\,p_y(0)}{A}\right]. \tag{5.95g}$$

These formulas provide the solution of the initial-value problem for the time evolution associated with the Hamiltonian (5.87). This motion takes place on a semicircle in the upper half-plane ($y > 0$), starting (in an asymptotic sense, at $t = -\infty$) infinitely slowly at one end of this semicircle, picking up some speed along the way, and becoming again infinitely slow as it approaches the other end of the semicircle, where it arrives (in an asymptotic sense) at $t = +\infty$. Of course this motion, considered as taking place on the surface of constant negative curvature defined by the metric (5.88), travels along a geodesic with constant velocity (corresponding to free motion), as it is plain from the formula [82]

$$\frac{\dot{x}^2 + \dot{y}^2}{y^2} = 2\,[H(0) + C]\,. \tag{5.96}$$

After this terse review of the standard treatment of the Hamiltonian model describing free motion in the Poincaré half-plane, let us introduce the ω-modified version of this Hamiltonian (5.87) via the formula (5.20),

$$\tilde{H}(\vec{p}, \vec{r}; \omega) = [1 + i\,\omega\,\Theta(\vec{p}, \vec{r})]\;H(\vec{p}, \vec{r})\,, \tag{5.97a}$$

of course with $H(\vec{p}, \vec{r})$ respectively $\Theta(\vec{p}, \vec{r})$ defined in terms of the canonical variables and momenta by (5.87) respectively (5.91), so that

$$\tilde{H}(\vec{p}, \vec{r}; \omega) = \frac{1}{2}\,y\,\left(p_x^2 + p_y^2\right)^{1/2} \left\{y\,\left(p_x^2 + p_y^2\right)^{1/2} - i\,\omega\,\text{arctanh}\left[\frac{p_y}{\left(p_x^2 + p_y^2\right)^{1/2}}\right]\right\}$$
$$-C\left\{1 - \frac{i\,\omega}{y\,\left(p_x^2 + p_y^2\right)^{1/2}}\,\text{arctanh}\left[\frac{p_y}{\left(p_x^2 + p_y^2\right)^{1/2}}\right]\right\}. \tag{5.97b}$$

Let us now proceed and consider the time evolution yielded by this ω-modified Hamiltonian $\tilde{H}(\vec{p}, \vec{r}; \omega)$: of course, due to the presence of the imaginary unit i in the right-hand side of this formula (5.97), we must hereafter consider the canonical variables x, y and the momenta p_x, p_y as *complex* variables.

We leave it to the interested reader to display the equations of motion yielded by this ω-modified Hamiltonian (5.97). We rather report [82] directly the solution of the corresponding initial-value problem:

$$x(t) = x(0) + \frac{\tilde{A}(t)\tanh[\tilde{\varphi}(t)] - \tilde{A}(0)\tanh[\tilde{\varphi}(0)]}{p_x(0)}, \tag{5.98a}$$

$$y(t) = \frac{\tilde{A}(t)}{p_x(0)\cosh[\tilde{\varphi}(t)]} = y(0)\,\frac{\tilde{A}(t)\cosh[\tilde{\varphi}(0)]}{\tilde{A}(0)\cosh[\tilde{\varphi}(t)]}, \tag{5.98b}$$

$$p_x(t) = p_x(0), \tag{5.98c}$$

$$p_y(t) = -p_x(0)\sinh[\tilde{\varphi}(t)] = p_y(0)\,\frac{\sinh[\tilde{\varphi}(t)]}{\sinh[\tilde{\varphi}(0)]}, \tag{5.98d}$$

where

$$\tilde{A}(t) = \left\{[y(0)]^2\,\left[p_x^2(0) + p_y^2(0)\right]\,\exp(-i\,\omega\,t) + 2\,C\,[1 - \exp(-i\,\omega\,t)]\right\}^{1/2}, \tag{5.99}$$

$$\tilde{A}(t) = \{2\,[H(0)\exp(-i\,\omega\,t) + C]\}^{1/2}, \tag{5.100a}$$

$$\tilde{\varphi}(t) \equiv \tilde{A}(t)\left[\Theta_0\exp(i\,\omega\,t) + \frac{\exp(i\,\omega\,t) - 1}{i\,\omega}\right]. \tag{5.100b}$$

Here the quantity Θ_0 is again related to the initial values of the canonical quantities and momenta via (5.95g).

The fact that this formulas reduce to (5.95) when ω vanishes is plain. It is also clear that, when ω does *not* vanish, both $\tilde{A}(t)$ and $\tilde{\varphi}(t)$ are *periodic* with period $T = 2\,\pi\,/\,\omega$ if C does *not* vanish, and they are both *periodic* with (primitive) period $T\,/\,2 = \pi\,/\,\omega$ if C vanishes. Clearly the same periodicity is generically featured by the two canonical coordinates $x(t)$, $y(t)$ and by the canonical momentum $p_y(t)$, while the other canonical momentum, p_x, is time-independent. The only exceptional cases occur when the initial data are such that, for some *positive* value of t, the quantity $\tilde{\varphi}(t)$ takes the value $i\,\pi\,/\,2 \mod (i\,\pi)$, because then both $x(t)$ and $y(t)$ diverge at this value of t, signifying that the time evolution has run into a singularity. It is possible, but not very enlightening, to exhibit explicitly the necessary and sufficient conditions that the initial data must satisfy in order that this happen; they amount to the requirement that the cubic equation

$$\alpha\,z^3 + \beta\,z^2 + \gamma\,z + \delta = 0 \tag{5.101a}$$

where

$$\alpha = C\left(\Theta_0 + \frac{i}{\omega}\right)^2, \tag{5.101b}$$

$$\beta = \left(\Theta_0 + \frac{i}{\omega}\right)\left[H(0)\left(\Theta_0 + \frac{i}{\omega}\right) - \frac{2\,C}{i\,\omega}\right], \tag{5.101c}$$

$$\gamma = \frac{\pi^2}{8}(2n+1)^2 - \frac{2H(0)\left(\Theta_0 + \frac{i}{\omega}\right)}{i\omega} - \frac{C}{\omega^2}, \tag{5.101d}$$

$$\delta = -\frac{H(0)}{\omega^2} \tag{5.101e}$$

with n an arbitrary integer (positive, negative or zero), possesses a root z having unit modulus, $|z| = 1$.

In the particularly interesting case of initial data such that $H(0) = 0$, the quantity $\tilde{A}(t)$, see (5.100a), becomes time-independent, $\tilde{A}(t) = A = (2C)^{1/2}$ (see (5.95e)), and the above condition becomes simpler, inasmuch as the cubic equation (5.101a) with the condition $|z| = 1$ can now be replaced by the following more explicit condition on the initial data:

$$|C|^2 \left[\omega |\Theta_0|^2 + 2\,\mathrm{Im}\,\Theta_0\right] = \frac{(2n+1)\,\pi}{2}\left\{\mathrm{Re}\left[(2C)^{1/2}\right] + \frac{(2n+1)\,\pi\,\omega}{4}\right\}, \tag{5.102}$$

where n is again an arbitrary integer. A more explicit condition is also implied if the initial data entail $\Theta_0 = -i/\omega$. It clearly reads (see (5.101))

$$\left|\frac{\pi^2}{8}(2n+1)^2\,\omega^2 - C\right| = |H(0)|, \tag{5.103}$$

where n is again an arbitrary integer.

But in any case it is plain that the model characterized by the ω-modified Hamiltonian (5.97) is *entirely isochronous.*

Example 5.4-5 The last two examples we consider are characterized by (one-body one-dimensional) Hamiltonians *linear* in the canonical momentum—hence of a different type from the standard N-body Hamiltonian (5.5). Such an (unmodified) *Hamiltonian* reads

$$H(p,\, q) = f(q)\, p, \tag{5.104}$$

with $f(q)$ an *a priori arbitrary* function. Clearly this *Hamiltonian* yields the (*first-order*) equation of motion

$$\dot{q} = f(q), \tag{5.105}$$

and a function $\Theta \equiv \Theta(q)$ satisfying the basic condition (5.18) is

$$\Theta(q) = \int^q \frac{dz}{f(z)}. \tag{5.106}$$

Hence the ω-modified *Hamiltonian*

$$\tilde{H}(p,\, q) = \left[1 + i\omega \int^q \frac{dz}{f(z)}\right] f(q)\, p \tag{5.107}$$

(see (5.20)) is *isochronous*. The two examples we exhibit explicitly (without detailing the proofs of the results we report, which the diligent reader will easily verify) correspond to the following two assignments:

$$\text{case (i): } f\,(q) = a\,\exp\,(\gamma\,q)\,, \quad \tilde{H} = \left[a\,\exp\,(\gamma\,q) - \frac{i\,\omega}{\gamma}\right]\,p, \tag{5.108a}$$

$$\text{case (ii): } f\,(q) = a\,q^{\gamma}, \quad \tilde{H} = \left[a\,q^{\gamma} - \frac{i\,\omega\,q}{\gamma - 1}\right]\,p. \tag{5.108b}$$

In case (i) the solution of the initial-value problem reads

$$q\,(t) = -\frac{1}{\gamma}\,\log\left\{\exp\left[-\gamma\,q\,(0) + i\,\omega\,t\right] - a\,\gamma\,\frac{\exp\,(i\,\omega\,t) - 1}{i\,\omega}\right\}, \tag{5.109a}$$

and it is, clearly, *periodic* with period T, see (5.4), provided the initial datum $q\,(0)$ satisfies the inequality

$$\left|1 - \frac{i\,\omega}{a\,\gamma}\,\exp\left[-\gamma\,q\,(0)\right]\right| < 1. \tag{5.109b}$$

Likewise, in case (ii) the solution of the initial-value problem reads

$$q(t) = q(0)\left\{\exp\,(i\,\omega\,t) + a(1-\gamma)[q\,(0)]^{\gamma-1}\,\frac{\exp\,(i\,\omega\,t) - 1}{i\,\omega}\right\}^{1/(1-\gamma)}, \tag{5.110a}$$

and it is, clearly, *periodic* with period T provided the initial datum $q\,(0)$ satisfies the inequality

$$\left|1 + \frac{i\,\omega\,[q\,(0)]^{1-\gamma}}{a\,(1-\gamma)}\right| < 1. \tag{5.110b}$$

5.5 Yet another trick

In this section we describe another trick to *Ω-modify* a *Hamiltonian* so that the *Ω-modified Hamiltonian* thereby obtained is *entirely isochronous*. The basic period of *isochrony* shall now be

$$T = \frac{2\pi}{\Omega}; \tag{5.111}$$

note that we use here an upper case Ω, to emphasize the difference of this technique from that described above.

This technique is applicable to a *vast* category of *Hamiltonians*, again restricted essentially only by the availability of an explicitly known function $\Theta\,(\underline{p}, \underline{q})$ of the canonical variables satisfying the Poisson-commutation formula (5.18) [84]. But to be specific our treatment below is focussed on the *Hamiltonian* describing the standard *translation-invariant* nonrelativistic many-body problem; our presentation is for simplicity restricted to the equal-mass one-dimensional case, but the extension to arbitrary masses and arbitrary space dimensions is easy, as outlined below.

As will be clear from the following this new technique (especially in the version applicable to the standard *translation-invariant* nonrelativistic many-body problem) is more effective than that described above, on two counts: it transforms *real Hamiltonians* into *real Hamiltonians*, and it generally yields *entirely isochronous Hamiltonians*, namely Ω*-modified Hamiltonians* yielding motions *all* of which (in their *entire* natural phase space) are *completely periodic* with period T, see (5.111).

5.5.1 *Main results*

Let us begin with a terse review of the basic formulas describing the standard equal-mass one-dimensional *translation-invariant* nonrelativistic many-body problem. Its Hamiltonian (in self-evident notation) reads,

$$H\left(\underline{p},\underline{q}\right)=\frac{1}{2}\sum_{n=1}^{N}p_n^2+V\left(\underline{q}\right). \tag{5.112a}$$

Its *translation invariance* property corresponds to the requirement

$$V\left(\underline{q}+\underline{a}\right)=V\left(\underline{q}\right), \tag{5.112b}$$

where $\underline{a}$ denotes an *arbitrary* constant N-vector; when this vector is infinitesimal the corresponding formula reads of course

$$\sum_{n=1}^{N}\frac{\partial V\left(\underline{q}\right)}{\partial q_n}=0. \tag{5.112c}$$

As usual, we denote with P the total momentum, and with Q the (canonically-conjugate) center-of-mass coordinate:

$$P=\sum_{n=1}^{N}p_n,\quad Q=\frac{1}{N}\sum_{n=1}^{N}q_n,\quad [P,Q]=1. \tag{5.113}$$

Thanks to the translation invariance property (5.112c) there holds of course the Poisson-commutation property

$$[H,P]=0. \tag{5.114}$$

It is now convenient to introduce the "relative coordinates" x_n and the "relative momenta" y_n via the definitions

$$x_n=q_n-Q,\quad y_n=p_n-\frac{P}{N}. \tag{5.115a}$$

[*Warning*: this notation is different from that used above]. Note that these are *not* canonically conjugated quantities, since $[y_n, x_m] = \delta_{nm} - 1/N$, and they are *not* independent since they obviously satisfy the property

$$\sum_{n=1}^{N} x_n = \sum_{n=1}^{N} y_n = 0. \tag{5.115b}$$

Moreover, it is convenient to introduce the "relative-motion Hamiltonian" $h\left(\underline{y}, \underline{x}\right)$ via the formula

$$h\left(\underline{y}, \underline{x}\right) = \frac{1}{2} \sum_{n=1}^{N} y_n^2 + V\left(\underline{x}\right) = \frac{1}{4N} \sum_{n,m=1}^{N} \left(p_n - p_m\right)^2 + V\left(\underline{q}\right), \tag{5.116a}$$

so that

$$H\left(\underline{p}, \underline{q}\right) = h\left(\underline{y}, \underline{x}\right) + \frac{P^2}{2N}. \tag{5.116b}$$

Note that to write the two versions of (5.116a) we used the identification $V\left(\underline{x}\right) = V\left(\underline{q}\right)$ implied by (5.115a) and (5.112b). Note moreover that this definition, (5.116a), of the "relative-motion Hamiltonian" $h\left(\underline{y}, \underline{x}\right)$ entails that it Poisson-commutes with both P and Q:

$$[P, h] = [Q, h] = 0. \tag{5.116c}$$

For completeness and future reference let us also display the equations of motion implied by the Hamiltonian $H\left(\underline{p}, \underline{q}\right)$, see (5.112a):

$$q_n' = p_n, \quad p_n' = -\frac{\partial V\left(\underline{q}\right)}{\partial q_n}, \quad q_n'' = -\frac{\partial V\left(\underline{q}\right)}{\partial q_n}, \tag{5.117a}$$

where (for reasons that will be clear below) we denote as τ the independent variable corresponding to this Hamiltonian flow and with appended primes the differentiations with respect to these variable:

$$q_n \equiv q_n(\tau), \quad p_n \equiv p_n(\tau); \qquad q_n' \equiv \frac{\partial q_n(\tau)}{\partial \tau}, \quad p_n' \equiv \frac{\partial p_n(\tau)}{\partial \tau}. \tag{5.117b}$$

Hence, from (5.113) and (5.112c),

$$Q' = \frac{P}{N}, \quad P' = 0, \tag{5.118a}$$

yielding

$$Q(\tau) = Q(0) + \frac{P(0)}{N}\tau, \quad P(\tau) = P(0), \tag{5.118b}$$

and from (5.115a) and (5.117a)

$$x_n' = y_n = \frac{\partial h\left(\underline{y}, \underline{x}\right)}{\partial y_n}, \quad y_n' = -\frac{\partial V\left(\underline{x}\right)}{\partial x_n} = -\frac{\partial h\left(\underline{y}, \underline{x}\right)}{\partial x_n}. \tag{5.119}$$

Note that these equations of motion have the standard *Hamiltonian* look, in spite of the *noncanonical* character of the variables x_n and y_n.

The Ω-modified *Hamiltonian* $\tilde{H}\left(\underline{\tilde{p}}, \underline{\tilde{q}}; \Omega\right)$ is now defined by the formula

$$\tilde{H}\left(\underline{\tilde{p}}, \underline{\tilde{q}}; \Omega\right) = \frac{1}{2}\left\{\left[\tilde{P} + \frac{\tilde{h}\left(\underline{\tilde{y}}, \underline{\tilde{x}}\right)}{b}\right]^2 + \Omega^2 \tilde{Q}^2\right\}, \qquad (5.120a)$$

where b is an *arbitrary* constant (introduced for dimensional reasons: it has the dimensions of a momentum), Ω is of course a *positive* constant and we introduced (perhaps unnecessarily) the notation

$$\tilde{h}\left(\underline{\tilde{y}}, \underline{\tilde{x}}\right) \equiv h\left(\underline{\tilde{y}}, \underline{\tilde{x}}\right) = \frac{1}{2}\sum_{n=1}^{N} \tilde{y}_n^2 + V\left(\underline{\tilde{x}}\right). \qquad (5.120b)$$

Here the superimposed tildes over the canonical coordinates and momenta, $\tilde{q}_n \equiv \tilde{q}_n\left(t\right)$ and $\tilde{p}_n \equiv \tilde{p}_n\left(t\right)$, are introduced to emphasize that these variables evolve now according to the Ω-modified Hamiltonian $\tilde{H}\left(\underline{\tilde{p}}, \underline{\tilde{q}}; \Omega\right)$. [*Warning*: this notational distinction was not practiced in previous chapters, where sometimes the apposition of superimposed tildes instead had quite different meanings]. Likewise, the total momentum $\tilde{P} \equiv \tilde{P}\left(t\right)$ and the center-of-mass coordinate $\tilde{Q} \equiv \tilde{Q}\left(t\right)$, as well as the relative coordinates and momenta $\tilde{x}_n \equiv \tilde{x}_n\left(t\right)$ and $\tilde{y}_n \equiv \tilde{y}_n\left(t\right)$, are defined by formulas analogous to the previous ones:

$$\tilde{P} = \sum_{n=1}^{N} \tilde{p}_n, \quad \tilde{Q} = \frac{1}{N}\sum_{n=1}^{N} \tilde{q}_n, \qquad (5.121a)$$

$$\tilde{x}_n = \tilde{q}_n - \tilde{Q}, \quad \tilde{y}_n = \tilde{p}_n - \frac{\tilde{P}}{N}; \qquad (5.121b)$$

and, as already indicated above, the corresponding independent variable is now denoted as t ("time"), and differentiations with respect to this variable will be denoted, as usual, by superimposed dots.

It is now easily seen that there hold the following Poisson-commutation formulas:

$$\left[\tilde{H}, \tilde{Q}\right] = \tilde{P} + \frac{\tilde{h}\left(\underline{\tilde{y}}, \underline{\tilde{x}}\right)}{b}, \quad \left[\tilde{H}, \tilde{P}\right] = -\Omega^2 \tilde{Q}, \quad \left[\tilde{H}, \tilde{h}\right] = 0, \qquad (5.122)$$

so that the quantities $\tilde{Q}$, $\tilde{P}$ and $\tilde{h}$ evolve as follows under the flow induced by the Ω-modified Hamiltonian $\tilde{H}\left(\underline{\tilde{p}}, \underline{\tilde{q}}; \Omega\right)$, see (5.120):

$$\dot{\tilde{Q}} = \tilde{P} + \frac{\tilde{h}\left(\underline{\tilde{y}}, \underline{\tilde{x}}\right)}{b}, \quad \dot{\tilde{P}} = -\Omega^2 \tilde{Q}, \quad \dot{\tilde{h}} = 0, \qquad (5.123)$$

entailing

$$\tilde{Q}(t) = \tilde{Q}(0)\cos(\Omega t) + \dot{\tilde{Q}}(0)\frac{\sin(\Omega t)}{\Omega}, \tag{5.124a}$$

$$\tilde{P}(t) = \tilde{P}(0)\cos(\Omega t) + \dot{\tilde{P}}(0)\frac{\sin(\Omega t)}{\Omega} + \frac{\tilde{h}\left[\underline{\tilde{y}}(0), \underline{\tilde{x}}(0)\right]}{b}\left[\cos(\Omega t) - 1\right], \tag{5.124b}$$

$$\tilde{h}\left[\underline{\tilde{y}}(t), \underline{\tilde{x}}(t)\right] = \tilde{h}\left[\underline{\tilde{y}}(0), \underline{\tilde{x}}(0)\right]. \tag{5.124c}$$

It is moreover plain that the total momentum $\tilde{P}$ and the center-of-mass coordinate $\tilde{Q}$ Poisson-commute with the relative-motion momenta and coordinates $\tilde{y}_n$ and $\tilde{x}_n$, hence as well with any function of these variables,

$$\left[\tilde{P}, \tilde{x}_n\right] = \left[\tilde{P}, \tilde{y}_n\right] = \left[\tilde{P}, \tilde{h}\right] = \left[\tilde{Q}, \tilde{x}_n\right] = \left[\tilde{Q}, \tilde{y}_n\right] = \left[\tilde{Q}, \tilde{h}\right] = 0. \tag{5.125}$$

Hence, the evolution equations of the relative-motion coordinates and momenta $\tilde{x}_n$ and $\tilde{y}_n$ under the flow induced by the Ω-modified Hamiltonian $\tilde{H}\left(\underline{\tilde{p}}, \underline{\tilde{q}}; \Omega\right)$ read

$$\dot{\tilde{x}}_n = \frac{1}{b}\left[\tilde{P} + \frac{\tilde{h}\left(\underline{\tilde{y}}, \underline{\tilde{x}}\right)}{b}\right]\frac{\partial \tilde{h}\left(\underline{\tilde{y}}, \underline{\tilde{x}}\right)}{\partial \tilde{y}_n} = \frac{\dot{\tilde{Q}}}{b}\frac{\partial \tilde{h}\left(\underline{\tilde{y}}, \underline{\tilde{x}}\right)}{\partial \tilde{y}_n}, \tag{5.126a}$$

$$\dot{\tilde{y}}_n = -\frac{1}{b}\left[\tilde{P} + \frac{\tilde{h}\left(\underline{\tilde{y}}, \underline{\tilde{x}}\right)}{b}\right]\frac{\partial \tilde{h}\left(\underline{\tilde{y}}, \underline{\tilde{x}}\right)}{\partial \tilde{x}_n} = -\frac{\dot{\tilde{Q}}}{b}\frac{\partial \tilde{h}\left(\underline{\tilde{y}}, \underline{\tilde{x}}\right)}{\partial \tilde{x}_n}, \tag{5.126b}$$

namely, via (5.124) and (5.126),

$$\dot{\tilde{x}}_n = C\cos\left[\Omega\left(t - t_0\right)\right]\frac{\partial \tilde{h}\left(\underline{\tilde{y}}, \underline{\tilde{x}}\right)}{\partial \tilde{y}_n}, \tag{5.127a}$$

$$\dot{\tilde{y}}_n = -C\cos\left[\Omega\left(t - t_0\right)\right]\frac{\partial \tilde{h}\left(\underline{\tilde{y}}, \underline{\tilde{x}}\right)}{\partial \tilde{x}_n}, \tag{5.127b}$$

where

$$C = \frac{\sqrt{2\tilde{H}}}{b}, \quad \sin\left(\Omega t_0\right) = -\frac{\Omega \tilde{Q}(0)}{bC} = -\frac{\Omega \tilde{Q}(0)}{\sqrt{2\tilde{H}}}. \tag{5.127c}$$

It is now crucial to observe—by comparing these evolution equations with (5.119)—that it is justified to set

$$\tilde{x}_n(t) = x_n\left(\tilde{t}\right), \quad \tilde{y}_n(t) = y_n\left(\tilde{t}\right), \tag{5.128a}$$

with

$$\tilde{t} = C\frac{\sin\left[\Omega\left(t - t_0\right)\right]}{\Omega}. \tag{5.128b}$$

Here the coordinates and momenta $x_n\left(\tilde{t}\right)$ and $y_n\left(\tilde{t}\right)$ evolve according to the original flow yielded by the Hamiltonian $H\left(\underline{p},\underline{q}\right)$, see (5.119), and are uniquely identified by "initial" data assigned at the time

$$\tilde{t}_0=\frac{C\tilde{Q}\left(0\right)}{\sqrt{2\tilde{H}}}\tag{5.128c}$$

according to the following prescription:

$$x_n\left(\tilde{t}_0\right)=\tilde{x}_n\left(0\right),\quad y_n\left(\tilde{t}_0\right)=\tilde{y}_n\left(0\right),\tag{5.128d}$$

where $\tilde{x}_n\left(0\right)$ and $\tilde{y}_n\left(0\right)$ are the initial data for the relative-motion dynamics induced by the Ω-modified Hamiltonian $\tilde{H}\left(\underline{\tilde{p}},\underline{\tilde{q}};\Omega\right)$. In this manner the dynamics of the canonical coordinate and momenta $\tilde{q}_n\left(t\right)$ and $\tilde{p}_n\left(t\right)$ is finally obtained via (5.121b) and (5.124). And it is now plain that this dynamics is *isochronous* with period T, see (5.111), (5.124), and (5.128): note that the time evolution of the coordinates $x_n\left(\tilde{t}\right)$ and $y_n\left(\tilde{t}\right)$ is uniquely well-defined for all *real* time, see (5.119)—unless it runs into singularities, which should not be the case for physically sound models, and in any case should only happen exceptionally.

It is plain that, when Ω vanishes, the Ω-modified Hamiltonian $\tilde{H}\left(\underline{\tilde{p}},\underline{\tilde{q}};\Omega\right)$, see (5.120), does not quite reduce to the unmodified Hamiltonian $H\left(\underline{p},\underline{q}\right)$, see (5.112); but it is also clear that the dynamics yielded by the Hamiltonian $\tilde{H}\left(\underline{\tilde{p}},\underline{\tilde{q}};0\right)$ differs only marginally from that yielded by the original Hamiltonian $H\left(\underline{p},\underline{q}\right)$. To illustrate this point we now display the version that the most relevant formulas written above take when Ω vanishes, $\Omega=0$.

Let us consider first the evolution (see (5.124) and (5.123)) of the center-of-mass $\tilde{Q}$ and the total momentum $\tilde{P}$ yielded by the Hamiltonian $\tilde{H}\left(\underline{\tilde{p}},\underline{\tilde{q}};0\right)$:

$$\tilde{Q}\left(t\right)=\tilde{Q}\left(0\right)+\left[\tilde{P}\left(0\right)+\frac{\tilde{h}\left[\underline{\tilde{y}}\left(0\right),\underline{\tilde{x}}\left(0\right)\right]}{b}\right]t,\quad \tilde{P}\left(t\right)=\tilde{P}\left(0\right),\tag{5.129a}$$

to be compared with the analogous evolution yielded by the Hamiltonian $H\left(\underline{p},\underline{q}\right)$,

$$Q\left(t\right)=Q\left(0\right)+\frac{P\left(0\right)}{N}t,\quad P\left(t\right)=P\left(0\right).\tag{5.129b}$$

(Let us note that getting rid of the $1/N$ discrepancy among the right-hand sides of (5.129a) and (5.129b) could have been easily achieved via a slight change in the definition (5.120a) of the Ω-modified Hamiltonian $\tilde{H}\left(\underline{\tilde{p}},\underline{\tilde{q}};\Omega\right)$—which we preferred not to make for notational simplicity).

Next, let us compare the evolution of the relative-motion variables. The relevant formula is of course always (5.128a), but now with (5.128b) replaced by (see (5.127c))

$$\tilde{t} = Ct + \frac{\tilde{Q}\,(0)}{b}. \tag{5.130}$$

These formulas confirm the assertion that the dynamics yielded by the Hamiltonian $\tilde{H}\left(\underline{\tilde{p}}, \underline{\tilde{q}}; 0\right)$ differs only marginally from that yielded by the original Hamiltonian $H\left(\underline{p}, \underline{q}\right)$.

In fact an analogous relationship—entailing that $\tilde{t}$ on a sufficiently short timescale varies linearly in t—holds generally, since in the neighborhood of any time $\bar{t}$, except when $\dot{\tilde{Q}}(\bar{t})$ vanishes,

$$\tilde{t}(t) = \frac{\tilde{Q}(\bar{t}) + \dot{\tilde{Q}}(\bar{t})(t - \bar{t})}{b} + O\left[\Omega^2\,(t - \bar{t})^2\right]. \tag{5.131}$$

One therefore finds that throughout the time evolution the Ω-modified dynamics differs from the unmodified one solely by a time rescaling—by a possibly negative coefficient—and by a time shift. The coefficient and the shift are time-independent over a timescale much smaller than the *isochrony* period $T = 2\pi/\Omega$, but vary periodically with period T, see (5.128b). A peculiar state of affairs arises, however, whenever $\dot{\tilde{Q}}(\bar{t})$ vanishes, namely when $d\tilde{t}/dt$ changes its sign: this of course happens twice within every time period T, see (5.128b), this being in fact a consequence of the periodicity of $\tilde{t}\,(t)$, which itself is the cause of the *isochrony*. These aspects are apparent in the two figures displayed below.

All the formulas written above are appropriate to the one-dimensional N-body problem with equal mass particles. These restrictions have been imposed merely for the sake of simplicity: the alert reader will have no difficulty to extend the treatment—if need be—to many-body problems with particles having different masses and/or moving in higher-dimensional space. Of course in the latter case there is an ample choice of the collective variables playing an analogous role to that played above by the total momentum and the center-of-mass coordinate: indeed, in a multidimensional context, any component of the vectors $\vec{P}$ and $\vec{Q}$ representing the total momentum and the center-of-mass coordinate could be used, or, not to spoil rotation invariance, one could replace, in the definition (5.120a) of the Ω-modified *Hamiltonian*, P with $\vec{P}\cdot\vec{Q}/\sqrt{\vec{Q}\cdot\vec{Q}}$ and Q^2 with $\vec{Q}\cdot\vec{Q}$.

5.5.2 *Transient chaos*

It is interesting to speculate on the application of this Ω-*modification* technique to any Hamiltonian describing a translation-invariant many-body problem featuring, in its center-of-mass system, *chaotic* motions with a natural timescale T_C. Then—provided the constant Ω is assigned so that the *isochrony* period T, see (5.111), is much larger than this timescale, $T >> T_C$—the Ω-modified problem shall exhibit a *chaotic* behavior for quite some time before the *isochronous* character of all its motions takes over. This phenomenology—qualitative rather than quantitative as it necessarily is, since a precise definition of *chaoticity* requires generally that a system displaying it be observed for *infinite* time—is nevertheless remarkable, justifying the title of this short section and further investigations. Its origin and explanation is, of course, in the fact that the Ω*-modification* technique yields an Ω*-modified* dynamics evolving more or less according to the original, *unmodified* dynamics but in terms of a variable, say $\tilde{t}$ (playing the role of time) that is itself a periodic function of the "physical time" t (see (5.128b)).

5.5.3 *A simple example*

In this section we exhibit a very simple example, reporting, with minimal comments, the findings obtained by applying our Ω-modification technique to the very simple Hamiltonian describing a couple of equal-mass one-dimensional particles interacting pairwise with a force proportional to their mutual distance; when this force is attractive this model corresponds of course, in the center-of-mass system, to the standard "harmonic" oscillator model. (We write "harmonic" under inverted commas to emphasize that the *entirely isochronous* Ω-modified Hamiltonians yielded by our technique *all* yield motions deserving to be called *harmonic*, inasmuch as they are characterized by just a single frequency of oscillation: in the case of many-body problems the term "nonlinear harmonic oscillators" [75] is perhaps the most appropriate to describe the corresponding dynamics...). In spite of its simplicity this example is adequate to illustrate the remarkable fact that, for most of the time, an Ω*-modified* system may well behave as the *unmodified* system from which it has been obtained, yet eventually exhibit the periodicity with period $T = 2\pi/\Omega$ implied by its *entirely isochronous* character.

The original Hamiltonian:

$$H(p_1, p_2; q_1, q_2) = \frac{1}{2}\left(p_1^2 + p_2^2\right) + \frac{\omega^2}{4}\left(q_1 - q_2\right)^2 . \tag{5.132}$$

Some standard definitions and related formulas:

$$Q = \frac{q_1 + q_2}{2}, \qquad P = p_1 + p_2; \tag{5.133}$$

$$x_1 = q_1 - Q = \frac{q_1 - q_2}{2}, \quad x_2 = q_2 - Q = \frac{q_2 - q_1}{2}, \quad x_1 + x_2 = 0, \tag{5.134a}$$

$$y_1 = p_1 - \frac{P}{2} = \frac{p_1 - p_2}{2}, \quad y_2 = p_2 - \frac{P}{2} = \frac{p_2 - p_1}{2}, \quad y_1 + y_2 = 0; \tag{5.134b}$$

$$h(y_1, y_2; x_1, x_2) = \frac{1}{2}\left(y_1^2 + y_2^2\right) + \frac{\omega^2}{4}(x_1 - x_2)^2, \tag{5.135a}$$

$$H(p_1, p_2; q_1, q_2) = \frac{1}{4}P^2 + h(y_1, y_2; x_1, x_2); \tag{5.135b}$$

$$Q' = \frac{1}{2}P, \quad P' = 0, \tag{5.136a}$$

$$Q(\tau) = Q(\tau_0) + \frac{1}{2}P(\tau_0)(\tau - \tau_0), \quad P(\tau) = P(\tau_0); \tag{5.136b}$$

$$x_n' = y_n, \quad y_n' = -\frac{\omega^2}{2}(x_n - x_{n+1}), \quad n = 1, 2 \mod(2), \tag{5.137a}$$

$$x_n(\tau) = x_n(\tau_0)\cos[\omega(\tau - \tau_0)] + y_n(\tau_0)\frac{\sin[\omega(\tau - \tau_0)]}{\omega}, \quad n = 1, 2, \tag{5.137b}$$

$$y_n(\tau) = y_n(\tau_0)\cos[\omega(\tau - \tau_0)] - \omega x_n(\tau_0)\sin[\omega(\tau - \tau_0)], \quad n = 1, 2. \tag{5.137c}$$

The Ω-modified Hamiltonian:

$$\tilde{H}(\tilde{p}_1, \tilde{p}_2; \tilde{q}_1, \tilde{q}_2; \Omega) = \frac{1}{2}\left\{\left[\tilde{P} + \frac{\tilde{h}(\tilde{y}_1, \tilde{y}_2; \tilde{x}_1, \tilde{x}_2)}{b}\right]^2 + \Omega^2\tilde{Q}^2\right\}; \tag{5.138}$$

$$\tilde{Q} = \frac{\tilde{q}_1 + \tilde{q}_2}{2}, \quad \tilde{P} = \tilde{p}_1 + \tilde{p}_2; \tag{5.139}$$

$$\tilde{x}_1 = \tilde{q}_1 - \tilde{Q} = \frac{\tilde{q}_1 - \tilde{q}_2}{2}, \quad \tilde{x}_2 = \tilde{q}_2 - \tilde{Q} = \frac{\tilde{q}_2 - \tilde{q}_1}{2}, \quad \tilde{x}_1 + \tilde{x}_2 = 0, \tag{5.140a}$$

$$\tilde{y}_1 = \tilde{p}_1 - \frac{\tilde{P}}{2} = \frac{\tilde{p}_1 - \tilde{p}_2}{2}, \quad \tilde{y}_2 = \tilde{p}_2 - \frac{\tilde{P}}{2} = \frac{\tilde{p}_2 - \tilde{p}_1}{2}, \quad \tilde{y}_1 + \tilde{y}_2 = 0; \tag{5.140b}$$

$$\tilde{h}(\tilde{y}_1, \tilde{y}_2; \tilde{x}_1, \tilde{x}_2) = \frac{1}{2}\left(\tilde{y}_1^2 + \tilde{y}_2^2\right) + \frac{\omega^2}{4}(\tilde{x}_1 - \tilde{x}_2)^2; \tag{5.141a}$$

$$\tilde{H}(\tilde{p}_1, \tilde{p}_2; \tilde{q}_1, \tilde{q}_2; \Omega) = \frac{1}{2}\left\{\left[\tilde{p}_1 + \tilde{p}_2 + \frac{(\tilde{p}_1 - \tilde{p}_2)^2}{4b}\right]^2 + \frac{\omega^2}{2b}\left[\tilde{p}_1 + \tilde{p}_2 + \frac{(\tilde{p}_1 - \tilde{p}_2)^2}{4b}\right] (\tilde{q}_1 - \tilde{q}_2)^2 + \left(\frac{\omega^2}{4b}\right)^2(\tilde{q}_1 - \tilde{q}_2)^4 + \frac{\Omega^2}{4}(\tilde{q}_1 + \tilde{q}_2)^2\right\}. \tag{5.141b}$$

The Ω*-modified equations of motions*: see (5.123) and (5.126) yielding

$$\dot{\tilde{x}}_n = \frac{1}{b}\left[\tilde{P} + \frac{\tilde{h}\left(\tilde{y}_1, \tilde{y}_2; \tilde{x}_1, \tilde{x}_2\right)}{b}\right]\tilde{y}_n, \quad n = 1, 2, \tag{5.142a}$$

$$\dot{\tilde{y}}_n = -\frac{\omega^2}{2b}\left[\tilde{P} + \frac{\tilde{h}\left(\tilde{y}_1, \tilde{y}_2; \tilde{x}_1, \tilde{x}_2\right)}{b}\right]\left(\tilde{x}_n - \tilde{x}_n\right), \quad n = 1, 2 \mod(2). \tag{5.142b}$$

The isochronous *motions yielded by the* Ω*-modified Hamiltonian* $\tilde{H}\left(\tilde{p}_1, \tilde{p}_2; \tilde{q}_1, \tilde{q}_2; \Omega\right)$: see (5.124), and (5.128) yielding (5.137b) and (5.137c) namely (see (5.128b))

$$\tilde{x}_n\left(t\right) = \tilde{x}_n\left(0\right)\cos\left\{\omega[\tau(t) - \tau(0)]\right\} + \tilde{y}_n\left(0\right)\frac{\sin\left\{\omega[\tau(t) - \tau(0)]\right\}}{\omega}, \quad n = 1, 2, \tag{5.143a}$$

$$\tilde{y}_n\left(t\right) = \tilde{y}_n\left(0\right)\cos\left\{\omega[\tau(t) - \tau(0)]\right\} - \omega\tilde{x}_n\left(0\right)\sin\left\{\omega[\tau(t) - \tau(0)]\right\}, \quad n = 1, 2, \tag{5.143b}$$

$$\tau(t) = C\frac{\sin[\Omega(t - t_0)]}{\Omega}, \tag{5.143c}$$

with C and t_0 defined in terms of the initial data by (5.127c).

The *entirely isochronous* character of this motion, with period $T = 2\pi/\Omega$, is evident. And note that this outcome obtains even if ω is purely imaginary, $\omega = i\alpha$; in which case, the original Hamiltonian H and the Ω-modified Hamiltonian $\tilde{H}$, and as well of course these solutions, see (5.124) and the formulas written just above, are nevertheless all *real*. Figures 5.1 and 5.2 display examples of these motions.

5.5.4 *Quantization: equispaced spectrum*

In this section we tersely show why, in a *quantal* context, the *Hamiltonian* $\tilde{H}\left(\underline{\tilde{p}}, \underline{\tilde{q}}; \Omega\right)$, see (5.120), features an (infinitely degenerate) *equispaced* spectrum with spacing $\hbar\Omega$.

This spectrum consists of the eigenvalues E_k of the stationary Schrödinger equation

$$\frac{1}{2}\left\{\left[-i\hbar\frac{\partial}{\partial Z} + \frac{\lambda}{b}\right]^2 + \Omega^2 Z^2\right\}\psi_k\left(Z; \lambda\right) = E_k\psi_k\left(Z; \lambda\right), \tag{5.144}$$

obtained from the expression (5.120) of the Hamiltonian $\tilde{H}\left(\underline{\tilde{p}}, \underline{\tilde{q}}; \Omega\right)$ via the standard quantization rule

$$\tilde{P} \Longrightarrow -i\hbar\frac{\partial}{\partial Z}, \quad \tilde{Q} \Longrightarrow Z, \tag{5.145}$$

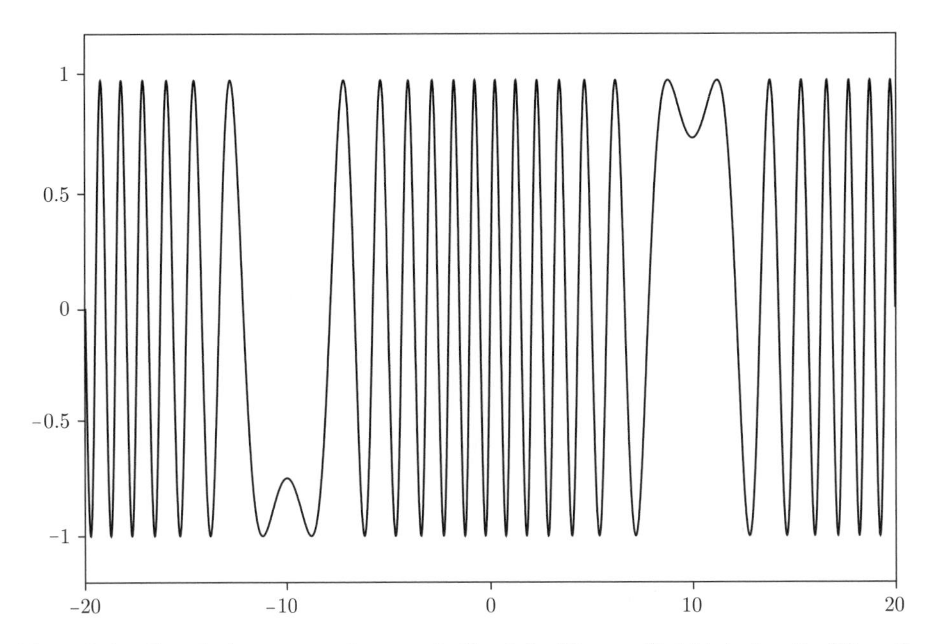

FIG. 5.1. Graph (over one time period) of $\tilde{x}_n(t)$, see (5.144a), for $\tilde{x}_n(0) = 0$, $\tilde{y}_n(0) = \omega = 40\Omega = 2\pi$, $\Omega = \pi/20$, $C = 1$, $t_0 = 0$. Note the overall periodicity with period $T = 2\pi/\Omega = 40$, the large regions where the behavior is nearly periodic with the original period $2\pi/\omega = 1$ of the solutions of the unmodified Hamiltonian (5.132), and the transition regions around the times $t = 10$ and $t = 30$ when $\dot{\tilde{t}}(t)$ vanishes. (Reprinted from [83].)

and by identifying λ as an eigenvalue of the quantized version of the relative-motion Hamiltonian $\tilde{h}\left(\underline{\tilde{y}}, \underline{\tilde{x}}\right)$, see (5.116a). Indeed, this Schrödinger equation is obtained by assuming that the eigenfunctions of the quantized version of the Hamiltonian $\tilde{H}\left(\underline{\tilde{p}}, \underline{\tilde{q}}; \Omega\right)$ factor into the product of an eigenfunction, $\psi_k(Z; \lambda)$, depending on the variable Z, on which acts the differential operator $\partial/\partial Z$ (see (5.145)), and of the eigenfunction corresponding to the eigenvalue λ of the quantized version of the relative-motion Hamiltonian $\tilde{h}\left(\underline{\tilde{y}}, \underline{\tilde{x}}\right)$. The justification for this factorization is in the commutativity of the operators representing the quantal version of the canonical variables $\tilde{P}$ and $\tilde{Q}$ with the operator representing the quantal version of the relative-motion Hamiltonian $\tilde{h}\left(\underline{\tilde{y}}, \underline{\tilde{x}}\right)$—a commutativity reflecting the Poisson-commutativity of the corresponding quantities in the classical context, see (5.125).

It is now plain that the Schrödinger equation (5.144) features the spectrum

$$E_k = \hbar\Omega\left(k + \frac{1}{2}\right), \quad k = 0, 1, 2, ..., \tag{5.146a}$$

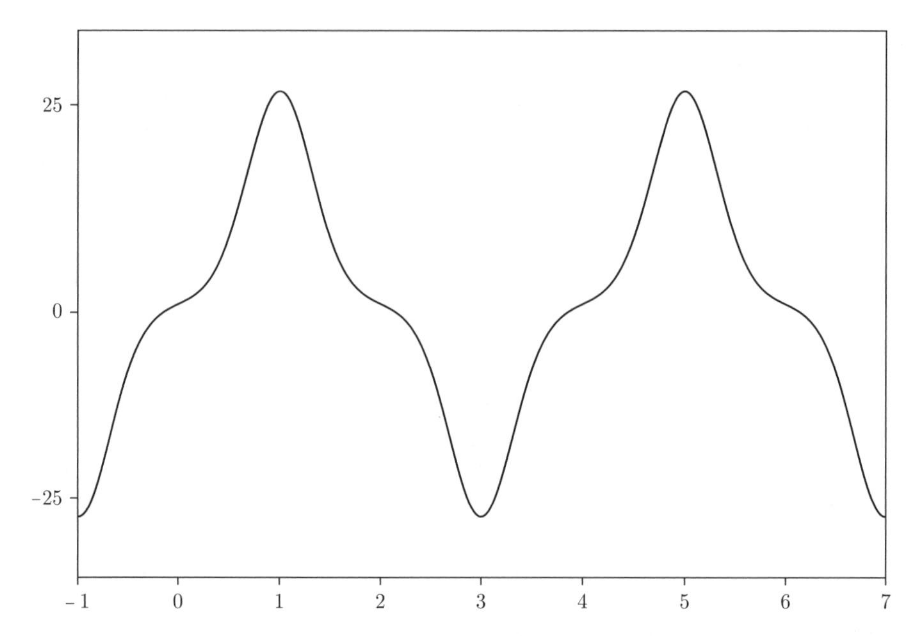

FIG. 5.2. Graph (over two time periods) of $\tilde{x}_n(t)$, see (5.143a), for $\tilde{x}_n(0) = 0$, $\tilde{y}_n(0) = 2\pi$, $\omega = 4i\Omega = 2\pi i$, $\Omega = \pi/2$, $C = 1$, $t_0 = 0$. Note the overall periodicity with period $T = 2\pi/\Omega = 4$, the regions where the time evolution resembles the original $\sin(\omega t)/i = \sinh(2\pi t)$ behavior of the corresponding solution of the unmodified Hamiltonian (5.132), and the transition regions around the times $t = -1, 1, 3, 5, 7$ when $\dot{\tilde{t}}(t)$ vanishes. (Reprinted from [83].)

with the corresponding eigenfunctions reading

$$\psi_k(Z;\lambda) = \exp\left(\frac{i\lambda z}{b\sqrt{\hbar\Omega}} - \frac{z^2}{2}\right) H_k(z), \quad z = Z\sqrt{\frac{\Omega}{\hbar}}, \tag{5.146b}$$

where $H_k(z)$ denotes the standard Hermite polynomial of order k.

This spectrum, (5.146a), is of course *equispaced* with spacing $\hbar\Omega$, and it is *infinitely degenerate* inasmuch as it does not feature any dependence on the eigenvalues λ.

5.N Notes to Chapter 5

The presentation in this chapter is largely based (often *verbatim*) on recent joint papers with François Leyvraz: in particular, for the material in Sections 5.1-3 see [80], for **Example 5.4-4** see [82], and for the material in Section 5.5 see [84]. Further results (also numerical) on the *transient chaos* phenomenology outlined in Section 5.5.2 (and in the last paragraph of the introductory Chapter 1) are in preparation [85]. For the applicability of the Ω-modification technique of

Section 5.5 to more general Hamiltonians than that of the nonrelativistic N-body problem see [84].

The trick described in Section 5.1 has the disadvantage that the ω-modified Hamiltonians it yields are generally *complex*, see (5.20). A variant of this trick that does not have this defect is characterized by the replacement of (5.20b) with

$$u(\underline{p},\, \underline{q}; \omega) = 1 + \omega^2 \left[\Theta(\underline{p},\, \underline{q})\right] . \tag{5.147}$$

This variant, however, often yields (via (5.20a)) ω-modified Hamiltonians generating motions which seem *periodic* but are in fact *singular* (in the same sense as the function $\tan(\omega t)$ is periodic but singular). This is, however, not always the case: for instance, no singularity emerges when this new kind of ω-modification is applied to the Hamiltonian model treated in **Example 5.4-4** [82]. For additional investigations (including several examples) of this technique see [83]. We omitted to describe this technique in this monograph—except for this terse mention here—because its main merit—to yield *real* Hamiltonians—is also possessed by the more powerful technique reported in Section 5.5 and its subsections.

The results presented in this Chapter 5 open the possibility to investigate the *quantization* of many *isochronous* systems, including the exploration of the limits of validity of the natural hunch that such systems feature, in an *appropriately quantized* context, *equispaced* spectra. This topic exceeds the scope of the present monograph—although we did present a glimpse of developments in this direction in Section 5.5.4. The interested reader will find additional results on this topic in recent joint papers with François Leyvraz [81] [82] [83] [84] (and some results in this direction also in the previous papers [74] [44] [45]).

6

ASYMPTOTICALLY ISOCHRONOUS SYSTEMS

In this chapter we outline an amusing idea that has been recently revived (see [71], whose presentation is closely followed, often *verbatim*, in this chapter): dynamical systems whose *generic* solutions approach *asymptotically* (at large time) *isochronous* evolutions, namely *all* their dependent variables tend *asymptotically* to functions *periodic* with the *same* fixed period. The *definition* of such dynamical systems is provided by the simultaneous validity of the two formulas

$$\lim_{t\to+\infty}\left[z_n(t)-\tilde{z}_n(t)\right]=0,\quad n=1,...,N, \tag{6.1a}$$

$$\tilde{z}_n\left(t+\tilde{T}\right)=\tilde{z}_n(t),\quad n=1,...,N. \tag{6.1b}$$

Notation: the N (generally *complex*; but see below) numbers $z_n(t)$ denote the N dependent variables of the dynamical system under consideration; we restrict consideration to the case when N is a *finite* positive integer; the *real* variable t denotes the time; the N functions $\tilde{z}_n(t)$ characterize the asymptotic behavior of the dynamical system via (6.1a) and the periodicity requirement (6.1b) they satisfy characterizes the property of *asymptotic isochrony.* This property is supposed to hold in an *open* (hence *fully dimensional*) region of the phase space of the dynamical system under consideration (possibly coinciding with its *entire* natural phase space): hence the dependent variables $z_n(t)$ denote here (the N components of) a *generic* solution of the dynamical system evolving (at least for sufficiently large time) within that region, while the functions $\tilde{z}_n(t)$, which shall generally be different for different solutions $z_n(t)$, are required to satisfy the periodicity property (6.1b) with the *fixed* period $\tilde{T}$ (the *same* for all the solutions in the phase space region under consideration). Of course the formula (6.1a) does not define uniquely—for a given N-vector $\underline{z}(t)$—a corresponding N-vector $\underline{\tilde{z}}(t)$: the time-dependent N-vector $\underline{\tilde{z}}(t)$ is only identified by (6.1a) up to arbitrary corrections whose effects disappear in the asymptotic limit $t\to+\infty$. The property of *asymptotic isochrony* is guaranteed provided there does exist an N-vector $\underline{\tilde{z}}(t)$ satisfying *both* relations (6.1), for every *generic* solution $\underline{z}(t)$ in an *open, fully dimensional*, region of phase space—namely, for every solution $\underline{z}(t)$ in that region of phase space, except possibly for some *exceptional*, generally *singular*, solutions belonging to a *lower dimensional* sector of that phase space region.

The elementary idea underlying the identification of large classes of such *asymptotically isochronous* dynamical systems is to start from *isochronous* systems and then modify them by introducing a deformation whose effects are significant through the time evolution but disappear at large time: so that the modified systems lose their *isochronous* character (at finite times) but in some sense retain it (at large times) as the dominant feature characterizing their *asymptotic* behavior.

There are several possible ways to implement this strategy in order to manufacture *asymptotically isochronous* systems: some are rather trivial, some less so. This kind of judgment is, of course, subjective: for instance we tend to think that an important requirement for such systems to be deemed "interesting" is that they be *autonomous*—because the interest of dynamical systems is also related to their potential usefulness in order to model natural phenomena, which are generally described by *autonomous* evolution equations—and moreover because the freedom to introduce, instead, an *explicit* time dependence in the equations of motion of a dynamical system would provide too easy a way to influence more or less at will the asymptotic behavior of such a system. But, of course, the difference between *autonomous* and *nonautonomous* systems is not fundamental, since any *nonautonomous* system can be made *autonomous* by treating time itself as an additional dependent variable.

Below we focus on two mechanisms yielding (autonomous) *asymptotically isochronous* systems, and illustrate each of them via a representative example. The first example (see Section 6.1) belongs to a recently discovered class of *integrable* indeed *solvable* many-body problems, described above as the third item in **Example 4.2.2-10**; while in this case we focus on as simple and specific an example as possible, we trust our presentation is nevertheless adequate to illustrate the generality of the approach. In this case the periodic behavior prevailing asymptotically corresponds to a special solution of the dynamical system under consideration belonging to a region of phase space with *positive* codimension—albeit *not* an *isolated* solution of this system, so not quite identifiable as a *limit cycle*. Hence this model might be considered a representative example of a phenomenology characterized by the presence of some kind of friction. The second example (see Section 6.2) consists of a broad class of models obtained by deforming appropriately the well-known *integrable* and *isochronous* one-dimensional many-body problem with inverse-cube two-body forces and a one-body linear ("harmonic oscillator") force (see the first model described in Chapter 1); the alert reader will again appreciate the generality of the approach, even though we illustrate it by focusing on a specific model (also restricting consideration to *real* dependent variables). In this second case the time-dependent N-vector to which the solutions of the model tend *asymptotically* is *not* restricted to be in a sector of phase space with *positive* codimension and is generally *not* itself a solution of the *asymptotically isochronous* N-body model, so this phenomenology does not correspond to what is generally referred to as a *limit cycle* behavior. In each of these two cases we back the qualitative understanding of the origin of the relevant phenomenology with a *proof* of its actual emergence, see (6.1).

6.1 An asymptotically isochronous class of solvable many-body problems

A particular mechanism to manufacture *integrable*, indeed *solvable*, dynamical systems interpretable as many-body problems inasmuch as they are characterized by *Newtonian* equations of motion ("acceleration equal force") was introduced about three decades ago [29] and has been subsequently exploited to identify and investigate several such systems; it has been amply described in Section 4.2.2. Let us recall here the main idea: to exploit the *nonlinear* relation among the N coefficients $c_m(t)$ and the N zeros $z_n(t)$ of a (for definiteness, monic) time-dependent polynomial of degree N,

$$\psi(z,t) = z^N + \sum_{m=1}^{N} c_m(t)\, z^{N-m} = \prod_{n=1}^{N} [z - z_n(t)] \,. \tag{6.2}$$

A class of such systems is characterized by the fact that the N coefficients $c_m(t)$ evolve in time according to a system of *linear* second-order constant-coefficent ODEs, the solution of which is a purely *algebraic* task (requiring essentially the diagonalization of an explicitly known matrix of order N). The determination of the corresponding time evolution of the N zeros $z_n(t)$ is therefore as well a purely *algebraic* task: computing the N zeros of a known polynomial. While it so happens that in many cases this time evolution is indeed interpretable as that characterizing a *Newtonian* N-body problem—a *solvable* such problem, since its solution can be achieved by purely *algebraic* means.

Indeed the solution $z_n(t)$ of such a model is generally obtained by finding the N zeros of a polynomial of degree N in the (*complex*) variable z, see (6.2), whose coefficients $c_m(t)$ generally evolve exponentially in time, typically

$$c_m(t) = \sum_{\ell=1}^{N} \left\{ \gamma^{(\ell,+)} u_m^{(\ell,+)} \exp\left[\lambda^{(\ell,+)} t\right] + \gamma^{(\ell,-)} u_m^{(\ell,-)} \exp\left[\lambda^{(\ell,-)} t\right] \right\}, \tag{6.3}$$

where the $2N$ constants $\gamma^{(\ell,\pm)}$ are arbitrary (to be determined by the initial data $z_n(0)\,, \dot{z}_n(0)$ in the context of the initial-value problem for the N-body system) and the $2N$ numbers $\lambda^{(\ell,\pm)}$ respectively the quantities $u_m^{(\ell,\pm)}$ are the *eigenvalues* respectively the (components of the) *eigenvectors* of the matrix eigenvalue problem characterizing, as explained above, the corresponding dynamics of the coefficients $c_m(t)$. Note that these *eigenvalues* and *eigenvectors* are associated to the dynamical problem under consideration: they do *not* depend on the initial data identifying a particular solution, namely they are the *same* for all the solutions of the system.

If the $2N$ eigenvalues $\lambda^{(\ell,\pm)}$ are all *integer* multiples of a single *imaginary* number $i\omega$ (with $\omega > 0$), $\lambda^{(\ell,\pm)} = ik^{(\ell,\pm)}\omega$ with the $2N$ numbers $k_\ell^{(\pm)}$ arbitrary *integers* (positive or negative, but *not* vanishing, and with $\lambda^{(\ell,+)} \neq \lambda^{(\ell,-)}$), then the polynomial $\psi(z,t)$ is clearly *periodic* with the (possibly nonprimitive) period

$$T = \frac{2\pi}{\omega}, \tag{6.4a}$$

$$\psi(z, t+T) = \psi(z,t)\,; \tag{6.4b}$$

hence all its zeros $z_n(t)$ are as well *periodic* with this same period or possibly with a (generally small [100]) integer multiple p of this period, $\tilde{T} = pT$, due to the possibility that they exchange their role through the time evolution. Hence the corresponding N-body problem is *isochronous.*

And it is as well plain that if, out of the $2N$ eigenvalues $\lambda^{(\ell,\pm)}$, only a (nonempty) subset have the property indicated above while *all* the others feature a *negative* real part, then the many-body problem in question is *asymptotically isochronous.* This observation is not new, see for instance Section 4.2.3 of [37] (entitled "Some special cases: models with a limit cycle, models with confined and periodic motions, Hamiltonian models, translation-invariant models, models featuring equilibrium and spiraling configurations, models featuring only completely periodic motions"); but, to the best of our knowledge, this mechanism yielding *asymptotically isochronous* many-body problems was never analyzed in explicit detail until quite recently [71]: in the following section we tersely report these findings.

6.1.1 *A specific example*

The N-body problem on which we focus is the third one described above in the context of **Example 4.2.2-10**, see (4.184): hence the evolution of the N "particle coordinates" $z_n(t)$—taking generally place in the *complex* z-plane—coincides with the evolution of the N zeros of a monic polynomial of degree N in the variable z analogous to $\psi(z,t)$, see (6.2), but more specifically reading as follows (see (4.185)):

$$\psi(z,t) = \pi_N(z) + \sum_{m=1}^{N-3} \left[c_m(t)\,\pi_{N-m}(z)\right] + c_N(t), \tag{6.5a}$$

$$\pi_m(z) = z^m - \varepsilon_m \frac{m}{2} z^2 - \varepsilon_{m+1} m z, \quad m = 0, 1, ..., N, \tag{6.5b}$$

$$\varepsilon_m = 1 \text{ if } m \text{ is } \textit{even}\,, \quad \varepsilon_m = 0 \text{ if } m \text{ is } \textit{odd}. \tag{6.5c}$$

And the coefficients $c_m(t)$ evolve indeed according to formulas analogous to (6.3), but more specifically reading as follows (see (4.186)):

$$c_m(t) = \sum_{\ell=1;\ \ell\neq N-1,N-2}^{N} \left\{\gamma^{(\ell,+)} u_m^{(\ell,+)} \exp\left[\lambda^{(\ell,+)} t\right] + \gamma^{(\ell,-)} u_m^{(\ell,-)} \exp\left[\lambda^{(\ell,-)} t\right]\right\},$$
$$m = 1, ..., N-3 \ \text{ and } \ m = N, \tag{6.6a}$$

$$\lambda^{(\ell,\pm)} = \frac{-a_1 \pm \Delta_\ell}{2}, \quad \Delta_\ell^2 = a_1^2 + 4\ell\left[a_2 + (2N-\ell-3)\,a_4\right],$$
$$\ell = 1, ..., N-3, N. \tag{6.6b}$$

We now restrict attention to the $N = 3$ case, since this is sufficient, indeed convenient, for exhibiting quite explicitly an *asymptotically isochronous* model. Then the only relevant coefficient (see (6.6a)) is

$$c_3(t) = \gamma_+ \exp(\lambda_+ t) + \gamma_- \exp(\lambda_- t), \tag{6.7a}$$

$$\lambda_\pm = \frac{-a_1 \pm \Delta}{2}, \quad \Delta^2 = a_1^2 + 12 a_2, \tag{6.7b}$$

where the somewhat simplified notation we are now using is we trust self-explanatory (and note that in this case with $N = 3$ the eigenvalues $\lambda_{\pm}$ only depend on the two coupling constants a_1 and a_2). Correspondingly, the positions of the three moving particles are the three zeros $z_n(t)$ of the third-degree polynomial

$$\psi(z,t) = \pi_3(z) + c_3(t) = z^3 - 3z + c_3(t) = \prod_{n=1}^{3} [z - z_n(t)]. \qquad (6.7c)$$

Note that these three zeros automatically satisfy the requirements (4.184c), which corresponds [70] to the condition that the partial derivative of $\psi(z,t)$ with respect to z vanish at $z = \pm 1$, $\psi_z(\pm 1, t) = 0$.

Assume now that the two coupling constants a_1 and a_2 entail, via (6.7b),

$$\lambda_+ = i\omega, \quad \lambda_- = -\alpha + i\beta, \qquad (6.8a)$$

with α *positive*, $\alpha > 0$, ω also *positive*, $\omega > 0$ (for definiteness), and β *real* but otherwise *arbitrary*. This indeed happens provided

$$a_1 = \alpha - i(\beta + \omega), \quad a_2 = \frac{\omega(\beta + i\alpha)}{3}. \qquad (6.8b)$$

It is now plain that the asymptotic condition (6.1a) holds now with $\tilde{z}_n(t)$ being the three roots of the polynomial $z^3 - 3z + \gamma_+ \exp(i\omega t)$,

$$z^3 - 3z + \gamma_+ \exp(i\omega t) = \prod_{n=1}^{3} [z - \tilde{z}_n(t)]. \qquad (6.9)$$

And it is as well plain that the time evolution of this polynomial is periodic with period T, see (3.8b), hence the corresponding evolution of each of its three zeros is clearly periodic with periods T or $2T$ or $3T$ (for an explicit display of this evolution see [71]).

6.2 A (generally nonintegrable) class of asymptotically isochronous many-body models

In this section we consider a class of *asymptotically isochronous* models obtained by deforming the well-known *integrable* N-body problem with two-body inverse cube forces and a one-body linear force, which is of course *isochronous* when no deformation is present (see the first N-body problem, (1.1), discussed in Chapter 1). In particular we focus on the following equations of motion:

$$\ddot{x}_n + \frac{1}{4}\omega^2 x_n = g^2 \sum_{m=1, m\neq n}^{N} (x_n - x_m)^{-3} + F(w, \underline{x}, \underline{\dot{x}}), \quad n = 1, ..., N, \qquad (6.10a)$$

$$\dot{w} = w[\alpha \log w - f(w, \underline{x}, \underline{\dot{x}})], \quad \alpha > 0, \quad 0 < w(0) < 1. \qquad (6.10b)$$

Notation: N is an arbitrary positive integer ($N \geq 2$); the N dependent variables $x_n \equiv x_n(t)$ may be interpreted as the coordinates of N particles evolving according to the *Newtonian* ("acceleration equal force") equations of motion (6.10a); these variables x_n are now assumed to be all *real* (until we mention below to what extend the results change if the variables x_n are allowed to be *complex*), and $\underline{x}$ denotes of course the N-vector with components x_n; likewise the (also *real*) "auxiliary" dependent variable $w \equiv w(t)$ evolves according to the first-order ODE (6.10b) (but clearly, see below, replacing this first-order ODE with an *appropriate* second-order "Newtonian" ODE would make little difference); t denotes of course the (*real*) independent variable ("time": ranging for simplicity from the *initial* time $t = 0$ to the *asymptotic* time $t = +\infty$), and superimposed dots denote as usual differentiations with respect to this variable; ω, g^2 and α are three *positive* (but otherwise *arbitrary*) constants.

The main restriction on the, otherwise *arbitrary*, function $F(w, \underline{x}, \underline{v})$ is that it vanish when w vanishes,

$$F(0, \underline{x}, \underline{v}) = 0, \tag{6.10c}$$

and the main restriction on the function $f(w, \underline{x}, \underline{v})$ is that it entail via (6.10b) a (very fast: see below) asymptotic vanishing (as $t \to +\infty$) of the auxiliary variable $w(t)$,

$$w(+\infty) = 0. \tag{6.10d}$$

A condition generally sufficient (but by no means necessary) to cause this is clearly (see (6.10b) and below) the requirement that $f(w, \underline{z}, \underline{v})$ be *finite* and *nonnegative*,

$$0 \leq f(w, \underline{z}, \underline{v}) \leq a^2, \tag{6.10e}$$

for all (*real*) values of w, $\underline{z}$ and $\underline{v}$; it is indeed plain (for a proof, see below) that these conditions together with (6.10b) entail the inequalities

$$0 < w(t) \leq [w(0)]^{\exp(\alpha t)}, \tag{6.10f}$$

hence that the auxiliary variable $w(t)$ is always *positive* and vanishes asymptotically *faster than exponentially*,

$$\lim_{t \to +\infty} [w(t) \exp(bt)] = 0, \tag{6.10g}$$

with b any arbitrary constant. Restrictions on the dependence of the function $F(w, \underline{x}, \underline{v})$ upon the N-vectors $\underline{x}$ and $\underline{v}$ are also required: a simple sufficient (but of course not necessary) condition, also encompassing (6.10c), is that there exist a (*finite*) constant C and a *positive* number β such that

$$|F(w, \underline{z}, \underline{v})| \leq C |w|^\beta, \quad \beta > 0, \tag{6.10h}$$

for all (*real*) values of w, $\underline{z}$ and $\underline{v}$ (for instance, functions satisfying these conditions are $F(w,\underline{x},\underline{v}) = Cw^{\beta}/\left[1+\sum_{n=1}^{N}\left(A_n^2 x_n^2 + B_n^2 v_n^2\right)\right]$ or $F(w,\underline{z},\underline{v}) = Cw^{\beta}$ $\exp\left[-\sum_{n=1}^{N}\left(A_n^2 x_n^2 + B_n^2 v_n^2\right)\right]$ with A_n and B_n arbitrary *real* constants).

Our main result states that, for *every* (N-vector) solution $\underline{x}(t)$ of this dynamical system, an (N-vector) $\tilde{\underline{x}}(t)$ characterizing its asymptotic behavior (as $t \to +\infty$) via the formula (6.1a) (exists and) has the property to be *completely periodic* (i.e., *periodic* with the *same* period in *all* its components), see (6.1b) with $\tilde{T} = T$, see (6.4a). Of course this asymptotic N-vector $\tilde{\underline{x}}(t)$ will depend on the solution $\underline{x}(t)$ under consideration—in particular, it will depend on the initial data, $\underline{x}(0)$ and $\dot{\underline{x}}(0)$, determining that solution in the context of the initial-value problem for the N-body problem (6.10): but let us re-emphasize that, for any arbitrary choice of these data (of course, satisfying the condition $x_n(0) \neq x_m(0)$ for $n \neq m$, see (6.10a)) it shall feature the property (6.1), namely *all* solutions $\underline{x}(t)$ of the system (6.10) shall feature the property of *complete asymptotic isochrony* (6.1) (with $\tilde{T} = T$, see (6.4a)).

This result is a natural consequence of the well-known fact (see Chapter 1) that *all* solutions of the system of *Newtonian* equations (6.10a) *without* the F term in the right-hand side are *completely periodic* with period T, see (6.4a), namely they *all* feature themselves the property (6.1b) with $\tilde{T} = T$. It stands therefore to reason that, if the function $F(w,\underline{x},\dot{\underline{x}})$ vanishes (fast enough) when w vanishes, see (6.10c), and if the time evolution (6.10b) of the auxiliary variable $w(t)$ entails that this dependent variable indeed vanishes *asymptotically*, see (6.10d), fast enough (see (6.10g)), then *asymptotically* all solutions of our model (6.10) shall behave as the solutions of the same model *without* the F term, entailing the *asymptotic* phenomenology (6.1) with $\tilde{T} = T$, see (6.4a).

To turn this hunch into a theorem a *proof* must be provided. This we do in the following subsection. Then in Section 6.3 we tersely discuss, again in the same qualitative vein as done above, to what extent the phenomenology described in this paper, and shown to occur in a specific, representative model, can be expected to occur in more general contexts.

6.2.1 *A theorem and its proof*

Theorem. The conditions (6.10e) and (6.10h) are sufficient to guarantee that *every* solution of the N-body problem (6.10) with the three constants ω, g^2 and α all *positive* yield the outcomes (6.10d) and (6.1) with $\tilde{T} = T$, see (6.4a); in particular, they guarantee that there exist, corresponding to *every* solution $\underline{x}(t)$ of the N-body problem (6.10), an N-vector $\tilde{\underline{x}}(t)$ satisfying both formulas (6.1), i.e.

$$\lim_{t\to+\infty}\left[x_n(t) - \tilde{x}_n(t)\right] = 0, \quad n = 1, ..., N, \tag{6.11a}$$

$$\tilde{x}_n\left(t+\tilde{T}\right) = \tilde{x}_n(t), \quad n = 1, ..., N. \; \boxdot \tag{6.11b}$$

Proof. First of all let us prove the inequalities (6.10f), obvious as they are. To this end we set

$$w\left(t\right)=\left[w\left(0\right)\right]^{\exp\left[\varphi\left(t\right)\right]}, \tag{6.12a}$$

so that

$$\varphi\left(0\right)=0 \tag{6.12b}$$

and (from (6.10b))

$$\dot{\varphi}\left(t\right)=\alpha+f\left[w\left(t\right),\underline{x}\left(t\right),\dot{\underline{x}}\left(t\right)\right]\exp\left[-\varphi\left(t\right)\right]\left|\log\left[w\left(0\right)\right]\right|^{-1}, \tag{6.12c}$$

where we used the fact that $\log\left[w\left(0\right)\right]=-\left|\log\left[w\left(0\right)\right]\right|$, see (6.10b). This ODE, together with the initial datum (6.12b) and the inequalities (6.10e), clearly imply that $\varphi\left(t\right)$ is *positive* and *finite* for $0\leq t<\infty$, indeed validity of the inequalities

$$\alpha t<\varphi\left(t\right)<\infty,\quad 0\leq t<\infty, \tag{6.12d}$$

which, via (6.12a) and (6.10b), yield (6.10f).

Next, let us introduce the counterpart of the Newtonian equations of motion (6.10a), but without the F term in the right-hand side:

$$\ddot{\tilde{x}}_n+\frac{1}{4}\omega^2\,\tilde{x}_n=g^2\sum_{m=1,m\neq n}^{N}\left(\tilde{x}_n-\tilde{x}_m\right)^{-3},\quad n=1,...,N. \tag{6.13}$$

Here it is justified to use the notation $\tilde{x}_n\equiv\tilde{x}_n\left(t\right)$ for the dependent variables, since it is well known (see the discussion in Chapter 1 following (1.1)) that *all* the solutions of this Newtonian N-body problem are *completely periodic* with period T, see (6.4a), consistently with (6.11b).

Let us now remark that, due to the strict positivity of g^2, this system of ODEs entails that

$$\left|\tilde{x}_n\left(t\right)-\tilde{x}_m\left(t\right)\right|>\tilde{c}^2,\quad \tilde{c}^2>0,\quad n\neq m, \tag{6.14a}$$

where $\tilde{c}^2$ is a time-independent constant that generally depends on the particular solution under consideration but is certainly strictly positive, $\tilde{c}^2>0$. Likewise, again due to the strict positivity of g^2, the system of ODEs (6.10a) with (6.10h) and (6.10f) (entailing $\left|F\left(w,\underline{z},\underline{v}\right)\right|\leq D,\ D=C\left|w\left(0\right)\right|^{\beta}$) implies that

$$\left|x_n\left(t\right)-x_m\left(t\right)\right|>c^2,\quad c^2>0,\quad n\neq m, \tag{6.14b}$$

where c^2 is again a time-independent constant that generally depends on the particular solution under consideration but is certainly strictly positive, $c^2>0$. Moreover the systems of ODEs (6.13) and (6.10) clearly imply that, for all (finite, positive) time, the functions $\tilde{x}_n\left(t\right)$ and $x_n\left(t\right)$ are finite.

Let us now set

$$\xi_n(t) = x_n(t) - \tilde{x}_n(t) . \tag{6.15}$$

These functions $\xi_n(t)$ satisfy—as implied by subtracting (6.13) from (6.10a)—the system of ODEs

$$\ddot{\xi}_n + \frac{1}{4}\omega^2 \xi_n + g^2 \sum_{m=1, m \neq n}^{N} \left[(\xi_n - \xi_m) \varphi_{nm}(\underline{x}, \underline{\tilde{x}}) \right] = F(w, \underline{x}, \underline{\dot{x}}) \tag{6.16a}$$

with

$$\varphi_{nm}(\underline{x}, \underline{\tilde{x}}) = \frac{(x_n - x_m)^2 + (x_n - x_m)(\tilde{x}_n - \tilde{x}_m) + (\tilde{x}_n - \tilde{x}_m)^2}{(x_n - x_m)^3 (\tilde{x}_n - \tilde{x}_m)^3} . \tag{6.16b}$$

Note that the above bounds, (6.14), as well as the finiteness of x_n and $\tilde{x}_n$ for all (positive) time, guarantee that these functions $\varphi_{nm}(\underline{x}, \underline{\tilde{x}})$ remain *finite* for all time, namely that there always exist time-independent *finite* upper and lower bounds $\varphi_\pm$ satisfied by them for all time,

$$\varphi_- \leq \varphi_{nm}(\underline{x}, \underline{\tilde{x}}) \leq \varphi_+ . \tag{6.16c}$$

These bounds depend of course on the particular solutions $\underline{x}$ and $\underline{\tilde{x}}$ under consideration, but let us re-emphasize that, for any such solutions, they are *finite.*

It is now clear that the theorem is proven if we can show that these ODEs, (6.16), admit a solution satisfying the asymptotic condition

$$\lim_{t \to +\infty} [\xi_n(t)] = 0, \quad n = 1, ..., N \tag{6.17}$$

(see (6.11a) and (6.15)). As can be easily verified such a solution of (6.16) is provided by the formula

$$\xi_n(t) = \int_t^\infty dt' F[w(t'), \underline{x}(t'), \underline{\dot{x}}(t')] \, G_n(t, t') , \quad n = 1, ..., N, \tag{6.18a}$$

where the functions $G_n(t, t')$ are the Green's functions associated with the left-hand side of the system of ODEs (6.16), namely the solutions of the system of ODEs

$$\frac{\partial^2 G_n(t, t')}{\partial t^2} + \frac{1}{4}\omega^2 G_n(t, t')$$
$$+ g^2 \sum_{m=1, m \neq n}^{N} [G_n(t, t') - G_m(t, t')] \varphi_{nm}[\underline{x}(t), \underline{\tilde{x}}(t)] = 0, \quad t \leq t',$$
$$G_n(t, t) = 0, \quad \left. \frac{\partial G_n(t, t')}{\partial t} \right|_{t'=t} = -1, \quad n = 1, ..., N. \tag{6.18b}$$

Indeed, while these Green functions cannot be computed explicitly (since we do not know the N-vectors $\underline{x}(t')$ and $\underline{\tilde{x}}(t')$), it is plain—from the linear character of this system of ODEs and from the bounds (6.16c)—that these Green functions can grow (in modulus) at most exponentially as $t \to +\infty$ and/or $t' \to +\infty$; so that the *faster than exponential* asymptotic vanishing of $F\left[w\left(t'\right), \underline{x}\left(t'\right), \dot{\underline{x}}\left(t'\right)\right]$ as $t' \to +\infty$ implied by (6.10h) with (6.10g) entails that the integral in the right-hand side of the solution formula (6.18a) vanishes asymptotically (as $t \to +\infty$). Q. E. D. ⊡

Remark. It is clear how this example could have been made more general by allowing the function F appearing in the right-hand side of (6.10a) to depend on the index n, and/or by replacing the single auxiliary variable $w(t)$ by a J-vector $\underline{w}(t)$ with J an arbitrary positive integer, and so on; without invalidating our conclusion, but complicating our proof. Let us also re-emphasize that the hypotheses made above to prove this *theorem* are *sufficient* but by no means *necessary* for its validity. More specific, and possibly considerably less stringent, conditions yielding an analogous conclusion can and will be introduced whenever this kind of result shall be considered in specific (possibly applicative) contexts. Our motivation to assume here quite simple (hence overly stringent) hypotheses is because we are just interested to show that the main idea presented above does indeed work. ⊡

6.3 Some additional considerations

Clearly the kind of approaches illustrated above via the detailed treatment of two specific examples can be applied much more widely: it will be particularly interesting to do so in specific applicative contexts.

A natural point of departure for such applications are *isochronous* systems, namely models whose *generic* solutions—in their *entire* natural phase space, or in *open*, hence *fully dimensional*, regions of it—are *completely periodic* (i.e. periodic in *all* their degrees of freedom) with the *same fixed* period (independent of the initial data, provided they stay within the *isochrony* region). The reader of this monograph now knows that quite a lot of dynamical systems can be modified so that they become *isochronous*, entailing the conclusion that *isochronous systems are not rare*. Each of these *isochronous* systems can then be further extended—along the lines obviously suggested by the treatment detailed above, see in particular the specific case treated in Section 6.2—in order to generate classes of *asymptotically isochronous* systems, namely systems featuring *open*, hence *fully dimensional*, regions in their natural phase space (possibly including all of it) in which *all* (or *almost all*) their solutions display *asymptotically* a *completely periodic* behavior with the *same fixed* period, see (6.1). The technique to manufacture such generalized systems is clearly suggested by the examples treated above: of course these systems could be *autonomous,* as the examples treated above, or they might feature an *explicit* time-dependence, as could have been included in the system treated in Section 6.2 by assuming the functions F and f to also feature an *explicit* time dependence (but *autonomous* systems are generally more interesting than *nonautonomous* ones).

As the attentive reader of this monograph now knows, often (although not always, see for instance Section 5.5) the natural context to investigate *isochronous* systems is in the *complex* rather than the *real*—although of course every system with *complex* dependent variables can of course be reformulated as a system with twice as many *real* dependent variables. Hence it may be of interest to mention how the findings detailed in Section 6.2 would be affected if the dependent variables x_n in the model (6.10) were allowed to be *complex*—keeping of course *real* the time t and *positive* the constant ω, while the constant g^2 could now also be *complex*. It is then well known (see Chapter 1) that the *isochronous* character of the motions still prevails for the *integrable* many-body problem (6.10a) *without* the F term (i.e. with an identically vanishing F: see (1.1))—describing motions taking place in the *complex* z-plane rather than on the *real* line. But the phase space is then divided into sectors separated by lower dimensional manifolds characterized by solutions which hit a *singularity* at a finite time due to a particle collision—an event forbidden in the *real* case with *positive* g^2, when the particles move on the *real* axis and the two-body force, *singular* at zero separation, is *repulsive*, see (6.10a)—but which can happen in the *complex* case, although not for *generic* initial data. In the different sectors the motion is still *completely periodic*, but with different periods, characterizing each sector and being (generally rather small [100]) *integer* multiples of the basic period T, see (6.4a). Accordingly, the *generic* solution of the (generally *nonintegrable*) generalized model (6.10) will be *nonsingular* throughout its time evolution and it shall eventually settle within a sector, approaching asymptotically one of the *completely periodic* solutions in that sector of the model (6.10a) with identically vanishing F, namely of the *integrable* model (1.1).

A somewhat analogous outcome obtains for the model analogous to (6.10) but with (6.10a) replaced by

$$\ddot{z}_n + \frac{1}{4}\omega^2 z_n = \sum_{m=1, m\neq n}^{N} \left[g_{nm}^2 \left(z_n - z_m\right)^{-3} \right] + F\left(w, \underline{z}, \underline{\dot{z}}\right), \quad n = 1, ..., N, \tag{6.19}$$

featuring $N(N-1)$ different coupling constants g_{nm}^2 acting among every particle pair. In this case the model without F is generally *not integrable*, yet (if considered in the *complex*, namely without restricting the dependent variables z_n—nor, for that matter, the coupling constants g_{nm}^2—to be *real*) it still does feature an *open*, hence *fully dimensional*, region in its phase space where *all* solutions are *completely periodic* with the same period T; while in other regions of its phase space it might also be *periodic* but with periods $\tilde{T} = pT$ where the numbers p are *integers* but might be very large; or it might even display an *aperiodic*, quite complicated (in some sense *chaotic*) behavior (as described in Chapter 1, after (1.4)). It then stands to reason that the solutions of the generalized model (6.10) with (6.10a) replaced by (6.19) (and of course $\underline{x}$ in (6.10b) replaced by $\underline{z}$) shall again approach asymptotically solutions—including, from certain *open* regions of initial data, *completely periodic* ones—of the model (6.19) *without* F: entailing a remarkable, and quite rich, phenomenology. Clearly our motivation to mention this specific model is because of its prototypical role: indeed, the main aspects of this phenomenology shall also characterize the large class of *isochronous* (but, by no means, necessarily *integrable*) systems that can now be manufactured (as explained throughout this monograph), once they are extended by adding to their equations of motion other, fairly general, terms having the property to disappear asymptotically (as $t \to +\infty$) sufficiently fast, as a consequence of the very dynamics implied by these extended equations of motion.

In conclusion, let us re-emphasize that these results (as indeed all mathematically correct findings) might well be deemed remarkable or trivial, depending on the level of understanding of the reader. Once their foundation is understood, it becomes obvious how they can be extended to many other models—suggesting an ample applicative potential.

7

ISOCHRONOUS PDEs

In the preceding part of this monograph we only considered time evolutions described by (systems of) nonlinear Ordinary Differential Equations (ODEs: "dynamical systems"), with the time playing the role of sole independent variable; and in that context we have shown how a large class of *isochronous* evolutions could be identified and investigated via the trick (first introduced in Section 2.1). In this chapter we indicate how this approach can be extended to treat time evolutions described by Partial Differential Equations (PDEs: "nonlinear evolution equations"). Our presentation is essentially limited to displaying a list of *ω-modified isochronous* PDEs, characterized by the property to possess lots of *completely periodic* solutions, living in open regions of their phase space. Only in a few cases some such solutions are exhibited.

In the following Section 7.1 an extended version of the trick appropriate to PDEs is introduced and certain, mainly notational, preliminaries are presented, the main purpose of which is to minimize subsequent repetitions. The nonlinear evolution PDEs are then listed, with minimal comments, in Section 7.2, generally both in their unmodified *avatars* (for appropriate—if inevitably incomplete—references see Section 7.N) as well as in their ω-modified *isochronous* versions. In Section 7.3 the possibility is outlined to manufacture PDEs featuring lots of solutions which are *periodic* not only in *time* but also in *space*, and this possibility is illustrated by the exhibition of a single example.

Let us conclude these introductory words by emphasizing that purpose and scope of the following presentation is to apply the trick to a number of well-known, *autonomous*, evolution PDEs, which thereby yield (with few exceptions, see below) also *autonomous* (at least as regards the dependence on the time variable) ω-modified PDEs, thereby demonstrating the wide applicability of this approach, and the fact that, in some sense, also *isochronous PDEs are not rare.* We believe the ω-modified PDEs obtained below deserve to be exhibited and—at least some of them—warrant further investigation inasmuch as they are likely to constitute useful theoretical tools, in view of the remarkable phenomenology featured by their solutions. Of course the PDEs displayed below are merely a (hopefully representative) sample of the vast class of *isochronous* PDEs that can be manufactured: the alert reader will have no difficulty to manufacture many more, just for the fun of it or for applications.

7.1 The trick for PDEs

This section is mainly devoted to notation and preliminaries, whose purpose is to streamline the presentation, in the following Section 7.2, of the list of

PDEs constituting the core of this Chapter 7. A generalized version of the trick suitable for PDEs is presented, but no elaboration is provided to explain why the ω-modified PDEs yielded by its application must be expected to possess lots of *completely periodic* solutions with a *fixed* period, namely to be *isochronous*; we believe this should be clear to any reader who absorbed the essence of the findings presented in the preceding part of this monograph.

The independent variables of the *unmodified* evolution PDE are denoted as $\underline{\xi} \equiv (\xi_1, \ldots, \xi_N)$ and τ; the (scalar) dependent variable of the *unmodified* evolution PDE is denoted as $w\left(\underline{\xi};\, \tau\right) \equiv w\left(\xi_1, \ldots, \xi_n;\, \tau\right)$, and if a second (scalar) dependent variable also enters, it is denoted as $\tilde{w}\left(\underline{\xi};\, \tau\right) \equiv \tilde{w}\left(\xi_1, \ldots, \xi_n;\, \tau\right)$; upper case letters, $W\left(\underline{\xi};\, \tau\right) \equiv W\left(\xi_1, \ldots, \xi_n;\, \tau\right)$, $\tilde{W}\left(\underline{\xi}; \tau\right) \equiv \tilde{W}\left(\xi_1, \ldots, \xi_n\ \tau\right)$, are used for *matrices*. The independent variables of the ω-*modified* evolution PDE are denoted as $\underline{x} \equiv (x_1, \ldots, x_n)$ and t; the (scalar) dependent variable of the ω-*modified* evolution PDE is denoted as $u\left(\underline{x};\, t\right) \equiv u\left(x_1, \ldots, x_n;\, t\right)$, and if a second (scalar) dependent variable also enters, it is denoted as $\tilde{u}\left(\underline{x};\, t\right) \equiv \tilde{u}\left(x_1, \ldots, x_n;\, t\right)$; and again upper case letters, $U\left(\underline{x};\, t\right) \equiv U\left(x_1, \ldots, x_n;\, t\right)$, $\tilde{U}\left(\underline{x};\, t\right) \equiv \tilde{U}\left(x_1, \ldots, x_n;\, t\right)$, are used for *matrices*. The relation among the (independent and dependent) variables of the *unmodified* evolution PDE and the ω-*modified* evolution PDE are given by the following formulas (*the trick*):

$$\tau = \frac{\exp\left(i\,\omega\, t\right) - 1}{i\,\omega}, \tag{7.1a}$$

$$\xi_n = \xi_n(t) = x_n \exp\left(i\,\mu_n\,\omega\, t\right), \qquad n = 1, \ldots, N, \tag{7.1b}$$

$$u\left(\underline{x};\, t\right) = \exp\left(i\,\lambda\,\omega\, t\right)\, w\left(\underline{\xi};\, \tau\right), \tag{7.1c}$$

$$\tilde{u}\left(\underline{x};\, t\right) = \exp\left(i\,\tilde{\lambda}\,\omega\, t\right)\, \tilde{w}\left(\underline{\xi};\, \tau\right), \tag{7.1d}$$

with analogous formulas, see (7.1c) and (7.1d), in the matrix case. Here and hereafter constants such as μ_n, λ, $\tilde{\lambda}$, α, β (Greek letters) generally denote *rational* numbers (not necessarily positive), which whenever necessary shall be properly assigned, while Latin letters such as a, b, c denote *complex* (or, as the case may be, *real*) constants (sometimes we keep such constants even when they could be eliminated by trivial rescaling transformations; and of course in many cases by such transformations additional constants might instead be introduced). When $N = 1$ we drop the index n, namely we write ξ instead of ξ_1, x instead of x_1, μ instead of μ_1, and for $N = 2$ we also, to simplify the notation, write η instead of ξ_2, y instead of x_2, ν instead of μ_2. Of course the *isochrony* property of the ω-*modified* equations obtained via this trick (7.1) is characterized by the usual time period,

$$T = \frac{2\,\pi}{\omega}. \tag{7.2}$$

Note that this transformation, (7.1), entails that, at the initial time, $\tau = t = 0$, the change of variables disappears altogether:

$$\underline{\xi}(0) = \underline{x}, \qquad u\left(\underline{x};\, 0\right) = w\left(\underline{\xi};\, 0\right), \qquad \tilde{u}\left(\underline{\xi};\, 0\right) = \tilde{w}\left(\underline{\xi};\, 0\right). \tag{7.3}$$

Hereafter subscripted variables denote partial differentiations, $w_\tau \equiv \partial w(\underline{\xi};\, \tau)/\partial \tau$, $u_{x_n} \equiv \partial u\left(\underline{x};\, t\right)/\partial x_n$ and so on.

Let us emphasize—obvious as this may be—that, since the transition from an (*unmodified*) PDE satisfied by $w\left(\underline{\xi};\, \tau\right)$ to the corresponding (ω-*modified*) PDE satisfied by $u\left(\underline{x};\, t\right)$ is performed via the explicit change of variables (7.1), properties such as *integrability* or *solvability*, if possessed by the *unmodified* PDE satisfied by $w\left(\underline{\xi};\, \tau\right)$, carry over to the corresponding (ω-*modified*) PDE satisfied by $u\left(\underline{x};\, t\right)$—which generally has, in addition, the property of *isochrony*. And let us emphasize that the property of *isochrony* of the ω-*modified* evolution PDE does not require that the original, *unmodified* PDE from which it has emerged be *integrable*.

Let us end this Section 6.1 by pointing out that, in most cases, the ω-*modified* PDEs are *complex*; and even in the few cases when an ω-*modified* PDE is *real*, generally its property of *isochrony* is only valid in the *complex*, indeed the initial data yielding *isochronous* solutions are generally *not real*. Of course these ω-modified PDEs can be rewritten in *real* form by introducing the *real* and *imaginary* parts of all the quantities that enter in these PDEs, and by then considering the two, generally coupled, *real* PDEs that obtain from each *complex* PDE by considering separately its *real* and *imaginary* parts.

7.2 A list of isochronous PDEs

In this section we display, with minimal commentary, a list of nonlinear evolution PDEs, generally firstly in their *unmodified avatars*, then in their ω-*modified* versions. It is remarkable that so many "well-known" *autonomous* evolution PDEs possess ω-modified versions which are as well *autonomous*, at least as regards their time dependence; in several cases this constitutes a minor miracle, inasmuch as the number of available parameters λ, μ, ν, $\tilde{\lambda}, \ldots$ is smaller than the equations they are required to satisfy in order to guarantee the *autonomous* character of the ω-modified equations, yet nontrivial parameters satisfying all these conditions do exist. All the ω-modified evolution PDEs displayed below possess many *isochronous* solutions—although quite a few of these solutions—but certainly not all of them—feature singularities at some (*real*) values of the independent variables. But only in a few cases we exhibit examples of these solutions. A discussion of each of these nonlinear evolution PDEs would indeed require much more space. Let us also re-emphasize that the list reported below includes only some kind of *representative* instances of this phenomenology; obviously many more examples can be easily manufactured.

The list is arranged in a *user-friendly* manner, being ordered (up to a few exceptions) according to the following taxonomic rules: of primary importance is the number of independent variables (with priority given, of course, to smaller numbers); next, the number of dependent variables; next, the order of the differential equation, giving precedence to the *time* variable over the *space* variables; finally, the type of nonlinearity (except when an equation is presented as a special case of a more general equation, see for instance below (7.11) and (7.42)).

Example 7.2-1 The following unmodified (1+1)-dimensional "generalized shock-type" PDE is *integrable*, indeed *solvable*:

$$w_\tau = a\, w^\alpha\, w_\xi . \tag{7.4a}$$

By setting $\mu = 1 - \alpha\,\lambda$ one gets the corresponding ω-modified evolution PDE

$$u_t - i\,\lambda\,\omega\,u + i\,(\alpha\,\lambda - 1)\,\omega\,x\,u_x = a\,u^\alpha\,u_x . \tag{7.4b}$$

The *general* solution of the initial-value problem for this PDE, (7.4b), is given, in *implicit* form, by the following formula:

$$u(x;t) = \exp\left(i\,\lambda\,\omega\,t\right)\,u_0\left(\exp\left[i(1-\alpha\,\lambda)\,\omega\,t\right]\,\left\{x + a\,\frac{1-\exp\left(-i\,\omega\,t\right)}{i\,\omega}\,\left[u(x;t)\right]^{\alpha}\right\}\right), \tag{7.4c}$$

where of course $u_0(x) = u(x;0)$.

Note that for $\lambda = 1\,/\,\alpha$ the ω-modified PDE (7.4b) is *autonomous* also as regards the space variable x.

Example 7.2-2 The following unmodified $(1+1)$-dimensional "generalized Burgers–Hopf" PDE reads

$$w_\tau = a\,w\,w_\xi + b\,\left(w^{\,\alpha}\,w_\xi\right)_\xi . \tag{7.5a}$$

By setting $\lambda = 1/\,(2-\alpha)$ and $\mu = (1-\alpha)\,/\,(2-\alpha)$ one gets the corresponding ω-modified evolution PDE

$$u_t + i\,\frac{1}{\alpha-2}\,\omega\,u + i\,\frac{1-\alpha}{\alpha-2}\,\omega\,x\,u_x = a\,u\,u_x + b\,\left(u^{\,\alpha}\,u_x\right)_x . \tag{7.5b}$$

Note that, for $\alpha = 1$, this PDE becomes *autonomous* also with respect to the space variable x, while for $\alpha = 0$ the PDE (7.5a) becomes the standard (*solvable*) Burgers–Hopf equation.

Example 7.2-3 The following unmodified $(1+1)$-dimensional dispersive KdV-like PDE is *integrable* indeed *solvable*:

$$w_\tau = w_{\xi\xi\xi} + 3\,\left[w_{\xi\xi}\,w^{\,2} + 3\,(w_\xi)^{\,2}\right] + 3\,w_\xi\,w^{\,4} . \tag{7.6a}$$

By setting $\lambda = 1\,/\,6$, $\mu = 1\,/\,3$ and (for notational simplicity) $\Omega = \omega\,/\,6$ one gets the corresponding ω-modified evolution PDE

$$u_t - i\,\Omega\,u - 2\,i\,\Omega\,x\,u_x = u_{xxx} + 3\,\left(u_{xx}\,u^{\,2} + 3\,u_x^2\right) + 3\,u_x\,u^{\,4} . \tag{7.6b}$$

Example 7.2-4 The symmetry properties, and some explicit solutions, of the following unmodified $(1+1)$-dimensional "generalized KdV equation" (reducing to the KdV or the modified KdV equation for $\alpha = 1$ and $\beta = 2$ or $\beta = 3$) have been investigated recently:

$$w_\tau = a\,(w^{\alpha})_{\xi\xi\xi} + b\,\left(w^{\beta}\right)_{\xi}. \tag{7.7a}$$

By setting $\lambda = 2/(3\beta - \alpha - 2)$, $\mu = (\beta - \alpha)/(3\beta - \alpha - 2)$ one gets the ω-modified evolution PDE

$$u_t - \frac{2\,i\,\omega}{3\,\beta - \alpha - 2}\,u - \frac{i(\beta - \alpha)\,\omega}{3\,\beta - \alpha - 2}\,x\,u_x = a\,(u^{\alpha})_{xxx} + b\,\left(u^{\beta}\right)_x. \tag{7.7b}$$

We assume here of course that $3\beta - \alpha - 2 \neq 0$. Particularly interesting is the case with $\alpha = \beta$, when this ω-modified PDE becomes *autonomous* also in the space variable x. In the even more special case with $\alpha = \beta = 2$, and by setting $u = u_1 + i\,u_2$, $a = c_1 + i\,c_2$, $b = c_3 + i\,c_4$, we re-write this evolution PDE as a system of two coupled PDEs

$$\begin{aligned} u_{1_t} + \omega\,u_2 &= \left[c_1\,\left(u_1^2 - u_2^2\right) - 2\,c_2\,u_1\,u_2\right]_x \\ &\quad + \left[c_3\,\left(u_1^2 - u_2^2\right) - 2\,c_4\,u_1\,u_2\right]_{xxx}, \\ u_{2_t} - \omega\,u_1 &= \left[c_2\,\left(u_1^2 - u_2^2\right) + 2\,c_1\,u_1\,u_2\right]_x \\ &\quad + \left[c_4\,\left(u_1^2 - u_2^2\right) + 2\,c_3\,u_1\,u_2\right]_{xxx}. \end{aligned} \tag{7.7c}$$

Here we assume of course that the two dependent variables $u_1 \equiv u_1(x,\,t)$, $u_2 \equiv u_2(x,t)$ are *real*, and that as well *real* are the four arbitrary constants c_1, c_2, c_3, c_4.

Example 7.2-5 The following unmodified (1+1)-dimensional "Schwarzian KdV" equation is *integrable*:

$$w_\tau = w_{\xi\xi\xi} + a\,\frac{w_{\xi\xi}^2}{w_\xi}. \tag{7.8a}$$

By setting $\mu = 1/3$ one gets the corresponding ω-modified evolution PDE

$$u_t - i\,\lambda\,\omega\,u - i\,\frac{\omega}{3}\,x\,u_x = u_{xxx} + a\,\frac{u_{xx}^2}{w_x}. \tag{7.8b}$$

Example 7.2-6 The following unmodified $(1+1)$-dimensional "Cavalcante–Tenenblat" equation is *integrable*:

$$w_\tau = a\,\left[(w_\xi)^{-1/2}\right]_{\xi\xi} + b\,(w_\xi)^{3/2}. \tag{7.9a}$$

By setting $\mu = 1/3$ one gets the corresponding ω-modified evolution PDE

$$u_t - i\,\lambda\,\omega\,u - i\,\frac{\omega}{3}\,x\,u_x = a\,\left[(w_\xi)^{-1/2}\right]_{xx} + b\,(u_x)^{3/2}. \tag{7.9b}$$

Example 7.2-7 The KdV class of unmodified (1 + 1)-dimensional *integrable* evolution PDEs reads

$$w_\tau = \Lambda^m \, w_\xi, \quad m = 1, 2, \ldots, \tag{7.10a}$$

where Λ is the integrodifferential operator (depending on the dependent variable $w(\xi;\tau)$) that acts on a generic (twice-differentiable, and integrable at infinity) function $\phi(\xi)$ as follows:

$$\Lambda \, \phi \, (\xi) = \phi_{\xi\xi}(\xi) - 4 \, w(\xi;\tau) \, \phi(\xi) + 2 \, w_\xi(\xi;\tau) \int_\xi^\infty \phi \, (\xi') \, d\xi'. \tag{7.10b}$$

By setting $\lambda = 2 / \, (2 \, m + 1)$, $\mu = 1 / \, (2 \, m + 1)$ and (for notational simplicity) $\Omega_m = \omega / \, (2 \, m + 1)$ one gets the corresponding class of Ω_m-modified evolution PDEs

$$u_t - i \, \Omega_m \, (2 \, u + x \, u_x) = L^m \, u_x, \quad m = 1, 2, \ldots, \tag{7.10c}$$

where L is the integrodifferential operator (depending on the dependent variable $u(x;t)$) analogous to Λ, namely the operator that acts on a generic (twice-differentiable, and integrable at infinity) function $f(x)$ as follows:

$$L \, f \, (x) = f_{xx}(x) - 4 \, u(x;t) \, f(x) + 2 \, u_x(x;t) \int_x^\infty f \, (x') \, dx'. \tag{7.10d}$$

For $m = 1$ the PDE (7.10a) becomes the well-known KdV equation

$$w_\tau = w_{\xi\xi\xi} - 6 \, w \, w_\xi, \tag{7.11a}$$

and the corresponding ω-modified equation reads

$$u_t - i \, \frac{\omega}{3} \, (2 \, u + x \, u_x) = u_{xxx} - 6 \, u \, u_x. \tag{7.11b}$$

Example 7.2-8 The unmodified (1+1)-dimensional "Monge–Ampere" *integrable* PDE reads

$$w_{\tau\tau} \, w_{\xi\xi} - w_{\xi\tau}^2 = 0. \tag{7.12a}$$

The corresponding ω-modified PDE is in this case t-*autonomous* for *any* choice of λ and μ:

$$\begin{aligned} &u_{tt} \, u_{xx} - (u_{tx})^2 + i \, \omega \, \left[-(2 \, \lambda + 1) u_t \, u_{xx} + 2 \, (\lambda + \mu) \, u_{tx} \, u_x\right] \\ &+\omega^2 \left[-\lambda \, (\lambda + 1) \, u \, u_{xx} + \mu \, (\mu + 1) \, x \, u_x \, u_{xx} + (\lambda + \mu)^2 \, (u_x)^2\right] = 0. \end{aligned} \tag{7.12b}$$

The *general* solution of this PDE (7.12b) can be obtained via the following two steps: first, for an *arbitrary* function $F(r)$, one should find the function $r(x;t)$

solving the *nondifferential* equation

$$r(x;t) = x \exp\left[i\,(\mu - 1)\,\omega\,t\right] - \exp\left(-i\,\omega\,t\right)\,F\left[r(x;t)\right]; \tag{7.12c}$$

then, for an *arbitrary* function $G(r)$ and an arbitrary constant a, one should evaluate the solution:

$$u(x;t) = \exp\left(i\,\lambda\,\omega\,t\right)\int_a^t dt'\,\exp\left(i\,\omega\,t\right)\,G\left[r(x;t')\right]. \tag{7.12d}$$

A class of *explicit* solutions of the ω-modified PDE (7.12b) is:

$$u(x;t) = \exp\left[i\,(\lambda + \mu)\,\omega\,t\right]\,x\,f\left\{\left[\frac{\sin\left(\frac{\omega\,t}{2}\right)\,\exp\left[i\,\left(\frac{1-2\,\mu}{2}\right)\,\omega\,t\right]}{x}\right]\right\}, \tag{7.12e}$$

where $f(z)$ is an *arbitrary* function.

Three particularly neat cases of the ω-*modified* evolution PDE (7.12b) are worth explicit display: for $\lambda = \mu = 0$,

$$u_{tt}\,u_{xx} - u_{tx}^2 - i\,\omega\,u_t\,u_{xx} = 0; \tag{7.12f}$$

for $\lambda = -1, \mu = 1$,

$$u_{tt}\,u_{xx} - u_{tx}^2 + i\,\omega\,u_t\,u_{xx} = 0; \tag{7.12g}$$

for $\lambda = -1\,/\,2, \mu = 1\,/\,2$,

$$u_{tt}\,u_{xx} - u_{tx}^2 + \left(\frac{\omega}{2}\right)^2\,(u - x\,u_x)\,u_{xx} = 0. \tag{7.12h}$$

Note that the first two of these three ω-modified PDEs are also x-*autonomous*, and that the last is *real*.

Example 7.2-9 The following unmodified (1 + 1)-dimensional *solvable* PDE reads

$$w_{\tau\tau}\,w_\xi - w_{\tau\xi}\,w_\tau = 0. \tag{7.13a}$$

The corresponding ω-modified PDE is again t-*autonomous* for *any* choice of λ and μ:

$$\begin{aligned} u_{tt}\,u_x - u_{tx}\,u_t + i\,\omega\,\left[(-\lambda + \mu + 1)\,u_t\,u_x + \lambda\,u\,u_{tx} - \mu\,x\,u_{tx}\,u_x + \mu\,x\,u_t\,u_{xx}\right] \\ + \omega^2\left[\lambda\,(\mu - 1)\,u\,u_x + (-\lambda + \mu - 2)\,\mu\,x\,(u_x)^2 + \lambda\,\mu\,x\,u\,u_{xx}\right] = 0. \end{aligned} \tag{7.13b}$$

The *general* solution of this PDE, (7.13b), reads

$$u(x;t) = \exp\left(i\,\lambda\,\omega\,t\right)\,f\left\{g\left[x\,\exp\left(i\,\mu\,\omega\,t\right)\right] + \exp\left(i\,\omega\,t\right)\right\}, \tag{7.13c}$$

where $f(z)$, $g(z)$ are two *arbitrary analytic* functions.

Note that for $\mu = 0$ this ω-modified PDE (7.13b) is x-*autonomous*.

Example 7.2-10 The following unmodified (1 + 1)-dimensional "Boussinesq" equation is *integrable*:

$$w_{\tau\tau} = \left(w_{\xi\xi\xi} + w\, w_{\xi} \right)_{\xi} . \tag{7.14a}$$

By setting $\lambda = 1$ and $\mu = 1/2$ one gets the corresponding ω-modified evolution PDE

$$u_{tt} - i\,\omega\, u - \frac{i\,\omega}{2}\, x\, u_x = \left(u_{xxx} - u\, u_x \right)_x . \tag{7.14b}$$

Example 7.2-11 The following unmodified (1+1)-dimensional "nonlinear wave equation" reads

$$w_{\tau\tau} = \left(w^{\alpha}\, w_{\xi} \right)_{\xi} . \tag{7.15a}$$

By setting $\mu = 1 - (\alpha\,\lambda/2)$ one gets the corresponding ω-modified evolution PDE

$$\begin{aligned} &u_{tt} - i\,(2\,\lambda + 1)\,\omega\, u_t - i\,\omega\,(2 - \alpha\,\lambda)\, x\, u_{t\,x} - \lambda\,(\lambda + 1)\,\omega^2\, u \\ &\quad - \frac{1}{4}\,(2 - \alpha\,\lambda)\,\left(4\,\lambda + 4 - \alpha\,\lambda\right)\,\omega^2\, x\, u_x - \frac{1}{4}\,(2 - \alpha\,\lambda)^2\, x^2\, u_{xx} = \left(u^{\alpha}\, u_x \right)_x . \end{aligned} \tag{7.15b}$$

Note that this ω-modified PDE is *x-autonomous for $\lambda = 2/\alpha$*. In particular two special cases warrant explicit display:

$$u_{tt} + \left(\frac{\omega}{2} \right)^2 u = \left(\frac{u_x}{u^4} \right)_x , \tag{7.15c}$$

corresponding to $\alpha = -4$, $\lambda = -1/2$;

$$u_{tt} - 5\, i\,\omega\, u_t - 6\,\omega^2\, u = \left(u\, u_x \right)_x , \tag{7.15d}$$

corresponding to $\alpha = 1$, $\lambda = 2$.

Example 7.2-12 Another class (out of many possible ones) of unmodified (1 + 1)-dimensional "nonlinear wave equations" reads

$$w_{\tau\tau} = \sum_k \frac{a_k}{w^{3+\alpha_k}} \left(\frac{\partial^{p_k} w}{\partial \xi^{p_k}} \right)^{\alpha_k} , \tag{7.16}$$

where the numbers p_k are *nonnegative integers*—or possibly just *integers*: with the standard definition

$$\frac{\partial^{-p}}{\partial\,\xi^{-p}}\, f(\xi) = \int^{\xi} d\xi_p \int^{\xi_p} d\xi_{p-1} \cdots \int^{\xi_2} d\xi_1\, f\,(\xi_1) \tag{7.17}$$

if p is a *positive integer*.

By setting $\lambda = -1/2, \mu = 0$ one gets the corresponding ω-modified evolution PDE

$$u_{t\,t} + \left(\frac{\omega}{2} \right)^2 u = \sum_k \frac{a_k}{u^{3+\alpha_k}} \left(\frac{\partial^{p_k} u}{\partial x^{p_k}} \right)^{\alpha_k} . \tag{7.18}$$

Example 7.2-13 Here is a longish list of "M. and N. Euler" ω-modified *integrable* PDEs in $1+1$ dimensions:

$$u_t + \frac{2}{3}i\omega u - \frac{1}{3}i\omega x u_x = \alpha_3 u_x^{-1} + \alpha_4 \left(u_{xxx} - \frac{3u_{xx}^2}{2u_x} \right); \tag{7.19a}$$

$$u_t - \frac{1}{3}i\omega x u_x = \alpha_2 u_x^3 + \alpha_4 \left(u_{xxx} - \frac{3u_{xx}^2}{2u_x} \right); \tag{7.19b}$$

$$u_t + i\omega u - \frac{1}{3}i\omega x u_x = \alpha_0 + \alpha_4 \left(u_{xxx} - \frac{3u_{xx}^2}{2u_x} \right); \tag{7.19c}$$

$$u_t - i\omega \left(\lambda u + \frac{1}{3} x u_x \right) = \alpha_4 \left(u_{xxx} - \frac{3}{2}\frac{u_{xx}^2}{u_x} \right); \tag{7.19d}$$

$$u_t + i\omega u - i\omega x u_x = \alpha_0 + \alpha_1 u_x + \alpha_2 u_x^3 + \frac{\alpha_3}{u_x}; \tag{7.19e}$$

$$u_t - i\omega x u_x = \frac{\alpha_3}{u_x}; \tag{7.19f}$$

$$u_t - i\lambda\omega u - i\frac{1-2\lambda}{3} x u_x = \alpha_2 u_x^3; \tag{7.19g}$$

$$u_t - i\lambda\omega u + (1+2\lambda) i\omega x u_x = \frac{\alpha_3}{u_x}; \tag{7.19h}$$

$$u_t + \frac{1}{3}i\omega u - \frac{1}{3}i\omega x u_x = \alpha_4 \left[u_{xxx} - \frac{3u_x u_{xxx}^2}{2\left(u_x^2 - c\right)} \right]; \tag{7.19i}$$

$$u_t + i\omega u - i\omega x u_x = \alpha_0 + \alpha_1 u_x + \alpha_2 u_x^3 + \alpha_3 \left(u_x^2 - c\right)^{3/2}; \tag{7.19j}$$

$$u_t - \frac{1}{3}i\omega u - \frac{1}{3}i\omega x u_x = \alpha_2 u_x^2 + \alpha_4 \left(u_{xxx} - \frac{3u_{xx}^2}{4u_x} \right); \tag{7.19k}$$

$$u_t - i\omega u - \frac{1}{3}i\omega x u_x = \alpha_3 u_x^{3/2} + \alpha_4 \left(u_{xxx} - \frac{3u_{xx}^2}{4u_x} \right); \tag{7.19l}$$

$$u_t - i\omega \left(\lambda u + \frac{1}{3} x u_x \right) = \alpha_4 \left(u_{xxx} - \frac{3}{4}\frac{u_{xx}^2}{u_x} \right); \tag{7.19m}$$

$$u_t + i\omega \left(u - \frac{1}{3} x u_x \right) = \alpha_0 + \alpha_4 \left(u_{xxx} - \frac{3}{4}\frac{u_{xx}^2}{u_x} \right); \tag{7.19n}$$

$$u_t + i\omega u - i\omega x u_x = \alpha_0 + \alpha_1 u_x + \alpha_2 u_x^2 + \alpha_3 u_x^{3/2}; \tag{7.19o}$$

$$u_t + i\omega u - \frac{1-\lambda}{2} i\omega x u_x = \alpha_2 u_x^2; \tag{7.19p}$$

$$u_t + i\omega u - \frac{2-\lambda}{3} i\omega x u_x = \alpha_3 u_x^{3/2}. \tag{7.19q}$$

In this list of equations λ denotes an arbitrary *rational* number, and the constants a_j are *arbitrary*. These PDEs have been displayed together because of their common origin; note however that some of them are of *third* order, while some are instead of *first* order, in the space derivative (and a few of the latter become x-*autonomous* for special choices of the *rational* number λ).

Example 7.2-14 The following unmodified $(1+1)$-dimensional system of two coupled PDEs is *integrable*:

$$w_\tau = w_{\xi\xi} + \tilde{w}^2, \quad \tilde{w}_\tau = w_{\xi\xi}. \tag{7.20a}$$

By setting $\lambda = \tilde{\lambda} = 1$, $\mu = 1/2$ one gets the corresponding ω-modified system:

$$\begin{aligned} u_t - i\,\omega\, u - \frac{i\,\omega}{2}\, x\, u_x &= u_{xx} + \tilde{u}^2, \\ \tilde{u}_\tau - i\,\omega\,\tilde{u} - \frac{i\,\omega}{2} x\,\tilde{u}_x &= u_{\xi\xi}. \end{aligned} \tag{7.20b}$$

Example 7.2-15 The following unmodified $(1+1)$-dimensional system of two coupled PDEs is *integrable*:

$$\begin{aligned} w_\tau &= a\,(w\,\tilde{w})_\xi\,, \\ \tilde{w}_\tau &= \left(b\,w + c\,\tilde{w}^2\right)_\xi. \end{aligned} \tag{7.21a}$$

By setting $\lambda = 2\,(1-\mu)$, $\tilde{\lambda} = 1-\mu$ one gets the corresponding ω-modified system:

$$\begin{aligned} u_t - 2\,i\,(1-\mu)\,\omega\,u - i\,\mu\,\omega\,x\,u_x &= a\,\left(u\,\tilde{u}\right)_x, \\ \tilde{u}_t - i\,(1-\mu)\,\omega\,\tilde{u} - i\,\mu\,\omega\,x\,\tilde{u}_x &= \left(b\,u + c\,\tilde{u}^2\right)_x. \end{aligned} \tag{7.21b}$$

Note that this ω-modified system is x-*autonomous* for $\mu = 0$.

Example 7.2-16 The following unmodified $(1+1)$-dimensional "Zakharov-Shabat" system of two coupled PDEs is *integrable*:

$$\begin{aligned} w_\tau + w_{\xi\xi} &= w^2\,\tilde{w}, \\ \tilde{w}_\tau - \tilde{w}_{\xi\xi} &= -\tilde{w}^2\,w. \end{aligned} \tag{7.22a}$$

By setting $\tilde{\lambda} = 1-\lambda, \mu = 1/2$ one gets the corresponding ω-modified system:

$$\begin{aligned} u_t - i\,\lambda\,\omega\,u - i\,\frac{\omega}{2}\,x\,u_x + u_{xx} &= u^2\,\tilde{u}, \\ \tilde{u}_t - i\,(1-\lambda)\,\omega\,\tilde{u} - i\,\frac{\omega}{2}\,x\,\tilde{u}_x - \tilde{u}_{xx} &= -u\,\tilde{u}^2. \end{aligned} \tag{7.22b}$$

Example 7.2-17 The following unmodified (1 + 1)-dimensional "Wadati–Konno–Ichikawa" system of two coupled PDEs is *integrable*:

$$w_\tau = a\left[w\,(1+w\,\tilde{w})^{-1/2}\right]_{\xi\xi},$$
$$\tilde{w}_\tau = b\left[\tilde{w}\,(1+w\,\tilde{w})^{-1/2}\right]_{\xi\xi}. \tag{7.23a}$$

By setting $\tilde{\lambda} = -\lambda$, $\mu = 1\,/\,2$ one gets the corresponding ω-modified system:

$$u_t - i\,\lambda\,\omega\,u - i\,\frac{\omega}{2}\,x\,u_x = a\left[u\,(1+u\,\tilde{u})^{-1/2}\right]_{xx},$$
$$\tilde{u}_t + i\,\lambda\,\omega\,\tilde{u} - i\,\frac{\omega}{2}\,x\,\tilde{u}_x = b\left[\tilde{u}\,(1+u\,\tilde{u})^{-1/2}\right]_{xx}. \tag{7.23b}$$

Example 7.2-18 The following unmodified (1 + 1)-dimensional "Landau–Lifshitz" system of two coupled PDEs is *integrable*:

$$w_\tau = -\sin(w)\,\tilde{w}_{\xi\xi} - 2\,\cos(w)\,w_\xi\,\tilde{w}_\xi$$
$$+(a-b)\,\sin(w)\,\cos(\tilde{w})\,\sin(\tilde{w}),$$
$$\tilde{w}_\tau = \frac{w_{\xi\,\xi}}{\sin(w)} - \cos(w)\,\left(\tilde{w}_\xi\right)^2$$
$$+\cos(w)\,\left[a\,\cos^2(\tilde{w}) + b\,\sin^2(\tilde{w}) + c\right]. \tag{7.24a}$$

By setting $\lambda = \tilde{\lambda} = 0$, $\mu = 1\,/\,2$ one gets the corresponding ω-modified system:

$$u_t - \frac{i\,\omega}{2}x\,u_x = -\sin(u)\,\tilde{u}_{xx} - 2\,\cos(u)\,u_x\,\tilde{u}_x$$
$$+(a-b)\,\sin(u)\,\cos(\tilde{u})\,\sin(\tilde{u}),$$
$$\tilde{u}_t - \frac{i\,\omega}{2}x\,\tilde{u}_x = \frac{u_{xx}}{[\sin(u)]} - \cos(u)\,\tilde{u}_x^2$$
$$+\cos(u)\,\left[a\,\cos^2(\tilde{u}) + b\,\sin^2(\tilde{u}) + c\right]. \tag{7.24b}$$

Note the simplification if $a = b$, and, moreover, if $c = -a$.

Example 7.2-19 The following unmodified (2 + 1)-dimensional PDE is *integrable*:

$$w_\tau = a\,w_\eta + b\,w\,w_\xi. \tag{7.25a}$$

By setting $\lambda = 0, \mu = \nu = 1$ one gets the corresponding ω-modified evolution PDE

$$u_t - i\,\omega\,(x\,u_x + y\,u_y) = a\,u_y + b\,u\,u_x. \tag{7.25b}$$

Example 7.2-20 The following unmodified (2 + 1)-dimensional PDE is *integrable*:

$$w_\tau = a\,w_\eta + b\,\left(w_\xi\right)^2. \tag{7.26a}$$

By setting $\lambda = 0$, $\mu = 1/2$, $\nu = 1$ one gets the corresponding ω-modified evolution PDE:

$$u_t - i\,\omega\,\left(\frac{x}{2}\,u_x + y\,u_y\right) = a\,u_y + b\,(u_x)^2. \tag{7.26b}$$

Example 7.2-21 The following unmodified $(2+1)$-dimensional PDE reads

$$w_\tau = a\,(w_\xi\,w_\eta - w\,w_{\xi\eta})^\alpha. \tag{7.27a}$$

By setting $\lambda = 1/\,(2\,\alpha - 1)\,,\ \mu = \nu = 0$ one gets the corresponding ω-modified evolution PDE:

$$u_t - \frac{i\,\omega}{2\,\alpha - 1}\,\omega\,u = a\,(u_x\,u_y - u\,u_{xy})^\alpha. \tag{7.27b}$$

A (rather trivial) separable solution of this PDE reads

$$u(x, y; t) = \exp\left(\frac{i\,\omega\,t}{2\,\alpha - 1}\right)\,f(x)\,g(y), \tag{7.27c}$$

where $f(x)$, $g(y)$ are two *arbitrary* functions.

Example 7.2-22 The following unmodified $(2+1)$-dimensional PDE reads

$$w_{\tau\tau} = a\,(w_\xi\,w_\eta - w\,w_{\xi\eta})^\alpha\,. \tag{7.28a}$$

By setting $\lambda = 2/\,(2\,\alpha - 1)\,,\ \mu = \nu = 0$ one gets the corresponding ω-modified evolution PDE:

$$u_{tt} - \frac{2\,\alpha + 3}{2\,\alpha - 1}\,i\,\omega\,u_t - \frac{2\,(2\,\alpha + 1)}{(2\,\alpha - 1)^2}\,\omega^2\,u = a\,(u_x\,u_y - u\,u_{xy})^\alpha\,. \tag{7.28b}$$

A (rather trivial) separable solution of this PDE reads

$$u(x, y; t) = \left[b\,\exp\left(\frac{2\,i\,\omega\,t}{2\,\alpha - 1}\right) + c\,\exp\left(\frac{2\,\alpha + 1}{2\,\alpha - 1}\,i\,\omega\,t\right)\right]\,f(x)\,g(y), \tag{7.28c}$$

where a, b are two *arbitrary* constants and $f(x)$, $g(y)$ are two *arbitrary* functions.

Example 7.2-23 The following unmodified $(2+1)$-dimensional system of two coupled PDEs is *integrable*:

$$w_\tau + w_{\xi\xi\xi} = (w\,\tilde{w})_\xi\,, \quad \tilde{w}_\eta = w_\xi. \tag{7.29a}$$

By setting $\lambda = \nu + (1/3)$, $\tilde{\lambda} = 2/3$, $\mu = 1/3$ and, for notational convenience, $\Omega = \omega/3$, one gets the corresponding Ω-modified system:

$$u_t + u_{xxx} - i\,(3\,\nu + 1)\,\Omega\,u - i\,\Omega\,x\,u_x - 3\,i\,\nu\,\Omega\,y\,u_y = (u\,\tilde{u})_x\,, \quad \tilde{u}_y = u_x. \tag{7.29b}$$

Example 7.2-24 The following unmodified (2+1)-dimensional "long-wave equation" system of two coupled PDEs is *integrable*:

$$w_{\tau\eta} + \tilde{w}_{\xi\xi} = \frac{1}{2}\left(w^2\right)_{\xi\eta}, \quad \tilde{w}_\tau + w_{\xi\xi} = \left(w\,\tilde{w} + w_{\xi\eta}\right)_\xi . \tag{7.30a}$$

By setting $\lambda = 1/2$, $\tilde{\lambda} = 0$, $\mu = 1/2$, $\nu = -1/2$ and, for notational convenience, $\Omega = \omega/2$, one gets the corresponding Ω-modified system:

$$u_{ty} - i\,\Omega\, u_y - i\,\Omega\,(x\,u_{xy} - y\,u_{yy}) + \tilde{u}_{xx} = \frac{1}{2}\left(u^2\right)_{xy},$$
$$\tilde{u}_t - i\,\Omega\,(x\,u_x - y\,u_y) + u_{xx} = \left(u\,\tilde{u} + u_{xy}\right)_x . \tag{7.30b}$$

Example 7.2-25 The following unmodified $(2+1)$-dimensional system of two coupled PDEs is *integrable*:

$$w_\tau + w_{\xi\xi\eta} = \left(w^2\right)_\eta + w_\xi\,\tilde{w}, \quad \tilde{w}_\xi = w_\eta . \tag{7.31a}$$

By setting $\tilde{\lambda} = 1 - (\lambda/2)$, $\mu = \lambda/2$, $\nu = 1 - \lambda$ one gets the corresponding ω-modified system:

$$u_t - i\,\lambda\,\omega\,u - \frac{i\,\lambda\,\omega}{2}\,x\,u_x - i\,(1-\lambda)\,\omega\,y\,u_y + u_{xxy} = \left(u^2\right)_y + u_x\,\tilde{u}, \quad \tilde{u}_x = u_y . \tag{7.31b}$$

A nontrivial family of solutions of this system, (7.31b), reads as follows:

$$u(x,y;t) = \frac{-4\,a^2\,\exp\left(i\,\lambda\,\omega\,t\right)}{\cosh^2\left\{X\,(x,\,y\,t)\right\}},$$
$$\tilde{u}(x,y;t) = -\frac{\exp\left(i\,\frac{\lambda}{2}\omega\,t\right)\,f'(t)}{a}$$
$$+\frac{\exp\left[i\,\left(1+\frac{\lambda}{2}\right)\,\omega\,t\right]\,g'\left[\eta\,(y,\,t)\right]\,\left\{i\,\omega\,b - 4\,a^2\,\tanh^2\left[X\,(x,\,y\,t)\right]\right\}}{a}, \tag{7.31c}$$

where

$$X\,(x,\,y\,t) = a\,x\,\exp\left(i\,\frac{\lambda}{2}\omega\,t\right) - f(t) - g\left[\eta\,(y,\,t)\right], \tag{7.31d}$$
$$\eta\,(y,\,t) = y\,\exp\left[i\,(1-\lambda)\,\omega\,t\right] - b\,\exp\left(i\,\omega\,t\right), \tag{7.31e}$$

a and b are two *arbitrary* constants and $f(t), g(z)$ are *arbitrary* functions (and, of course, $f'(t), g'(z)$ denote their derivatives). For conditions sufficient to guarantee that this solution be *isochronous* see [112].

Example 7.2-26 The following unmodified $(2+1)$-dimensional "Schwarzian KdV" system of two coupled PDEs is *integrable*:

$$w_\tau + \frac{1}{4}w_{\xi\xi\eta} - \frac{w_\xi\,w_{\xi\eta}}{2\,w} - \frac{w_{\xi\xi}\,w_\eta}{4\,w} + \frac{w_\xi^2\,w_\eta}{2\,w^2} - \frac{w_\xi\,\tilde{w}_\eta}{8} = 0, \quad \tilde{w}_\xi = \left(\frac{w_\xi}{w}\right)^2 . \tag{7.32a}$$

By setting $\tilde{\lambda} = \mu$, $\nu = 1 - \mu$ one gets the corresponding ω-modified system:

$$u_t - i\,\lambda\,\omega\,u - i\,\mu\,\omega\,x\,u_x + i\,(2\,\mu - 1)\,\omega\,y\,u_y + \frac{1}{4}u_{xxy} - \frac{u_x\,u_{xy}}{2\,u}$$
$$-\frac{u_{xx}\,u_y}{4\,u} + \frac{u_x^2\,u_y}{2\,u^2} - \frac{u_x\,\tilde{u}_y}{8} = 0,$$
$$\tilde{u}_x = \left(\frac{u_x}{u}\right)^2. \tag{7.32b}$$

Example 7.2-27 The following unmodified (2 + 1)-dimensional "*matrix* Kadomtsev–Petviashvili" system of two coupled *matrix* PDEs is *integrable*:

$$W_\tau + W_{\xi\xi\xi} - 3\,\tilde{W}_\eta = 3\,\left(W^2\right)_\xi + 3\,i\,\left[W,\,\tilde{W}\right], \quad \tilde{W}_\xi = W_\eta. \tag{7.33a}$$

Here $W \equiv W(\xi,\eta;\tau)$ and $\tilde{W} \equiv \tilde{W}(\xi,\eta;\tau)$ are (square) *matrices* (of course, of the same order), and the notation $\left[W,\,\tilde{W}\right]$ denotes their commutator. By setting $\lambda = 2\,/\,3$, $\tilde{\lambda} = 1$, $\mu = 1\,/\,3$, $\nu = 1$ one gets the corresponding ω-modified system:

$$U_t + U_{xxx} - 3\,\tilde{U}_y - \frac{2}{3}i\,\omega\,U - \frac{1}{3}i\,\omega\,x\,U_x - i\,\omega\,y\,U_y = 3\,\left(U^2\right)_x + 3\,i\,\left[U,\,\tilde{U}\right],$$
$$\tilde{U}_x = U_y. \tag{7.33b}$$

Example 7.2-28 The following class of *unmodified* $(N+1)$-dimensional PDEs,

$$\frac{\partial^{m+1} w}{\partial\tau\,\partial\xi^m} = f\,(w)\,, \tag{7.34a}$$

where m is a *positive integer* (or possibly just an *integer*) and $f(w)$ is an *arbitrary analytic* function, gets transformed, by setting $\lambda = 0$, $\mu = -1\,/\,m$, into the corresponding ω-modified evolution PDE:

$$\frac{\partial^{m+1} u}{\partial t\,\partial x^m} + i\,\omega\,\frac{\partial^m u}{\partial x^m} + i\,\frac{\omega}{m}\,x\,\frac{\partial^{m+1} u}{\partial x^{m+1}} = f\,(u)\,. \tag{7.34b}$$

For instance, for $m = 1$ and $f(w) = \exp(a\,w)$, (7.34a) becomes the *solvable* "Liouville" equation

$$w_{\tau\xi} = \exp(a\,w), \tag{7.35a}$$

and the corresponding ω-modified evolution PDE (7.34b) reads

$$u_{tx} + i\,\omega\,u_x + i\,\omega\,x\,u_{xx} = \exp\,(a\,u)\,. \tag{7.35b}$$

The *general* solution of this ω-modified "Liouville" equation reads

$$u(x;t) = g\left[x\,\exp\left(-i\,\omega\,t\right)\right] + f\left(t\right)$$
$$-\frac{2}{a}\ln\left\{b\,\exp\left(-i\,\omega\,t\right)\int_{x_0}^{x} dy\,\exp\left\{g\left[y\,\exp\left(-i\,\omega\,t\right)\right]\right\}\right.$$
$$\left.-\frac{a}{2\,b}\int_0^t ds\,\exp\left[i\,\omega\,s - a\,f\left(s\right)\right]\right\}, \qquad (7.35c)$$

where b is an arbitrary (nonvanishing) *complex* constant, x_0 is an arbitrary *real* constant, $g(x)$ is an *arbitrary analytic* function and $f(t)$ is as well an *arbitrary function* of the *real* "time" variable t, but it is of course *necessary* that it be *periodic* with period T, see (7.2), for the *isochrony* property

$$u(x;t+T) = u(x;t) \qquad (7.35d)$$

of this solution to hold; for conditions on this *general* solution *sufficient* to guarantee this see [112].

Example 7.2-29 The following unmodified $(N+1)$-dimensional PDE reads

$$\frac{\partial^{\,m+2}\,w}{\partial\tau^{\,2}\,\partial\xi^{\,m}} = f\left(w\right), \qquad (7.36a)$$

where m is an *integer* and $f(w)$ is an *arbitrary analytic* function. By setting $\lambda = 0$, $\mu = -2\,/\,m$ one gets the corresponding ω-modified evolution PDE:

$$\frac{\partial^{\,m+2}\,u}{\partial t^{\,2}\,\partial x^{\,m}} + 3\,i\,\omega\,\frac{\partial^{\,m+1}\,u}{\partial t\,\partial x^{\,m}} + 4\,i\,\omega\,x\,\frac{\partial^{\,m+2}\,u}{\partial t\,\partial x^{\,m+1}} - 2\,\omega^{\,2}\,\frac{\partial^{\,m}\,u}{\partial x^{\,m}}$$
$$-\frac{2}{m}\left(3+\frac{2}{m}\right)\omega^{\,2}\,x\,\frac{\partial^{\,m+1}\,u}{\partial x^{\,m+1}} - \left(\frac{2}{m}\right)^{2}\omega^{\,2}\,x^{\,2}\,\frac{\partial^{\,m+2}\,u}{\partial x^{\,m+2}} = f\left(u\right). \qquad (7.36b)$$

Example 7.2-30 The following unmodified $(N+1)$-dimensional PDE reads

$$w_{\tau\,\tau} = w_\tau\sum_{n=1}^{N}\left[a_n\,\left(w_{\xi_n}\right)^{\,\alpha_n}\right] + \sum_{n=1}^{N}\left[b_n\,\left(w_{\xi_n}\right)^{\,2\,(\beta\,\alpha_n-1)}\,w_{\xi_n\xi_n}\right]. \qquad (7.37a)$$

By setting $\lambda = \beta - 1$, $\mu_n = 1-\beta+(1\,/\,\alpha_n)$ one gets the corresponding ω-modified evolution PDE

$$u_{tt} - i\,\omega\left[\left(2\,\beta-1\right)u_t + \sum_{n=1}^{N}\left(1-\beta+\frac{1}{\alpha_n}\right)x_n\,u_{t\,x_n}\right]$$
$$-\,\omega^2\left[\beta\left(\beta-1\right)u + \beta\sum_{n=1}^{N}\left(1-\beta+\frac{1}{\alpha_n}\right)x_n\,u_{x_n}\right]$$

$$= \left\{ u_t - i\,\omega \left[(\beta - 1)\,u + \sum_{n=1}^{N} \left(1 - \beta + \frac{1}{\alpha_n} \right) x_n\, u_{x_n} \right] \right\} \sum_{n=1}^{N} \left[a_n\, (u_{x_n})^{\alpha_n} \right]$$
$$+ \sum_{n=1}^{N} \left[b_n\, (u_{x_n})^{2\,(\beta\,\alpha_n - 1)}\, u_{x_n x_n} \right]. \tag{7.37b}$$

Example 7.2-31 The following unmodified $(N+1)$-dimensional "nonlinear diffusion" PDE reads

$$w_\tau = w^\alpha \sum_{n=1}^{N} w_{\xi_n \xi_n}. \tag{7.38a}$$

By setting $\mu_n = (1 - \alpha\,\lambda)\,/\,2$ one gets the corresponding ω-modified evolution PDE:

$$u_t - i\,\lambda\,\omega\,u - \frac{1 - \alpha\,\lambda}{2}\, i\,\omega \sum_{n=1}^{N} x_n\, u_{x_n} = u^\alpha \sum_{n=1}^{N} u_{x_n\, x_n}. \tag{7.38b}$$

Note the simplification for $\lambda = 1\,/\,\alpha$: in this case this ω-modified PDE is *autonomous* also with respect to the *space* variables x_n.

Example 7.2-32 The following unmodified $(N+1)$-dimensional "nonlinear heat equation with a source" reads

$$w_\tau = \sum_{n=1}^{N} \left[a_n\, \left(w^{\alpha_n}\, w_{\xi_n} \right)_{\xi_n} \right] + b\, w^\beta. \tag{7.39a}$$

By setting $\lambda = 1\,/\,(\beta - 1)$, $\mu_n = (\beta - \alpha_n)\,/\,[2\,(\beta - 1)]$ one gets the corresponding ω-modified evolution PDE:

$$u_t - \frac{i\,\omega}{\beta - 1}\, u - \frac{(\beta - \alpha_n)\, i\,\omega}{2\,(\beta - 1)} \sum_{n=1}^{N} x_n\, u_{x_n} = \sum_{n=1}^{N} \left[a_n\, \left(u^{\alpha_n}\, u_{x_n} \right)_{x_n} \right] + b\, u^\beta. \tag{7.39b}$$

Example 7.2-33 The following unmodified $(N + 1)$-dimensional "Bateman" PDE is *solvable:*

$$\det \begin{pmatrix} 0 & w_\tau & w_{\xi_1} & \cdots & w_{\xi_n} \\ w_\tau & w_{\tau\,\tau} & w_{\xi_1\,\tau} & \cdots & w_{\tau\,\xi_n} \\ w_{\xi_1} & w_{\tau\,\xi_1} & w_{\xi_1\,\xi_1} & \cdots & w_{\xi_n\,\xi_1} \\ \vdots & \vdots & \vdots & \ddots & \vdots \\ w_{\xi_n} & w_{\tau\,\xi_n} & w_{\xi_1\,\xi_n} & \cdots & w_{\xi_n\,\xi_n} \end{pmatrix} = 0. \tag{7.40a}$$

The corresponding ω-modified PDE is in this case t-autonomous for any choice of λ and μ_n. For $\lambda = \mu_n = 0$ it reads

$$\det \begin{pmatrix} 0 & u_t & u_{x_1} & \cdots & u_{x_n} \\ u_t & u_{t\,t} & u_{x_1\,t} & \cdots & u_{x_n\,t} \\ u_{x_1} & u_{x_1\,t} & u_{x_1\,x_1} & \cdots & u_{x_n\,x_1} \\ \vdots & \vdots & \vdots & \ddots & \vdots \\ u_{x_n} & u_{t\,x_n} & u_{x_1\,x_n} & \cdots & u_{x_n\,x_n} \end{pmatrix}$$

$$-i\,\omega\,u_t \det \begin{pmatrix} 0 & u_{x_1} & \cdots & u_{x_n} \\ u_{x_1} & u_{x_1\,x_1} & \cdots & u_{x_n\,x_1} \\ \vdots & \vdots & \ddots & \vdots \\ u_{x_n} & u_{x_1\,x_n} & \cdots & u_{x_n\,x_n} \end{pmatrix} = 0. \tag{7.40b}$$

The general solution $u \equiv u(\underline{x};t)$ of this PDE is given by the implicit formula

$$[\exp(i\,\omega\,t) - 1]\; f_0\,(u) + \sum_{k=1}^{n} [x_k\, f_k\,(u)] = c, \tag{7.40c}$$

where the $N+1$ functions $f_k(z)$, $k = 0, 1, \ldots, N$ are *arbitrary.*

For $N = 1$, the unmodified $(1+1)$-dimensional "Bateman" equation reads

$$w_{\tau\tau}\;(w_\xi)^2 + w_{\xi\xi}\,w_\tau^2 - 2\,w_\xi\,w_\tau\,w_{\xi\tau} = 0, \tag{7.41a}$$

and its ω-modified version reads

$$u_{tt}\,u_x^2 + u_{xx}\,u_t^2 - 2\,u_{xt}\,u_x\,u_t + i\,\omega\,\left[2\,\lambda\,u\,(u_x\,u_{xt} - u_t\,u_{xx}) + (2\,\mu - 1)\,(u_x)^2\,u_t\right]$$
$$+\,\omega^2\left[\lambda^2 u\,(u_x)^2 - \lambda^2\,u^2\,u_{xx} + \lambda\,(2\,\mu - 1)\,u\,(u_x)^2 + \mu\,(\mu - 1)\,x\,(u_x)^3\right] = 0. \tag{7.41b}$$

The *general* solution of this equation reads (in implicit form)

$$[\exp(i\,\omega\,t) - 1]\; f\,[\exp(-i\,\omega\,t)\; u(x;t)] + x\,\exp(i\,\mu\,\omega\,t)$$
$$\times\; g\,[\exp(-i\,\lambda\,\omega\,t)\; u(x;t)] = c, \tag{7.41c}$$

with $f(z)$ and $g(z)$ two *arbitrary* functions which can be easily determined in terms of the initial data $u_0(x) = u(x;0)$ and $u_1(x) = u_t(x;0)$.

By setting $\lambda = 0$, $\mu = 1/2$ and (for notational convenience) $\Omega = \omega/2$ the Ω-modified $(1+1)$-dimensional Bateman equation (7.41b) takes the simple (*real*) form

$$u_{tt}\,u_x^2 + u_{xx}\,u_t^2 - 2\,u_{xt}\,u_x\,u_t = \Omega^2\,x\,u_x^3. \tag{7.42a}$$

Note that, if $u(x;t)$ is a solution of this PDE, $v(x;t) = f\,[u(a\,x;t-b)]$ is also a solution, with $f(z)$ an arbitrary function and a, b two arbitrary constants.

The initial-value problem for this equation is solved by the implicit formula

$$u(x;t) = u_0\left(\frac{\tan(\Omega\, t)}{\Omega}\frac{u_1\left(u_0^{(\text{inv})}\left[u(x;t)\right]\right)}{u_0'\left(u_0^{(\text{inv})}\left[u(x;t)\right]\right)} + \frac{x}{\cos[\Omega\, t]}\right), \tag{7.42b}$$

where of course $u_0^{(\text{inv})}(z)$ respectively $u_0'(z)$ are the inverse respectively the derivative of the function $u_0(z)$ (i.e., $u_0^{(\text{inv})}\left[u_0(x)\right] = x$, $u_0'(x) = du_0(x)\,/\,dx$). And two explicit solutions of this equation, (7.42b), read as follows:

$$u(x;t) = f\left\{\frac{c_1\, x + c_2\,\cos\left[\Omega\,(t-c_3)\right]}{c_4\, x + c_5\,\cos\left[\Omega\,(t-c_6)\right]}\right\}, \tag{7.42c}$$

$$u(x;t) = f\left\{\frac{c_1\,\cos\left[\Omega\,(t-c_2)\right]}{\cos\left[2\,\Omega\,(t-c_3)\right] + \cos\left[2\,\Omega(c_2-c_3)\right]}\, x\right.$$
$$\left. + c_4\,\tan\left[\Omega\,(t+c_2-2\,c_3)\right]\right\}. \tag{7.42d}$$

Here $f(z)$ denotes an *arbitrary* (twice differentiable) function, and the constants c_k are *arbitrary* as well. Clearly these solutions are *real* if the function $f(z)$ and the constants c_k are themselves *real*; and the conditions that the function $f(z)$ and the constants c_k must satisfy in order to guarantee the *isochrony* of these solutions are rather obvious.

Particularly interesting are the two cases with $\mu = 0$ and $\mu = 1$, when the ω-modified PDE (7.41b) becomes x-*autonomous*, reading

$$u_{tt}\, u_x^2 + u_{xx}\, u_t^2 - 2\, u_{xt}\, u_x\, u_t + i\,\omega\left[2\,\lambda\, u\,\left(u_x\, u_{xt} - u_t\, u_{xx}\right) \mp \left(u_x\right)^2\, u_t\right]$$
$$+ \lambda\,\omega^2\, u\left[\lambda\,\left(u_x\right)^2 - \lambda\, u\, u_{xx} \mp \left(u_x\right)^2\right] = 0. \tag{7.43}$$

Example 7.2-34 The following unmodified $(N+1)$-dimensional PDE reads

$$w_{\tau\tau} = w_\tau^2\, f\left(w, w_{\xi_1}, \ldots, w_{\xi_N}, w_{\xi_1\,\xi_1}, w_{\xi_1\,\xi_2}, \ldots, w_{\xi_N\,\xi_N}, \ldots\right), \tag{7.44a}$$

where $f\left(w, w_{x_1}, \ldots, w_{\xi_N}, w_{\xi_1\,\xi_1}, \ldots, w_{\xi_N\,\xi_N}\ldots\right)$ is an *arbitrary analytic* function. By setting $\lambda = \mu_n = 0$ one gets the corresponding ω-modified evolution PDE

$$u_{t\,t} - i\,\omega\, u_t = u_t^2\, f\left(u, u_{x_1}, \ldots, u_{x_N}, u_{x_1\,x_1}, u_{x_1\,x_2}, \ldots, u_{x_N\,x_N}, \ldots\right). \tag{7.44b}$$

Example 7.2-35 A class of *C-integrable* PDEs in multidimensions was identified some time ago, and recently the *isochronous* versions of these PDEs have been investigated. We report here only two examples of these (systems of) PDEs, for simplicity only in their *isochronous* ω-modified version (the unmodified version is of course obtained by setting ω to zero).

An *isochronous C-integrable* system of N *nonlinear* evolution PDEs "of Schrödinger ($\gamma = i$) or diffusion ($\gamma = 1$) type" reads as follows:

$$\gamma\,\psi_{n,t} - \Delta\,\psi_n + \lambda\,\omega\,\psi_n + \frac{\omega}{2}\,\vec{r}\cdot\vec{\nabla}\,\psi_n = 2 \sum_{m=1,\,m\neq n}^{N} \frac{a - \left(\vec{\nabla}\,\psi_n\right)\cdot\left(\vec{\nabla}\,\psi_m\right)}{\psi_n - \psi_m}, \quad (7.45)$$

where λ is an *arbitrary* (*real, rational*) parameter if $a = 0$, while $\lambda = -1\,/\,2$ if a is an *arbitrary* constant.

An *isochronous C-integrable* system of N *nonlinear* evolution PDEs "of Klein–Gordon type" reads as follows:

$$\psi_{n,tt} - \Delta\,\psi_n - 2\,i\,\omega\left(\vec{r}\cdot\vec{\nabla}\right)\psi_{n,t} - i\,(2\,\lambda + 1)\,\omega\,\psi_{n,t} - \lambda\,(\lambda + 1)\,\omega^2\,\psi_n$$
$$-2\,(\lambda + 1)\,\omega^2\left(\vec{r}\cdot\vec{\nabla}\right)\psi_n - \omega^2\left(\vec{r}\cdot\vec{\nabla}\right)^2\psi_n$$
$$= 2 \sum_{m=1,\,m\neq n}^{N} (\psi_n - \psi_m)^{-1}\left\{a - \left(\vec{\nabla}\,\psi_n\right)\cdot\left(\vec{\nabla}\,\psi_m\right)\right.$$
$$+ \left[\psi_{n,t} - i\,\lambda\,\omega\,\psi_n - \frac{1}{2}\,\omega\left(\vec{r}\cdot\vec{\nabla}\right)\psi_n\right]$$
$$\left.\left[\psi_{m,t} - i\,\lambda\,\omega\,\psi_m - \frac{1}{2}\,\omega\left(\vec{r}\cdot\vec{\nabla}\right)\psi_m\right]\right\}, \quad (7.46)$$

where λ is an *arbitrary* (*real, rational*) parameter if $a = 0$, while $\lambda = -1$ if a is an *arbitrary* constant.

In both these (systems of) evolution PDEs N is an arbitrary integer ($N \geq 2$), the dimensionality of the independent (space) variable $\vec{r}$ is *arbitrary*, and $\vec{\nabla}$ respectively $\Delta = \vec{\nabla}\cdot\vec{\nabla}$ denote the standard *gradient* respectively *Laplacian* operators.

7.3 PDEs with lots of solutions periodic in time and in space

In the preceding part of this Chapter 7 we focussed on ω-*modified* PDEs that feature many *isochronous* solutions. By a variant of the trick (7.1) it is in some cases possible to generate equations that feature many solutions which are *periodic* not only in the time variable t, but as well in the space variable x. For the sake of simplicity we restrict attention here to the $(1 + 1)$-dimensional case and we only report a *single* example, obtained from the *unmodified* PDE (7.4a) via the following generalized version of the trick, corresponding to the following change of dependent and independent variables:

$$\tau = \frac{\exp\,(i\,\omega\,t) - 1}{i\,\omega}, \quad (7.47a)$$

$$\xi = \frac{\exp\,(i\,k\,x) - 1}{i\,k}, \quad (7.47b)$$

$$u(x;t) = \exp\left[i\,\left(\lambda\,\omega\,t + \rho\,k\,x\right)\right]\,w(\xi;\tau). \quad (7.47c)$$

To apply this transformation to the *unmodified* PDE (7.4a), we set $\lambda = -\rho = 1 / \alpha$, while ω and k are of course two *real* (indeed, without loss of generality, *positive*) constants, that determine the basic period T in the (*real*) time variable t, see (7.2), as well as the basic period,

$$L = \frac{2\,\pi}{k}, \tag{7.48}$$

in the (also *real*) space variable x. The modified PDE corresponding to (7.4a) reads then

$$u_t - \frac{i\,\omega}{\alpha}\,u = a\,u^{\alpha}\left(u_x + \frac{i\,k}{\alpha}\,u\right), \tag{7.49a}$$

and its *general* solution, in implicit form, reads

$$\begin{aligned} u(x;t) &= \exp\left(\frac{i\,\omega\,t}{\alpha}\right)\left\{1 + \frac{a\,k}{\omega}\,[1 - \exp(-i\,\omega\,t)]\,[u(x;t)]^{\alpha}\right\}^{1/\alpha} \cdot \\ &\cdot u_0\left(x - \frac{i}{k}\log\left\{1 + \frac{a\,k}{\omega}\,[1 - \exp(-i\,\omega\,t)]\,[u(x;t)]^{\alpha}\right\}\right), \end{aligned} \tag{7.49b}$$

where of course the initial datum, $u_0(x) = u(x;0)$ should be itself *periodic* with period L (or some appropriate integer multiple, or fraction, of L), in order for this solution to be for *all* time *periodic* in space with period L (or some appropriate integer multiple, or fraction, of L)—in addition of course to being *periodic* in t with period T (or with some integer multiple, or fraction, of T)—depending on the analyticity properties of $u_0(z)$ as a function of the *complex* variable z. A special case of this *implicit* equation (corresponding to $u_0(x) = \exp(i\,\gamma\,x)$) reads

$$\begin{aligned} u(x;t) &= b\,\exp\left[i\left(\frac{\omega\,t}{\alpha} + \gamma\,x\right)\right] \cdot \\ &\cdot \left\{1 + \frac{a\,k}{\omega}\,[1 - \exp(-i\,\omega\,t)]\,[u(x;t)]^{\alpha}\right\}^{(k+\alpha\,\gamma)/(\alpha\,\gamma)}, \end{aligned} \tag{7.49c}$$

yielding, for $k + \alpha\,\gamma = 0$, the trivial solution of (7.49a)

$$u(x;t) = b\,\exp\left[i\,\frac{(\omega\,t - k\,x)}{\alpha}\right]. \tag{7.49d}$$

For $\gamma = k / \alpha$ one obtains instead from (7.49c) the following explicit solution of (7.49a):

$$\begin{aligned} u(x;t) &= \exp\left[\frac{i(\omega\,t - k\,x)}{\alpha}\right]\left\{\frac{\omega}{2\,a^2\,b\,k^2\,[\exp(i\,\omega\,t) - 1]^2} \cdot \right. \\ &\cdot [\omega - 2\,a\,b\,\exp(i\,k\,x)\,[\exp(i\,\omega\,t) - 1] \cdot \\ &\left. \cdot \left[- \left\{\omega^2 - 4\,a\,b\,k\,\omega\,\exp(i\,k\,x)\,[\exp(i\,\omega\,t) - 1]\right\}^{1/2}\right]\right\}^{1/\alpha}. \end{aligned} \tag{7.49e}$$

Of course many other explicit solutions could be exhibited, for instance in the special case with $\gamma = -k$ and $\alpha = 2$

$$u(x;t) = \pm \left[\frac{\omega \pm \left\{ \omega^2 - 4\, a\, b^2\, k\, \omega\, \exp\left(-2\, i\, k\, x\right)\, \left[1 - \exp\left(i\, \omega\, t\right)\right] \right\}^{1/2}}{2\, a\, k\, \left[1 - \exp\left(i\, \omega\, t\right)\right]} \right]^{1/2} . \tag{7.49f}$$

Finally, we exhibit below the *real* version of the ω-modified PDE (7.49a) with $\alpha = 1$, setting $u = u_1 + i\, u_2$, $a = c_1 + i\, c_2$ where the two dependent variables $u_1 \equiv u_1(x;t)$ and $u_2 \equiv u_2(x;t)$, as well as the two constants c_1, c_2, are of course now *real*:

$$\begin{aligned} u_{1,t} + \omega\, u_2 &= c_1\, \left[u_1\, u_{1,x} + u_2\, u_{2,x} - 2\, k\, u_1\, u_2\right] \\ &\quad - c_2\, \left[u_1\, u_{2,x} + u_2\, u_{1,x} + k\, \left(u_1^2 - u_2^2\right)\right] , \\ u_{2,t} - \omega\, u_1 &= c_2\, \left[u_1\, u_{1,x} + u_2\, u_{2,x} - 2\, k\, u_1\, u_2\right] \\ &\quad + c_1\, \left[u_1\, u_{2,x} + u_2\, u_{1,x} + k\, \left(u_1^2 - u_2^2\right)\right] . \end{aligned} \tag{7.50}$$

7.N Notes to Chapter 7

The presentation in this chapter follows closely a joint paper with Mauro Mariani [113], and also the last section of [51]. See also [112].

For the list of PDEs reported in Section 7.2—most of which have been selected because of the "named" status of their *unmodified* version—we report below, whenever possible, some key reference identifying its first introduction or some extended treatment of it.

For the unmodified PDE (7.4a) of **Example 7.2-1** see, for instance, [107].

For the unmodified PDE (7.5a) of **Example 7.2-2** see [104].

For the unmodified PDE (7.6a) of **Example 7.2-3** see, for instance, [31].

For a recent treatment of the unmodified PDE (7.7a) of **Example 7.2-4** see [12] [13].

For the unmodified PDE (7.8a) of **Example 7.2-5** see, for instance, [89].

For the unmodified PDE (7.9a) of **Example 7.2-6** see, for instance, [88].

For the unmodified PDE (7.10a) of **Example 7.2-7** see, for instance, [55].

For the unmodified PDE (7.12a) of **Example 7.2-8** see, for instance, [107] and [93].

For the unmodified PDE (7.14a) of **Example 7.2-10** see, for instance, [96].

For the unmodified PDE (7.15a) of **Example 7.2-11** see, for instance, [107]. Some solutions of the ω-modified PDE (7.15d) are exhibited in [112].

For the *integrable* ω-modified character of the PDEs of **Example 7.2-13**—as well as the display of some special (*isochronous*) solutions of these PDEs, and the possibility to manufacture *hierarchies* of nonlinear PDEs of which those displayed are first instances—see [60].

For the unmodified system of two coupled PDEs (7.20a) of **Example 7.2-14** see, for instance, [136].

For the unmodified system of two coupled PDEs (7.21a) of **Example 7.2-15** see, for instance, [1] and [136].

For the unmodified system of two coupled PDEs (7.22a) of **Example 7.2-16** see [144] and [2].

For the unmodified system of two coupled PDEs (7.23a) of **Example 7.2-17** see [134].

For the unmodified system of two coupled PDEs (7.24a) of **Example 7.2-18** see, for instance, [133] and [136].

For the unmodified PDEs (7.25a) and (7.26a) of **Examples 7.2-19** and **7.2-20** see for instance [107].

For the unmodified system of two coupled PDEs (7.29a) of **Example 7.2-23** see for instance [10].

For the unmodified PDE (7.30a) of **Example 7.2-24** see for instance [11] and [137].

For the unmodified system of two coupled PDEs (7.31a) of **Example 7.2-25** see for instance [27].

For the unmodified system of two coupled PDEs (7.32a) of **Example 7.2-26** see [131] and [126] (where a somewhat different notation is used). For a rather detailed discussion of its ω-modified version (7.32b), including the exhibition of several solutions, see [114]. Note that for some solutions of this ω-modified PDE, (7.32b), the property of periodicity might apply only to the first ("main") dependent variable, $u(x, y, t+T) = u(x, y, t)$, while the corresponding second ("auxiliary") dependent variable might instead satisfy a "shifted" periodicity property, say $\tilde{u}(x, y, t+T) = \tilde{u}(x, y, t) + v(x, y)$.

For the unmodified system of matrix PDEs (7.33a) of **Example 7.2-27** see for instance [110].

For the unmodified PDEs (7.38a) and (7.39a) of **Examples 7.2-31** and **7.2-32** see for instance [107].

For the unmodified PDE (7.40a) of **Example 7.2-33** see for instance [92].

In connection with **Example 7.2-35** let us recall that the term "*C-integrable*" is used to identify nonlinear evolution PDEs whose solution, via an appropriate (generally *nonlinear*) *Change of variables*, can be reduced to performing algebraic (nondifferential) operations and solving *linear* PDEs. For the systems of N *nonlinear* PDEs reported in this example the *Change of variables* in question is typically that relating the coefficients and the zeros of a polynomial of degree N, whose relevance to identify and investigate *nonlinear* dynamical systems (i. e., systems of ODEs) was originally highlighted in [29] and is discussed in considerable detail in Section 2.3 of [37]. The extension of this approach to (systems of) *nonlinear* evolution PDEs was originally done in [32]; it is mentioned as *Exercise 2.3.4.2-5* in [37] and is treated—including the *isochronous* versions reported in **Example 7.2-35**—in a recent electronic article coauthored with Matteo Sommacal [87], which also includes the display of solutions as animations. Extensions of these *C-integrable* PDEs in *multidimensional* space—including their *isochronous* versions—have been recently published, featuring coefficients involving *trigonometric* [67] and *elliptic* [68] functions.

For a more complete treatment of the topic of Section 7.3 see [112].

8

OUTLOOK

Five directions of future research are suggested by the findings reported in this monograph. They are tersely outlined in this final chapter, in what might well be an order of increasing importance.

(i) In Appendix C several *Diophantine* results and conjectures were reported. Other analogous findings might be obtained in analogous manners. And of course each conjecture should be proven (some already have been).

(ii) Some of the *isochronous* dynamical systems obtained in the first part of this monograph are *Hamiltonian*, and also *Hamiltonian* are, of course, *all* those obtained in Chapter 5. Moreover, the tricks described in Chapter 5 open the way to manufacturing many more *Hamiltonian* systems that are *isochronous*—or even *entirely isochronous*. The possibility offered by these techniques to manufacture so many dynamical systems that are both *Hamiltonian* and *isochronous*—even *entirely isochronous*—raises the issue of their quantization, to test the hunch that to the classical property of *isochrony* there correspond, in an appropriate quantal context, *equispaced* spectra. This problematique is also likely to shed light on the peculiarities of quantization.

(iii) Through the approach on which this monograph hinges, based on the application of the original trick (see Section 2.1), it is seen that the property of *isochrony* of an "ω-modified" system (regarding its evolution as function of the *real* independent variable t representing "time", and valid in a certain open, *fully dimensional*, region of its phase space) is in many cases connected with the property of solutions of a related "unmodified" dynamical system, considered as functions of the *complex* variable τ, to be *holomorphic* in a disk D of the *complex* τ-plane. Solutions of the ω-*modified* system corresponding to solutions of the *unmodified* system featuring singularities—typically *branch points*—within the disk D may instead display an *aperiodic* behavior, possibly so complicated to be considered an instance of a, possibly unorthodox, kind of "deterministic chaos". This phenomenology suggests a somewhat novel paradigm for understanding the genesis of certain deterministic motions displaying a very complicated behavior.

(iv) On the other hand, via the trick of Section 5.5 the way is open to transform a *real* Hamiltonian describing a *realistic* many-body system into an Ω-modified *real* and *entirely isochronous* Hamiltonian. Although the latter system is not characterized by a *normal* Hamiltonian (standard kinetic energy plus potential energy), the fact that its dynamics may be quite similar—over finite time periods—to that characterizing a standard many-body system, and yet it is—over longer time periods—*entirely isochronous*, is remarkable in view if the *irreversible* character—in the context of statistical mechanics and thermodynamics—of the time evolution of most *realistic* many-body systems. We tersely discussed this issue in Section 5.5.2 ("Transient chaos"). Although the origin of this phenomenology is to some extent obvious—a kind of reversal of the time evolution caused by the Ω-*modified* dynamics itself—a further study of it seems warranted: in particular an investigation of the behavior of such *entirely isochronous* Ω-modified systems in the context of *statistical mechanics* and *thermodynamics*, when their *isochrony* period is much larger than the natural timescale characterizing the (chaotic, or at least ergodic) evolution of the (for instance, "molecular") dynamics yielded by the corresponding *unmodified* Hamiltonians. And also of interest shall be the application of the Ω-*modification* technique to *general Hamiltonians*, as follow-up to the findings reported in [84].

(v) Finally most appealing, but possibly most elusive, is the prospect to employ *isochronous* models such as those developed in this monograph for the qualitative—or even quantitative—understanding of *isochronous* phenomena in *applicative* contexts, be it in physics, chemistry, biology, medicine, economy, ...

8.N Notes to Chapter 8

(i) In some cases the *Diophantine* findings arrived at via the investigation of *isochronous* systems in the neighborhood of their equilibrium configurations are related to the *remarkable matrices* treated in [37] (see in particular Section 2.4, Chapter 3 and Appendix D of [37]). For very recent developments—including the proofs of some of the *Diophantine* conjectures reported in Appendix C, and the investigation of three-term recursion relations generating orthogonal polynomials, some of which possess several, or even only, zeros given by neat formulas in terms of *integers*—see [22] [23] [24] [25] and future papers of this series.

(ii) Concerning the hunch that to the classical property of *isochrony* there correspond, in an appropriate quantal context, *equispaced* spectra, see [81] [82] [83] [84]. For some peculiarities of quantization—in some cases indeed related to *isochronous* systems—see [74] [44] [49] [56].

(iii) Research aimed at understanding irregular motions as travels on Riemann surfaces is now in progress, see [72] [101] [95] [73]. Its goal is to shed light on the nature of certain, possibly unorthodox, kinds of deterministic chaos and on techniques suitable for a better understanding of such time evolutions—to the extent this *oximoronic* goal, to *understand* a *chaotic* time evolution, is at all achievable.

(iv) Investigations of the *transient chaos* phenomenology, in collaboration with François Leyvraz, are in progress [85].

(v) *Isochronous* phenomena are of course rather *common* in the *real* world, see for instance [130].

APPENDIX A

SOME USEFUL IDENTITIES

In this Appendix—which is lifted essentially *verbatim* from [21]—we list certain useful relations; we consider their derivation sufficiently straightforward not to require any proof. Not all the formulas reported here are used in this monograph, but we thought it useful to report a rather extensive compilation as a convenient tool for future utilizations.

Let $\psi(z,t)$ be a *monic* polynomial of degree N in z, and let us indicate as $z_n \equiv z_n(t)$ its N zeros and as $c_m \equiv c_m(t)$ its N coefficients:

$$\psi(z,t) = \prod_{n=1}^{N} [z - z_n(t)] = z^N + \sum_{m=1}^{N} c_m(t)\, z^{N-m} = \sum_{m=0}^{N} c_m(t)\, z^{N-m}, \quad c_0 = 1. \tag{A.1}$$

Then clearly

$$\psi_z(z,t) = \psi(z,t) \sum_{n=1}^{N} [z - z_n(t)]^{-1}, \tag{A.2}$$

$$\psi_t(z,t) = \psi(z,t) \sum_{n=1}^{N} [z - z_n(t)]^{-1} [-\dot{z}_n(t)]. \tag{A.3}$$

Here (and throughout) subscripted variables denote partial differentiations, $\psi_z \equiv \partial\psi / \partial z$, $\psi_t \equiv \partial\psi / \partial t$, and so on.

To conveniently streamline the look of these two, and of all the following, formulas, we rewrite these two relations via the self-evident notation

$$\psi_z \Leftrightarrow 1, \tag{A.4}$$

$$\psi_t \Leftrightarrow -\dot{z}_n, \tag{A.5}$$

and we write accordingly the following relations:

$$z\,\psi_z - N\,\psi \Leftrightarrow z_n, \tag{A.6a}$$

$$z^2\,\psi_z + (c_1 - N\,z)\,\psi \Leftrightarrow z_n^2, \tag{A.6b}$$

$$z^3\,\psi_z - \left(N\,z^2 - c_1\,z + c_1^2 - 2\,c_2\right)\psi \Leftrightarrow z_n^3 \tag{A.6c}$$

$$\frac{1}{z}\left[\psi_z - \frac{c_{N-1}}{c_N}\,\psi\right] \Leftrightarrow \frac{1}{z_n}, \tag{A.6d}$$

$$z\,\psi_t - \dot{c}_1\,\psi \Leftrightarrow -\dot{z}_n\,z_n, \tag{A.7a}$$

$$\frac{1}{z}\left[\psi_t - \frac{\dot{c}_N}{c_N}\,\psi\right] \Leftrightarrow -\frac{\dot{z}_n}{z_n}, \tag{A.7b}$$

$$\psi_{zz} \Leftrightarrow \sum_{m=1,\,m\neq n}^{N} \frac{2}{z_n - z_m}, \tag{A.8a}$$

$$z\,\psi_{zz} - 2\,(N-1)\,\psi_z \Leftrightarrow \sum_{m=1,\,m\neq n}^{N} \frac{2\,z_m}{z_n - z_m}, \tag{A.8b}$$

$$z\,\psi_{zz} \Leftrightarrow \sum_{m=1,\,m\neq n}^{N} \frac{2\,z_n}{z_n - z_m}, \tag{A.8c}$$

$$z\,\psi_{zz} - (N-1)\,\psi_z \Leftrightarrow \sum_{m=1,\,m\neq n}^{N} \frac{z_n + z_m}{z_n - z_m}, \tag{A.8d}$$

$$z^2\,\psi_{zz} - N\,(N-1)\,\psi \Leftrightarrow \sum_{m=1,\,m\neq n}^{N} \frac{2\,z_n^2}{z_n - z_m}, \tag{A.8e}$$

$$z^2\,\psi_{zz} - 2\,[(N-2)\,z - c_1]\,\psi_z + N\,(N-3)\,\psi \Leftrightarrow \sum_{m=1,\,m\neq n}^{N} \frac{2\,z_m^2}{z_n - z_m}, \tag{A.8f}$$

$$z^2\,\psi_{zz} - [(N-2)\,z - c_1]\,\psi_z - N\,\psi \Leftrightarrow \sum_{m=1,\,m\neq n}^{N} \frac{z_n^2 + z_m^2}{z_n - z_m}, \tag{A.8g}$$

$$z^2\,\psi_{zz} - 2\,(N-1)\,z\,\psi_z + N\,(N-1)\,\psi \Leftrightarrow \sum_{m=1,\,m\neq n}^{N} \frac{2\,z_n\,z_m}{z_n - z_m}, \tag{A.8h}$$

$$z^3\,\psi_{zz} - N\,(N-1)\,z\,\psi + 2\,(N-1)\,c_1\,\psi \Leftrightarrow \sum_{m=1,\,m\neq n}^{N} \frac{2\,z_n^3}{z_n - z_m}, \tag{A.8i}$$

$$z\,\left[z^2\,\psi_{zz} - 2\,(N-1)\,z\,\psi_z + N\,(N-1)\,\psi\right] \Leftrightarrow \sum_{m=1,\,m\neq n}^{N} \frac{2\,z_n^2\,z_m}{z_n - z_m}, \tag{A.8j}$$

$$z^3\,\psi_{zz} - 2\,(N-2)\,z^2\,\psi_z + 2\,c_1\,z\,\psi_z + [N\,(N+1)\,z - 2\,(N-1)\,c_1]\,\psi$$
$$\Leftrightarrow \sum_{m=1,\,m\neq n}^{N} \frac{2\,z_n\,z_m^2}{z_n - z_m}, \tag{A.8k}$$

$$z^3\,\psi_{zz} - (2\,N-3)\;z^2\,\psi_z + c_1\,z\,\psi_z + \left[N^2\,z - (N-1)\;c_1\right]\,\psi$$
$$\Leftrightarrow \sum_{m=1,\,m\neq n}^{N} \frac{z_n\,z_m^2 + z_n^2\,z_m}{z_n - z_m}, \tag{A.8l}$$

$$z^4\,\psi_{zz} - \left[N\;(N-1)\;z^2 - 2\;(N-1)\;c_1\,z + 2\;(N-1)\;c_1{}^2 - 2\;(2\,N-3)\;c_2\right]\,\psi$$
$$\Leftrightarrow \sum_{m=1,\,m\neq n}^{N} \frac{2\,z_n^4}{z_n - z_m}, \tag{A.8m}$$

$$z^4\,\psi_{zz} - 2\,z^2\,[(N-2)\;z - c_1]\;\psi_z + \left[N\,(N-3)\,z^2 - 2\,(N-1)\,c_1\,z + 2\,c_2\right]\,\psi$$
$$\Leftrightarrow \sum_{m=1,\,m\neq n}^{N} \frac{2\,z_n^2\,z_m^2}{z_n - z_m}, \tag{A.8n}$$

$$\psi_{zt} \Leftrightarrow - \sum_{m=1,\,m\neq n}^{N} \frac{\dot{z}_n + \dot{z}_m}{z_n - z_m}, \tag{A.9a}$$

$$z\,\psi_{zt} \Leftrightarrow - \sum_{m=1,\,m\neq n}^{N} \frac{(\dot{z}_n + \dot{z}_m)\;z_n}{z_n - z_m}, \tag{A.9b}$$

$$z\,\psi_{zt} - (N-1)\;\psi_t \Leftrightarrow - \sum_{m=1,\,m\neq n}^{N} \frac{\dot{z}_n\,z_m + \dot{z}_m\,z_n}{z_n - z_m}, \tag{A.9c}$$

$$z\;[z\,\psi_{zt} - (N-1)\;\psi_t] \Leftrightarrow - \sum_{m=1,\,m\neq n}^{N} \frac{(\dot{z}_n\,z_m + \dot{z}_m\,z_n)\;z_n}{z_n - z_m}, \tag{A.9d}$$

$$z^2\,\psi_{zt} + [c_1 - (N-2)\;z]\;\psi_t - \dot{c}_1\,\psi$$
$$\Leftrightarrow - \sum_{m=1,\,m\neq n}^{N} \frac{\dot{z}_n\,z_m^2 + \dot{z}_m\,z_n^2}{z_n - z_m}, \tag{A.9e}$$

$$\psi_{tt} \Leftrightarrow -\ddot{z}_n(t) + \sum_{m=1,\,m\neq n}^{N} \frac{2\,\dot{z}_n\,\dot{z}_m}{z_n - z_m}. \tag{A.10}$$

To obtain some of the formulas written above we used the relations

$$\frac{c_{N-1}}{c_N} = -\sum_{n=1}^{N} \frac{1}{z_n}, \qquad \frac{\dot{c}_N}{c_N} = \sum_{n=1}^{N} \frac{\dot{z}_n}{z_n}, \tag{A.11}$$

which are obvious consequences of the formulas

$$c_N(t) = \prod_{n=1}^{N} [-z_n(t)] = \psi(0,t), \quad c_{N-1}(t) = \sum_{n=1}^{N} \prod_{m=1, m\neq n}^{N} [-z_m(t)] = \psi_z(0,t), \tag{A.12}$$

themselves direct consequences of (A.1).

Likewise, we introduce the following notation whereby in the formulas written below (which are also straightforward consequences of (A.1)) the expression in the right-hand side identifies the coefficient of z^{N-m} in the polynomial (of degree N or less) appearing in the left-hand side:

$$\psi \leftrightarrow c_m, \tag{A.13a}$$

$$\frac{\psi - c_N}{z} \leftrightarrow c_{m-1}, \tag{A.13b}$$

$$\psi_z \leftrightarrow (N-m+1)\, c_{m-1}, \tag{A.13c}$$

$$z\,\psi_z \leftrightarrow (N-m)\, c_m, \tag{A.13d}$$

$$\frac{\psi_z - c_{N-1}}{z} \leftrightarrow (N-m+2)\, c_{m-2}, \tag{A.13e}$$

$$\psi_{zz} \leftrightarrow (N-m+2)\,(N-m+1)\, c_{m-2}, \tag{A.13f}$$

$$z\,\psi_{zz} \leftrightarrow (N-m+1)\,(N-m)\, c_{m-1}, \tag{A.13g}$$

$$z^2\,\psi_{zz} \leftrightarrow (N-m)\,(N-m-1)\, c_m, \tag{A.13h}$$

$$\psi_t \leftrightarrow \dot{c}_m, \tag{A.13i}$$

$$z\,\psi_t \leftrightarrow \dot{c}_{m+1}, \tag{A.13j}$$

$$\frac{\psi_t - \dot{c}_N}{z} \leftrightarrow \dot{c}_{m-1}, \tag{A.13k}$$

$$\psi_{zt} \leftrightarrow (N-m+1)\, \dot{c}_{m-1}, \tag{A.13l}$$

$$z\,\psi_{zt} \leftrightarrow (N-m)\, \dot{c}_m, \tag{A.13m}$$

$$\psi_{tt} \leftrightarrow \ddot{c}_m; \tag{A.13n}$$

$$N\,\psi - z\,\psi_z \leftrightarrow m\, c_m, \tag{A.13o}$$

$$N^2\,\psi - (2\,N-1)\, z\,\psi_z + z^2\,\psi_{zz} \leftrightarrow m^2\, c_m, \tag{A.13p}$$

$$(N+1)\,\frac{\psi - c_N}{z} - \psi_z \leftrightarrow m\, c_{m-1}, \tag{A.13q}$$

$$(N+1)^2\,\frac{\psi - c_N}{z} - (2\,N+1)\,\psi_z + z\,\psi_{zz} \leftrightarrow m^2\, c_{m-1}, \tag{A.13r}$$

$$\frac{\psi - c_N - c_{N-1}\,z}{z^2} \leftrightarrow c_{m-2}, \tag{A.13s}$$

$$\frac{(N+2)\,\psi - z\,\psi_z - (N+2)\,c_N - (N+1)\,z\,c_{N-1}}{z^2} \leftrightarrow m\,c_{m-2}, \tag{A.13t}$$

$$\frac{(N+2)^2\,(\psi - c_N) - z\,\left[(2\,N+3)\,\psi_z + (N+1)^2\,c_{N-1}\right] + z^2\,\psi_{zz}}{z^2}$$
$$\leftrightarrow m^2\,c_{m-2}, \tag{A.13u}$$

$$(N+1)\,\frac{\psi_t - \dot{c}_N}{z} - \psi_{zt} \leftrightarrow m\,\dot{c}_{m-1}. \tag{A.13v}$$

APPENDIX B

TWO PROOFS

In this appendix we report (essentially from [47] respectively [76]) the proofs of *Propositions 4.2.2-4* respectively *4.2.2-8.*

The point of departure of both these proofs is of course the parameterization of the $N \times N$ matrix $U(t)$ in terms of its N eigenvalues $z_n(t)$ and of its diagonalizing $N \times N$ matrix $R(t)$:

$$U = R\, Z\, R^{-1}, \tag{B.1a}$$

$$Z = \text{diag}\,[z_n]\,; \tag{B.1b}$$

recall in this connection the treatment in Section 4.2.2, see (4.65) and the equations following it. Let us in particular recall the definition (4.67) of the matrix M,

$$M = R^{-1}\,\dot{R}, \tag{B.2}$$

as well as the fact that, because the formulas (B.1) define the matrix R only up to multiplication from the right by an *arbitrary* diagonal matrix D, this matrix M is defined only up to the "gauge transformation" corresponding to the replacement of M by

$$\tilde{M} = D^{-1}\, M\, D + D^{-1}\,\dot{D}, \tag{B.3a}$$

entailing

$$\tilde{M}_{nn} = M_{nn} + \frac{\dot{d}_n}{d_n}, \tag{B.3b}$$

$$\tilde{M}_{nm} = d_n^{-1}\, M_{nm}\, d_m, \quad n \neq m. \tag{B.3c}$$

Here of course $D \equiv D\,(t)$ is an essentially *arbitrary* $N \times N$ *diagonal* matrix, and the N quantities $d_n \equiv d_n\,(t)$ denote its (*diagonal*) elements. Hence, in our parameterization of the $N \times N$ matrix $U(t)$ (via (B.1) with (B.2)) the N^2 matrix elements of this matrix get replaced by the N elements $z_n(t)$ of the *diagonal* matrix $Z(t)$ (namely, by the N eigenvalues of the matrix $U(t)$: see (B.1)) and by the $N\,(N-1)$ off-diagonal elements $M_{nm}(t)$ (with $n \neq m$) of the $N \times N$ matrix $M(t)$, while the N diagonal elements $M_{nn}(t)$ can be arbitrarily adjusted by choosing appropriately the elements $d_n(t)$ of the diagonal matrix $D(t)$, see (B.3b) (of course, up to a corresponding adjustment of the corresponding off-diagonal elements, see (B.3c)).

The point of departure of our first proof is the $N \times N$ matrix equation (4.93), which is reported here to make this appendix self-contained:

$$\ddot{U} = a\left(\dot{U}\,U + U\,\dot{U}\right). \tag{B.4}$$

Let us as well report the *general* solution of this matrix evolution equation

$$U(t) = a^{-1}\left[\cos(A\,t) - B\,A^{-1}\,\sin(A\,t)\right]^{-1}\left[A\,\sin(A\,t) + B\,\cos(A\,t)\right], \tag{B.5a}$$

where A and B are two arbitrary *constant* $N \times N$ matrices. In terms of the initial-value problem for the matrix evolution equation (B.4), clearly (B.5a) entails

$$U(0) = a^{-1}\,B, \quad \dot{U}(0) = a^{-1}\left(A^2 + B^2\right), \tag{B.5b}$$

and these two matrix equations can be inverted to yield

$$A^2 = -a^2\,[U(0)]^2 + a\,\dot{U}(0), \quad B = a\,U(0). \tag{B.5c}$$

Note that the explicit expression (B.5a) entails that the matrix $U(t)$ is actually a function of the matrix A^2 rather than A.

Differentiation of (B.1a) with respect to the independent variable t yields, using (B.2),

$$\dot{U} = R\left\{\dot{Z} + [M,\,Z]\right\}R^{-1}, \tag{B.6a}$$

$$\ddot{U} = R\left\{\ddot{Z} + \left[\dot{M},\,Z\right] + 2\left[M,\,\dot{Z}\right] + [M,\,[M,\,Z]]\right\}R^{-1}. \tag{B.6b}$$

Here and hereafter we use the standard notation $[X,\,Y] \equiv X\,Y - Y\,X$ for the commutator of two matrices. These formulas coincide of course with (4.66).

Indeed, the treatment in this Appendix B has been up to now essentially identical to that of Section 4.2.2 (we reproduced it here to make this appendix self-contained). We now insert these formulas, (B.1) and (B.6), in the matrix evolution equation (B.4) and we thereby obtain the $N \times N$ matrix evolution equation

$$\begin{aligned}\ddot{Z} + \left[\dot{M},\,Z\right] + 2\left[M,\,\dot{Z}\right] + [M,\,[M,\,Z]]\\ = a\left\{\left(\dot{Z} + [M,\,Z]\right)Z + Z\left(\dot{Z} + [M,\,Z]\right)\right\},\end{aligned} \tag{B.7}$$

namely, by separating the diagonal and off-diagonal terms,

$$\ddot{z}_n = 2\,a\,\dot{z}_n\,z_n - 2\sum_{m=1,m\neq n}^{N}(z_n - z_m)\,M_{nm}\,M_{mn}, \tag{B.8a}$$

$$(z_n - z_m)\, \dot{M}_{nm} + 2\,(\dot{z}_n - \dot{z}_m)\, M_{nm}$$
$$= a\,\left(z_n^2 - z_m^2\right) M_{nm} + \sum_{\ell=1;\ell\neq n,m}^{N} (z_n + z_m - 2\,z_\ell)\, M_{n\ell}\, M_{\ell m}$$
$$- (z_n - z_m)\, M_{nm}\,(F_n - F_m)\,, \quad n \neq m, \tag{B.8b}$$

where we made the notational assignment

$$M_{nn}(t) \equiv \mu_n(t) \tag{B.9}$$

and, consistently with the observation made above, we retain the freedom to assign arbitrarily these N quantities μ_n.

Now we interpret the N eigenvalues $z_n \equiv z_n(t)$ as the coordinates identifying the positions of N moving particles, given the Newtonian look of the system of ODEs (B.8a); while the $N\,(N-1)$ "auxiliary variables" $M_{nm} \equiv M_{nm}(t)$ evolve according to the system of $N\,(N-1)$ first-order ODEs (B.8b), where the N "source terms" $\mu_n(t)$ can be assigned as arbitrary functions of time, or even of the other dependent variables $z_m(t)$ and $M_{m\ell}(t)$, without spoiling the *solvable* character of the model. Here and throughout indices like n, m, ℓ range of course from 1 to N.

We then introduce the *ansatz*

$$M_{nm}(t) = \frac{\left[\dot{z}_n(t)\, \dot{z}_m(t)\right]^{1/2}}{z_n(t) - z_m(t)}\, u_{nm}(t)\,. \tag{B.10}$$

The equations of motion (B.8) thereby take the form

$$\ddot{z}_n = 2\,a\,\dot{z}_n\, z_n + 2 \sum_{m=1,m\neq n}^{N} \frac{\dot{z}_n\, \dot{z}_m}{z_n - z_m}\, u_{nm}\, u_{mn}, \tag{B.11a}$$

$$\dot{u}_{nm} + \frac{\dot{z}_n - \dot{z}_m}{z_n - z_m}\, u_{nm}\,(1 - u_{nm}\, u_{mn})$$
$$= - \sum_{\ell=1;\ell\neq n,m}^{N} \dot{z}_\ell \left[\frac{u_{n\ell}\,(u_{\ell m} + u_{nm}\, u_{\ell n})}{z_n - z_\ell} + \frac{u_{\ell m}\,(u_{n\ell} + u_{nm}\, u_{m\ell})}{z_m - z_\ell}\right]$$
$$- u_{nm}\,(\mu_n - \mu_m)\,, \quad n \neq m. \tag{B.11b}$$

The auxiliary variables are now the $N\,(N-1)$ quantities $u_{nm}(t)$; and note that, while the (*Newtonian*) equations of motion (B.11a) that characterize the evolution of the "particle coordinates" $z_n(t)$ follow straightforwardly from (B.8a) via (B.10), in order to obtain the equations (B.11b) that characterize the evolution of the auxiliary variables $u_{nm}(t)$ from (B.8b) one had to use, in addition to (B.10), the equations of motion (B.11a). It is now clear that, for

$$\mu_n = M_{nn} = 0, \tag{B.12}$$

(or, equivalently, for $\mu_n = \mu$, see (B.11b)), the $N(N-1)$ evolution equations (B.11b) admit the (trivial) solution

$$u_{nm} = -1, \quad n \neq m, \tag{B.13}$$

entailing that the *Newtonian* equations of motion (B.11a) become then

$$\ddot{z}_n = 2\,a\,\dot{z}_n\,z_n + 2 \sum_{m=1,m\neq n}^{N} \frac{\dot{z}_n\,\dot{z}_m}{z_n - z_m}. \tag{B.14}$$

These system of ODEs coincides with the *Newtonian* equations of motion (4.95); we have thereby accomplished the main part of our proof.

To complete it, it is sufficient to note that we are free to assign the *initial* value $R(0)$ of the diagonalizing matrix $R(t)$—since this matrix is only constrained by the first-order matrix ODE (B.2). Clearly the most convenient choice is to set $R(0) = \mathbf{1}$, where of course $\mathbf{1}$ denotes the $N \times N$ unit matrix. This clearly entails (see (B.1))

$$U(0) = \text{diag}\left[z_n(0)\right], \tag{B.15}$$

as well as (via (B.6a) with (B.10), (B.13) and (B.12))

$$\dot{U}_{nm}(0) = \left[\dot{z}_n(0)\;\dot{z}_m(0)\right]^{1/2}. \tag{B.16}$$

And from these two formulas, via (B.5c), the proof of *Proposition 4.2.2-4* is completed.

The proof of *Proposition 4.2.2-8* is analogous, but somewhat more complicated, as to be expected given the formulation of this *Proposition.* Indeed it is instructive to frame the proof of this *Proposition 4.2.2-8* in a somewhat more general context, which incidentally also entails that we provide below once more also a proof of the *Proposition 4.2.2-4* we just proved—yet we considered worthwhile to also report separately, as done just above, a more straightforward proof of that simpler *Proposition 4.2.2-4.*

The starting point we adopt to eventually prove *Proposition 4.2.2-8* is to assume that the $N \times N$ matrix $U \equiv U(t)$ satisfy the first-order matrix ODE

$$\dot{U} = f(U) + C, \tag{B.17}$$

where C is a constant $N \times N$ matrix and the function $f(u)$ is scalar, namely the matrix $f(U)$ depends on no other matrix besides U, entailing that, for any invertible $N \times N$ matrix R,

$$R\,f(U)\,R^{-1} = f(R\,U\,R^{-1}). \tag{B.18}$$

A discussion of the *solvability* of this matrix ODE, (B.17), is postponed to the last part of this proof; of course the assignment $f(u) = a\,u^2$ corresponds to the treatment of **Example 4.2.2-6**. Also note that, although this matrix ODE, (B.17), with this assignment of $f(u)$ clearly entails (B.4), one is taking now a somewhat different point of departure than done above.

We now assume again that the $N \times N$ matrix $U(t)$ is diagonalized by the matrix $R(t)$, see (B.1), identifying again the N quantities $z_n \equiv z_n(t)$ as its N eigenvalues. And we also introduce the $N \times N$ matrix $\Gamma(t)$ by setting

$$C = R(t)\,\Gamma(t)\,[R(t)]^{-1}, \quad \Gamma(t) = [R(t)]^{-1}\,C\,R(t). \tag{B.19}$$

Clearly the relations (B.1) and (B.19) entail

$$\dot{U} = R\left\{\dot{Z} + [M,\ Z]\right\}R^{-1}, \quad \dot{C} = R\left\{\dot{\Gamma} + [M,\ \Gamma]\right\}R^{-1}, \tag{B.20}$$

where we introduced the $N \times N$ matrix $M(t)$ as above, see (B.2). Hence the original matrix ODE (B.17) entails, via (B.20) and (B.1),

$$\dot{Z} + [M,\ Z] = \Gamma + f(Z), \tag{B.21a}$$

and the time-independence of the matrix C entails, via (B.20) and (B.1),

$$\dot{\Gamma} + [M,\ \Gamma] = 0. \tag{B.21b}$$

Let us now write out separately, componentwise, the *diagonal* and *off-diagonal* parts of these two matrix equations (B.21), also indicating, for notational convenience, with γ_n respectively μ_n the diagonal elements of the matrices Γ respectively M, $\gamma_n \equiv \Gamma_{nn}$, $\mu_n \equiv M_{nn}$:

$$\dot{z}_n = \gamma_n + f(z_n), \tag{B.22a}$$

$$-M_{nm}\,(z_n - z_m) = \Gamma_{nm}, \quad n \neq m, \tag{B.22b}$$

$$\dot{\gamma}_n = \sum_{m=1,\,m\neq n}^{N} \left(\Gamma_{nm}\,M_{mn} - \Gamma_{mn}\,M_{nm}\right), \tag{B.22c}$$

$$\dot{\Gamma}_{nm} = -(\mu_n - \mu_m)\,\Gamma_{nm} + (\gamma_n - \gamma_m)\,M_{nm} + \sum_{\ell=1,\,\ell\neq m,n}^{N} \left(\Gamma_{n\ell}\,M_{\ell m} - \Gamma_{\ell m}\,M_{n\ell}\right), \quad n \neq m. \tag{B.22d}$$

The first *two* of these four equations can be immediately solved for γ_n respectively for M_{nm}; thereby the last *two* of these four equations become

$$\ddot{z}_n = \dot{z}_n\,f'(z_n) + 2\sum_{m=1,\,m\neq n}^{N} \frac{\Gamma_{nm}\,\Gamma_{mn}}{z_n - z_m}, \tag{B.23a}$$

$$\dot{\Gamma}_{nm} = -(\mu_n - \mu_m)\,\Gamma_{nm} - \frac{\{[\dot{z}_n - f(z_n)] - [\dot{z}_m - f(z_m)]\}\,\Gamma_{nm}}{z_n - z_m} + \sum_{\ell=1,\,\ell\neq m,n}^{N} \left\{\Gamma_{n\ell}\,\Gamma_{\ell m}\left[(z_n - z_\ell)^{-1} + (z_m - z_\ell)^{-1}\right]\right\}, \quad n \neq m. \tag{B.23b}$$

In (B.23a) and hereafter the prime appended to a function denotes differentiation with respect to its argument, e.g. $f'(z) \equiv df(z)\,/\,dz$.

Let us now introduce a new *ansatz* by setting

$$\Gamma_{nm} = \{[\dot{z}_n - g(z_n)]\;[\dot{z}_m - g(z_m)]\}^{1/2}\;\eta_{nm}, \tag{B.24}$$

where the quantities $\eta_{nm} \equiv \eta_{nm}(t)$ are new "auxiliary" dependent variables; note that we reserve the privilege to assign later the function $g(z)$. Insertion of this *ansatz* in (B.23a) yields

$$\ddot{z}_n = \dot{z}_n\, f'(z_n) + 2 \sum_{m=1,\, m\neq n}^{N} \frac{[\dot{z}_n - g(z_n)]\;[\dot{z}_m - g(z_m)]\;\eta_{nm}\,\eta_{mn}}{z_n - z_m}, \tag{B.25a}$$

and its insertion in (B.23b) yields (via (B.25a))

$$\frac{\dot{\eta}_{nm}}{\eta_{nm}} + \mu_n - \mu_m - \frac{f(z_n) - g\,(z_n) - [f(z_m) - g\,(z_m)]}{z_n - z_m} + \frac{\dot{z}_n\;[f'(z_n) - g'\,(z_n)]}{2\;[\dot{z}_n - g\,(z_n)]}$$
$$+ \frac{\dot{z}_m[f'(z_m) - g'(z_m)]}{2[\dot{z}_m - g(z_m)]} = \sum_{\ell=1,\ell\neq m,n}^{N} \left\{[\dot{z}_\ell - g\,(z_\ell)]\left(\frac{\eta_{n\ell}\eta_{\ell m}}{\eta_{nm}} - 1\right)\cdot\right.$$
$$\left.\cdot\left[(z_n - z_\ell)^{-1} + (z_m - z_\ell)^{-1}\right]\right\}, n \neq m. \tag{B.25b}$$

Again, the system of *second-order* ODEs (B.25a) is clearly interpretable as a *Newtonian* N-body problem, but it contains, in addition to the particle coordinates z_n, the auxiliary variables η_{nm}, which themselves evolve according to the system of *first-order* ODEs (B.25b). The question now—in order to identify systems of ODEs interpretable as a *Newtonian* N-body problem *without* additional variables—is to see how to get rid of the *auxiliary* variables η_{nm}. Clearly a possibility to do so, suggested by the structure of these two systems (B.25) of ODEs, is to set

$$\eta_{nm}(t) = \frac{\eta_n(t)}{\eta_m(t)}. \tag{B.26}$$

Through this assignment the auxiliary variables indeed disappear altogether from the first set (B.25a) of ODEs, which then read

$$\ddot{z}_n = \dot{z}_n\, f'(z_n) + 2 \sum_{m=1,\, m\neq n}^{N} \frac{[\dot{z}_n - g(z_n)]\;[\dot{z}_m - g(z_m)]}{z_n - z_m}, \tag{B.27a}$$

while the second set, (B.25b), of ODEs becomes

$$\frac{\dot{\eta}_n}{\eta_n} + \mu_n - \left(\frac{\dot{\eta}_m}{\eta_m} + \mu_m\right) = \frac{f(z_n) - g\,(z_n) - [f(z_m) - g\,(z_m)]}{z_n - z_m}$$
$$- \frac{\dot{z}_n\;[f'(z_n) - g'\,(z_n)]}{2\;[\dot{z}_n - g\,(z_n)]} - \frac{\dot{z}_m\;[f'(z_m) - g'\,(z_m)]}{2\;[\dot{z}_m - g\,(z_m)]}, \quad n \neq m. \tag{B.27b}$$

There now remains to see whether this second set, (B.27b), of ODEs can be satisfied.

We note first of all that the left-hand sides of these equations, (B.27b), are antisymmetric under the exchange of the two indices n and m, while the right-hand sides are symmetrical. Hence they must both vanish.

The vanishing of the left-hand side is easily achieved, by setting

$$\mu_n(t) = -\frac{\dot{\eta}_n(t)}{\eta_n(t)}, \tag{B.28}$$

an assignment which is permitted since we are free to chose at our convenience the diagonal elements $\mu_n(t)$ of the matrix $M(t)$.

To achieve the vanishing of the right-hand side of (B.27b) (which is quite overdetermined, especially because the functions f and g are required to be independent of the "velocities" $\dot{z}_n$) we can take advantage of our freedom to assign the function $g(z)$. This can be conveniently done in two quite different manners.

A first assignment is

$$g(z) = 0, \tag{B.29}$$

which has the merit to eliminate altogether from (B.27b) the presence of the "velocities" $\dot{z}_n$ and $\dot{z}_m$. The requirement that the right-hand sides of (B.27b) vanish yields then the following (apparently still overdetermined, but see below) set of ODEs for the function $f(z)$:

$$\frac{f(z_n) - f(z_m)}{z_n - z_m} = \frac{f'(z_n) + f'(z_m)}{2}, \quad n \neq m. \tag{B.30a}$$

It is easily seen that the general solution of this system of ODEs is

$$f(z) = a\, z^2 + b\, z + c, \tag{B.30b}$$

with a, b and c three *arbitrary* constants. The corresponding version of the original matrix evolution equation (B.17) reads

$$\dot{U} = a\, U^2 + b U + c + C. \tag{B.31a}$$

Without significant loss of generality this matrix ODE can be replaced by its simpler version

$$\dot{U} = a\, U^2 + C, \tag{B.31b}$$

which obtains by setting $b = c = 0$ in (B.31a), but clearly could as well be obtained from (B.31a) by shifting the two $N \times N$ matrices $U(t)$ and C by two appropriate *constant* multiples of the $N \times N$ *unit* matrix, resulting merely in a shift of all the eigenvalues $z_n(t)$ of the $N \times N$ matrix $U(t)$ by a *common constant*, and in a redefinition of the matrix C which in any case does not feature in the system of ODEs (B.27a). We conclude that the many-body problem yielded by this choice is characterized by the equations of motion

$$\ddot{z}_n = 2\,a\,\dot{z}_n\,z_n + 2 \sum_{m=1,\,m\neq n}^{N} \frac{\dot{z}_n\,\dot{z}_m}{z_n - z_m}, \tag{B.32}$$

yielded by (B.27a) with (B.29) and with (B.30b) which takes now the simpler form

$$f(z) = a\,z^2. \tag{B.33}$$

But this N-body model, (B.32), is *not* new: it is just the model (B.14) discussed in the first part of this Appendix B.

A second, and in fact more obvious, assignment of the function $g(z)$ which also achieves the vanishing of the right-hand side of (B.27b) is

$$g(z) = f(z). \tag{B.34}$$

Hence, by inserting this assignment in (B.27a), we conclude that the many-body system

$$\ddot{z}_n = \dot{z}_n\,f'(z_n) + 2 \sum_{m=1,\,m\neq n}^{N} \frac{[\dot{z}_n - f(z_n)]\,[\dot{z}_m - f(z_m)]}{z_n - z_m} \tag{B.35}$$

is *solvable*, provided the corresponding $N\times N$ matrix evolution ODE (B.17), with an *arbitrary* constant $N \times N$ matrix C, is itself *solvable*: indeed our treatment entails that the time evolution of the N particle coordinates $z_n(t)$ coincides then with the time evolution of the N eigenvalues of the $N \times N$ matrix $U(t)$ (with an appropriate assignment, in terms of the initial data $z_n(0)$, $\dot{z}_n(0)$, of the initial value $U(0)$ of the matrix $U(t)$ and of the constant matrix C: see below). Let us also point out that a more general choice of the function $g(z)$, also consistent with the vanishing of the right-hand side of (B.27b), would add an *arbitrary* constant c to the right-hand side of the formula (B.34); it is left to the private initiative of the alert reader to work out the marginal extension entailed by this possibility.

It is remarkable that, to obtain these Newtonian equations of motion, (B.35), no restriction has been required so far on the function $f(z)$. However, to the best of our knowledge, the *only* case in which the matrix ODE (B.17) is *solvable* is for the assignment (B.30b) of the function $f(z)$, and for the reasons explained above no significant loss of generality is then caused by restricting this assignment to the form (B.33). The corresponding N-body problem is then indeed characterized by the *Newtonian* equations of motion (4.117). The first part of our proof is thus completed.

The corresponding version of the matrix ODE is (B.31b), and it is easy to verify that the solution to the initial-value problem for this matrix ODE is again provided by the formulas (B.5a), but now with

$$A^2 = a\,C, \quad B = a\,U(0), \tag{B.36}$$

instead of (B.5c). We now note that, via (B.1) and (B.19), these two equations can be rewritten as follows:

$$A^2 = a\,\Gamma(0), \tag{B.37a}$$
$$B = a\,\text{diag}\,[z_n(0)]\,. \tag{B.37b}$$

Note that to write the first, (B.37a), of these two equations we took advantage of the time-independence of the matrix C in order to evaluate it, conveniently, at $t = 0$; and of course we used again, as above, the freedom to set $R(0) = \mathbf{1}$.

Hence, via (B.22a) and (B.24) with (B.34), (B.33) and (B.26), we get

$$\left(A^2\right)_{nm} = a\,\left[\dot{z}_n(0) - a\,z_n^2(0)\right] \quad \text{for } n = m, \tag{B.38a}$$
$$\left(A^2\right)_{nm} = a\,\left\{\left[\dot{z}_n - a\,z_n^2(0)\right]\,\left[\dot{z}_m - a\,z_m^2(0)\right]\right\}^{1/2}\,\frac{\eta_n(0)}{\eta_m(0)} \quad \text{for } n \neq m. \tag{B.38b}$$

We now note that we are also free to assign the initial values $\eta_n(0)$ of the quantities $\eta_n(t)$, hence we make hereafter the convenient choice $\eta_n(0) = 1$—any other assignment would make no significant difference, since clearly the presence of these numbers in the right-hand side of (B.38b) only entails a time-independent similarity transformation for the matrix $U(t)$, with no effect on its eigenvalues.

We thus see that the matrix A^2 is proportional to the *dyadic* matrix P, see (4.118f), (4.118g), (4.118h):

$$A^2 = \Omega^2\,P. \tag{B.39}$$

We now can use the following (rather obvious) property of any *dyadic* matrix W:

$$W_{nm} = v_n\,v_m \tag{B.40a}$$

entails

$$\varphi(X\,W) = \varphi(0) + \frac{\varphi(x) - \varphi(0)}{x}\,X\,W \tag{B.40b}$$

with

$$x = \sum_{n,m=1}^{N} v_n\,x_{nm}\,v_m. \tag{B.40c}$$

Here X is an *arbitrary* $N \times N$ matrix whose matrix elements are denoted as x_{nm} and $\varphi(x)$ is any scalar function for which these formulas make good sense. We actually use below only the special version of these formulas with $W = P$, see (4.118f), (4.118g), (4.118h), and with a *diagonal* matrix X, namely such that $x_{nm} = x_n\,\delta_{nm}$, in which case the definition (B.40c) takes the simpler form

$$x = \sum_{n=1}^{N} \left(\frac{\Omega_n^2}{\Omega^2}\right) x_n . \tag{B.40d}$$

In particular for $X = t^2\,\mathbf{1}$ and $\varphi(x) = \cos(x^{1/2})$, $\varphi(x) = x^{-1/2}\sin(x^{1/2})$ respectively $\varphi(x) = x^{1/2}\sin(x^{1/2})$ we get from (B.39) and (B.40) respectively

$$\cos(A\,t) = 1 + [\cos(\Omega\,t) - 1]\;P, \tag{B.41a}$$

$$A^{-1}\sin(A\,t) = t + \frac{\sin(\Omega\,t) - \Omega\,t}{\Omega}\;P, \tag{B.41b}$$

$$A\,\sin(A\,t) = \Omega\,\sin(\Omega\,t)\,P. \tag{B.41c}$$

The insertion in (B.5a) of these expressions, (B.41), and of (B.36), yields the formula

$$U(t) = \left\{1 + [\cos(\Omega\,t) - 1]\;P - a\,U(0)\left[t + \frac{\sin(\Omega\,t) - \Omega\,t}{\Omega}\;P\right]\right\}^{-1} \cdot \left\{U(0)\;(1 + [\cos(\Omega\,t) - 1]\;P) + \frac{\Omega\,\sin(\Omega\,t)}{a}\;P\right\}, \tag{B.42a}$$

which can be conveniently rewritten as follows:

$$U(t) = \left[1 - X^{(1)}(t)\;P\right]^{-1}\;[1 - a\,U(0)\,t]^{-1}\;U(0)\;\left[1 - X^{(2)}(t)\;P\right], \tag{B.42b}$$

with the two *diagonal* matrices $X^{(1)}(t)$ and $X^{(2)}(t)$ defined as follows:

$$X^{(1)}(t) = [1 - a\,U(0)\,t]^{-1}\left[1 - \cos(\Omega\,t) + \frac{\sin(\Omega\,t) - \Omega\,t}{\Omega}\;a\,U(0)\right], \tag{B.42c}$$

$$X^{(2)}(t) = 1 - \cos(\Omega\,t) - \Omega\,\sin(\Omega\,t)\;[a\,U(0)]^{-1}. \tag{B.42d}$$

Next we use again the formula (B.40b) with (B.40d) entailing

$$\left[1 - X^{(1)}(t)\;P\right]^{-1} = 1 + \frac{X^{(1)}(t)\;P}{1 - x^{(1)}(t)} \tag{B.43a}$$

with

$$x^{(1)}(t) = \sum_{n=1}^{N}\left(\frac{\Omega_n^2}{\Omega^2}\right)\frac{\Omega\;[1 - \cos(\Omega\,t)] + [\sin(\Omega\,t) - \Omega\,t]\;a\,z_n(0)}{\Omega\;[1 - a\,z_n(0)\,t]}, \tag{B.43b}$$

and we thereby rewrite (B.42b) as follows:

$$U(t) = X^{(0)}(t) + \tilde{X}^{(1)}(t)PX^{(0)}(t) - X^{(0)}(t)X^{(2)}(t)P \\ - \tilde{X}^{(1)}(t)PX^{(0)}(t)X^{(2)}(t)P, \tag{B.44a}$$

with the *diagonal* matrices $X^{(0)}(t)$ and $\tilde{X}^{(1)}(t)$ defined as follows:

$$X^{(0)}(t) = [1 - a\,U(0)\,t]^{-1}\;U(0), \tag{B.44b}$$

$$\tilde{X}^{(1)}(t) = \frac{X^{(1)}(t)}{1 - x^{(1)}(t)}. \tag{B.44c}$$

Finally we note that clearly

$$P\,X^{(0)}(t)\,X^{(2)}(t)\,P = \sum_{n=1}^{N} \left(\frac{\Omega_n^2}{\Omega^2}\right) \frac{[1 - \cos(\Omega\,t)]\;a\,z_n(0) - \Omega\,\sin(\Omega\,t)}{a\,[1 - a\,z_n(0)\,t]}\,P, \tag{B.45}$$

and by using this formula to evaluate the last term in the right-hand side of (B.44a) we get the expression (4.118) of $U(t)$. The *Proposition 4.2.2-8* is thereby finally proven.

APPENDIX C

DIOPHANTINE FINDINGS AND CONJECTURES

In this appendix we list several *Diophantine relations*—namely, equations whose solutions are *integer* numbers—which were discovered via the study of *isochronous* systems. As already mentioned, the general strategy to arrive at such findings and conjectures proceeds through the following steps: (i) to identify a multi-degrees-of-freedom *entirely isochronous* system whose *generic* solution is *completely periodic* with a fixed period—namely, *all* its degrees of freedom evolve *periodically* with the same period; (ii) to identify an equilibrium configuration of this system; (iii) to investigate the small oscillations of this system in the neighborhood of its equilibrium configuration. A matrix is thereby obtained whose eigenvalues provide the frequencies of these small oscillations. But since the *complete periodicity* of the motion must also apply to these small oscillations, one concludes that *all* these frequencies must be—up to a common scaling—*integer* numbers.

The identification of matrices the eigenvalues of which are *integer* numbers—or equivalently, of determinantal polynomials whose zeros are *all integers*—is therefore the typical outcome of this approach. Indeed these results are presented below in terms of the $2N$ roots w_j of the determinantal polynomial equation

$$\det\left[w^{2} + w\,A + B\right] = 0, \tag{C.1}$$

of degree $2N$ in the variable w, with the two $N \times N$ matrices A and B defined below case by case. These *Diophantine* findings are then confirmed by verifications which can be numerically performed (with the help of computers) for moderately small values of the positive integer N: and this generally leads to the formulation of *conjectures* for the actual values of the *integers* in question, presumably valid for *arbitrary* N.

Via this approach several findings were ascertained in the past, leading eventually to recognize the connection of results of this kind with the theory of Lagrangian interpolation. Related developments yielded general prescriptions to manufacture " remarkable matrices" with *integer* eigenvalues. These results have been treated in sufficient detail in the literature (see for instance [30] and especially Section 2.4 and Appendix D of [37]) to justify forsaking any additional review here. Accordingly, we report below without further commentary only *Diophantine* findings and conjectures obtained rather recently. The fact that—in spite of the availability of the machinery to identify *remarkable matrices* we just mentioned—these conjectures have not yet been all proven (nor disproved) is tantalizing and might appeal to some readers. Those wishing to know more about how these *Diophantine* findings were arrived at shall consult the original papers where they were obtained to begin with, as identified case by case below.

Notation. Unless otherwise indicated, below indices such as n, m run from 1 to N; δ_{nm} denotes the standard Kronecker symbol, $\delta_{nm} \equiv \delta_{n,m} = 1$ if $n = m$, $\delta_{nm} \equiv \delta_{n,m} = 0$ if $n \neq m$, and the binomial coefficients are defined as follows:

$$\binom{x}{y} = \frac{\Gamma(x+1)}{\Gamma(y+1)\,\Gamma\,(x-y+1)} \tag{C.2}$$

(also for noninteger arguments: Γ is of course the standard gamma function, so that $\Gamma\,(n+1) = n!$ if n is a *nonnegative integer*).

In [48] the following findings are obtained.

Proposition C-1.. Let the N numbers u_n be the N zeros of the generalized Laguerre polynomial $L_N^{(-2\,N-1)}(u)$ (see, for instance, [91]),

$$L_N^{(-2\,N-1)}(u_n) = 0, \tag{C.3}$$

and define in terms of these N numbers the following two $N \times N$ matrices defined componentwise as follows:

$$A_{nm} = -\delta_{nm} + (1-\delta_{nm})\,\frac{2\,u_n}{u_n - u_m}, \tag{C.4a}$$

$$B_{nm} = \delta_{nm}\,2\left[1 + u_n + \sum_{\ell=1,\ell\neq n}^{N} \frac{u_\ell^2}{(u_n - u_\ell)^2}\right] - (1-\delta_{nm})\,\frac{2\,u_n^2}{(u_n - u_m)^2}. \tag{C.4b}$$

Then the $2\,N$ roots w_j of (C.1) are *all integers.* ⊡

Conjecture C-1. With the two $N \times N$ matrices A and B defined in *Proposition C-1*

$$\det\left[w^2 + w\,A + B\right] = \prod_{k=1}^{N}\left[(w-2\,k)\,(w+2\,k-1)\right]. \;⊡ \tag{C.5}$$

Proposition C-2.. Let the following two $N \times N$ matrices be defined componentwise as follows:

$$A_{nm} = -\delta_{nm}\,(2\,n+1) - \delta_{n+1,m} + \delta_{1m}\,\frac{(N+n)!}{(N-n)!\,n!}, \tag{C.6a}$$

$$B_{nm} = -\delta_{nm}\,[N\,(N+1) - n\,(n+1)] + \delta_{n+1,m}\,(n+1) - \delta_{1m}\,\frac{(N+n)!}{(N-n)!\,n!}. \tag{C.6b}$$

Then the $2\,N$ roots w_j of (C.1) are *all integers.* Indeed, they coincide with those defined in *Proposition C-1.* ⊡

Conjecture C-2. With the two $N \times N$ matrices A and B defined in *Proposition C-2*

$$\det\left[w^{2} + w\,A + B\right] = \prod_{k=1}^{N} \left[(w - 2\,k)\,(w + 2\,k - 1)\right]. \boxdot \tag{C.7}$$

Of course the validity of this *Conjecture C-2* is entailed, via the last statement in *Proposition C-2*, by the eventual validity of *Conjecture C-1*; and viceversa.

In [20] the following findings are obtained.

Proposition C-3.. Let the N numbers u_n be the N zeros of the monic polynomial

$$P_N(u) = (1-\beta)\,u^N + \beta\left[(u-1)^N + N\,(u-1)^{N-1} + \frac{1}{2}N\,(N-1)\,(u-1)^{N-2}\right], \tag{C.8a}$$

where β is an *arbitrary* constant,

$$P_N\left(u_n\right) = 0, \tag{C.8b}$$

and let the two $N \times N$ matrices A and B be defined componentwise in terms of these N numbers as follows:

$$A_{nm} = \delta_{nm}\,(2\,u_n - 1) + 2\,\left(1 - \delta_{nm}\right)\,\frac{u_n\,\left(1 - u_n\right)}{u_n - u_m}, \tag{C.9a}$$

$$\begin{aligned} B_{nm} = {} & 2\,\delta_{nm} \sum_{m=1,\,m \neq n}^{N} \frac{u_n\,u_m\,\left(1-u_n\right)\,\left(1-u_m\right)}{\left(u_n - u_m\right)^2} \\ & - 2\,\left(1-\delta_{nm}\right)\,\frac{u_n\,\left(1-u_n\right)\,\left(u_m^2 - 2\,u_n\,u_m + u_n\right)}{\left(u_n - u_m\right)^2}. \end{aligned} \tag{C.9b}$$

Then the $2\,N$ roots w_j of (C.1) are *all integers.* $\boxdot$

Conjecture C-3. With the two $N \times N$ matrices A and B defined in *Proposition C-3*

$$\det\left[w^{2} + w\,A + B\right] = \prod_{j=1-N}^{N} (w - j)\,. \boxdot \tag{C.10}$$

Proposition C-4.. Let the N numbers α_n be defined by the *nonlinear* recursion relations

$$\begin{aligned} & -(m+2)\,(m-1)\,\alpha_{m+2} - (m+1)\,(N+1-2m)\,\alpha_{m+1} \\ & + m\,(N-m)\,\alpha_m = 0, \end{aligned} \tag{C.11a}$$

with the additional conditions

$$\alpha_1 = \alpha_2 = \alpha_{N+1} = \alpha_{N+2} = 0, \tag{C.11b}$$

and the two $N \times N$ matrices A and B be defined (componentwise) in terms of these numbers as follows:

$$A_{m,n} = (N - 2m)\,\delta_{m,n} + 2m\,\delta_{m+1,n}, \tag{C.12a}$$

$$B_{m,n} = -m\,(N-m)\,\delta_{m,n} + (m+1)\,(N+1-2m)\,\delta_{m+1,n} \\ + (m+2)(m-1)\delta_{m+2,n} - [2m\alpha_{m+1} + (N+1-2m)\alpha_m]\delta_{1,n} + 2\alpha_m\delta_{2,n}. \tag{C.12b}$$

Then the $2N$ roots w_j of (C.1) are *all integers.* Indeed, they coincide with those defined in *Proposition C-3.* ⊡

Conjecture C-4. With the two $N \times N$ matrices A and B defined in *Proposition C-4*

$$\det\left[w^2 + w\,A + B\right] = \prod_{j=1-N}^{N} (w - j)\,. \;⊡ \tag{C.13}$$

Of course the validity of this *Conjecture C-4* is entailed, via the last statement in *Proposition C-4*, by the eventual validity of *Conjecture C-3*; and viceversa.

In [21] the following findings are obtained.

Proposition C-5.. Let the two $N \times N$ matrices A and B be defined componentwise as follows:

$$A_{nm} = (2n - \nu)\,\delta_{n,m} + (2N + 1 - 2n)\,\delta_{n-1,m} + C_n\,\delta_{N,m}, \tag{C.14a}$$

$$B_{nm} = n\,(n-\nu)\,\delta_{n,m} - [N + (N+1-n)\,(\nu - 2n)]\,\delta_{n-1,m} \\ + (N-n)\,(N+2-n)\,\delta_{n-2,m} + C_n\,(\delta_{N-1,m} + N\,\delta_{N,m})\,, \tag{C.14b}$$

where

$$C_n = -(-)^{N-n}\,\frac{1}{1+f}\left[\binom{N}{n-1} + f\,\frac{N + (N-n)\,\nu}{(N-\nu)}\binom{N-\nu}{n-\nu-1} + g\binom{N-\nu-1}{n-\nu}\right], \tag{C.14c}$$

with ν a *positive integer* smaller than N and f, g two *arbitrary* numbers. Then the $2N$ roots w_j of (C.1) are *all integers.* ⊡

Conjecture C-5. With the two $N \times N$ matrices A and B defined in *Proposition C-5* where now ν, f and g are *three arbitrary* numbers

$$\det\left[w^2 + w\,A + B\right] = \prod_{k=1}^{N} \left[(w+k)\,(w - \nu - 1 + k)\right]. \;⊡ \tag{C.15}$$

Proposition C-6.. Let the two $N \times N$ matrices A and B be defined componentwise as follows:

$$A_{nm} = (2\,n - \nu - 1)\,\delta_{n,m} + (2\,N + 1 - 2\,n)\,\delta_{n-1,m} + C_n\,\delta_{N,m} \tag{C.16a}$$

$$B_{nm} = n\,(n - \nu - 1)\,\delta_{n,m} + [(N-n)\,(2\,n - \nu - 3) + N]\,\delta_{n-1,m}$$
$$+ (N-n)\,(N+2-n)\,\delta_{n-2,m} + C_n\,[\delta_{N-1,m} + (N - \nu - 1)\,\delta_{N,m}]\,,$$

where

$$C_n = (-)^{N-n}\,\frac{1}{1+f}\,\left\{\frac{g\,(N+1-n) - N}{N}\binom{N}{n-1}\right.$$
$$\left. + f\left[(g - \nu - 1)\binom{N-\nu-1}{n-\nu-1} - \binom{N-\nu-1}{n-\nu-2}\right]\right\}, \tag{C.16b}$$

with ν a *positive integer* smaller than N and f, g two *arbitrary* numbers. Then the $2\,N$ roots w_j of (C.1) are *all integers.* ⊡

Conjecture C-6. With the two $N \times N$ matrices A and B defined in *Proposition C-6* where now ν, f and g are *three arbitrary* numbers

$$\det\left[w^2 + w\,A + B\right] = (w+g)\left[\prod_{k=1}^{N-1}(w+k)\right]\left[\prod_{k=1}^{N}(w - \nu - 1 + k)\right]. \; ⊡ \tag{C.17}$$

In [79] the following findings are obtained.

Proposition C-7.. Let the two $N \times N$ matrices A and B be defined componentwise as follows:

$$A_{nm} = 2\,(n-1)\,\delta_{n+1,m} - (2\,n + 1 + 2\,c_1)\,\delta_{n,m} + 2\,c_n\,\delta_{1,m}, \tag{C.18a}$$

$$B_{nm} = (n+2)\,(n-3)\,\delta_{n+2,m} - 2\,(n-1)\,(n+1+c_1)\,\delta_{n+1,m}$$
$$+ \left[n\,(n+1) + 2\,(n-1)\,c_1 - 2\,c_1^2 + 6\,c_2\right]\,\delta_{n,m}$$
$$+ 2\,[-(n-1)\,c_{n+1} + (n - 1 - 2\,c_1)\,c_n]\,\delta_{1,m} + 6\,c_n\,\delta_{2,m}, \tag{C.18b}$$

with the numbers c_m defined as follows:

$$\text{for } \nu = 0, \quad c_m = (-)^m\binom{\mu}{m}, \tag{C.19a}$$

$$\text{for } \nu = 1, \quad c_m = \delta_{0m} + \delta_{1m} \quad \text{if } \mu = 1,$$
$$c_m = (-)^m\left[\binom{\mu-2}{m} - \binom{\mu-2}{m-2}\right] \quad \text{if } \mu > 1, \tag{C.19b}$$

$$\text{for } \nu = 3, \quad c_m = (-)^m\left[\binom{\mu-3}{m} + 6\binom{\mu-3}{m-1} + 14\binom{\mu-3}{m-2} + 14\binom{\mu-3}{m-3}\right], \tag{C.19c}$$

$$\text{for } \nu = 4\,, \quad c_m = (-)^m \sum_{k=0}^{4} \binom{\mu - 4}{m - k} \binom{5}{k}, \tag{C.19d}$$

$$\text{for } \nu = 5\,, \quad c_m = (-)^m \left[c \binom{\mu - 5}{m - 5} + \sum_{k=0}^{5} \binom{\mu - 5}{m - k} \binom{5}{k} \right], \quad c \text{ arbitrary.} \tag{C.19e}$$

As indicated above the parameter ν (the role of which here is to distinguish five different cases) can take any one of the five values $0, 1, 3, 4, 5$, while the parameter μ can take any *positive integer* value in the range $\nu \leq \mu \leq N$. Then the $2\,N$ roots w_j of (C.1) are *all integers.* ⊡

Conjecture C-7a. With the two $N \times N$ matrices A and B defined in *Proposition C-7*, with $\nu = 0, 1, 3, 4, 5$ and μ *integer* in the range $\nu \leq \mu \leq N$, there hold the following relations:

$$\text{for } \nu = 0, \quad \det\left[w^2 + A\,w + B\right] = \left\{ \prod_{k=1}^{N-\mu} \left[(w - k)\,(w - 1) \right] \right\} \left\{ \prod_{k=1}^{\mu} \left[(w + k)\,(w + k - 5) \right] \right\}, \tag{C.20a}$$

$$\text{for } \nu = 1, \quad \det\left[w^2 + A\,w + B\right] = (w + 1)\,(w - 4) \cdot \left\{ \prod_{k=1}^{N-\mu} \left[(w - k)\,(w - k - 5) \right] \right\} \left\{ \prod_{k=1}^{\mu-1} \left[(w + k)\,(w + k - 7) \right] \right\}, \tag{C.20b}$$

$$\text{for } \nu = 3, \quad \det\left[w^2 + A\,w + B\right] = (w + 1)\,(w - 4) \cdot \left\{ \prod_{k=1}^{N-\mu} \left[(w - k)\,(w - k + 5) \right] \right\} \left\{ \prod_{k=1}^{\mu-1} \left[(w + k)\,(w - k + \mu - 7) \right] \right\}, \tag{C.20c}$$

$$\text{for } \nu = 4, \quad \det\left[w^2 + A\,w + B\right] = (w + 1) \left[\prod_{k=1}^{3} (w - k - 1) \right] \cdot \left\{ \prod_{k=1}^{N-\mu} \left[(w - k)\,(w - k + 1) \right] \right\} \left[\prod_{k=1}^{\mu-4} (w + k) \right] \left[\prod_{n=1}^{\mu} (w + k + 1) \right], \tag{C.20d}$$

$$\text{for } \nu = 5, \quad \det\left[w^2 + A\,w + B\right] = \left\{ \prod_{k=1}^{N-\mu} \left[(w - k)\,(w - k - 1) \right] \right\} \left\{ \prod_{k=1}^{\mu} \left[(w + k)\,(w - k + \mu - 4) \right] \right\}. \tag{C.20e}$$

Here we use the standard convention according to which a product equals unity if the lower limit of the running index exceeds the upper limit. ⊡

Remark C-8.. For $\mu = \nu = 0$ this *Conjecture C-7a* is true [79]. $\boxdot$

Remark C-9.. This *Conjecture C-7a* can be proven by induction for all $N > \mu$ if one assumes its validity for $N = \mu$. $\boxdot$

The *Conjecture C-7a* only refers to integer values of the parameter μ in the range $\nu \leq \mu \leq N$. But computer-aided explorations suggest the validity, for *arbitrary* values of the parameter μ, of the following (to some extent more general) *conjecture* (which is only formulated below for sufficiently large values of N, to avoid less interesting complications).

Conjecture C-7b. With the two $N \times N$ matrices A and B defined in *Proposition C-7*, where $\nu = 0, 1, 3, 4, 5$ and μ is now an *arbitrary* number, the algebraic equation (C.1)—of order $2\,N$ in w—possesses the following $N-1$ solutions w_k:

$$2,\ 3,\ 4,\ 5-\mu,\ 6-\mu, \ldots,\ N-\mu, \quad \text{if } \nu = 0 \text{ or } \nu = 5 \text{ and } N \geq 5, \tag{C.21a}$$

$$2,\ 3,\ 4,\ 4-\mu, 5-\mu, \ldots, N-1-\mu, \quad \text{if } \nu = 4 \text{ and } N \geq 5, \tag{C.21b}$$

and the following $N-4$ solutions w_k

$$-1,\ 4,\ 6,\ 8-\mu, 9-\mu, \ldots, N-\mu, \quad \text{if } \nu = 1 \text{ and } N \geq 8, \tag{C.21c}$$

$$-1,\ 4,\ 6,\ 3-\mu, 4-\mu, \ldots, N-5-\mu, \quad \text{if } \nu = 3 \text{ and } N \geq 8. \ \boxdot \tag{C.21d}$$

Remark C-10.. This *Conjecture C-7b*—in contrast to the *Conjecture C-7a*—does not provide all the $2\,N$ roots of the algebraic equation (C.1). $\boxdot$

In [58] the following findings are obtained.

Proposition C-11.. Let the two *tridiagonal* $N \times N$ matrices A and B be defined componentwise as follows:

$$A_{n,n} = N, \quad A_{n,n-1} = N+1-n, \quad A_{n,n+1} = n, \tag{C.22a}$$

$$B_{n,n} = -2\,n^2 + (N+1)\,(2\,n-1) \equiv 2\,n\,(N+1-n) - (N+1)\,, \tag{C.22b}$$

$$B_{n,n-1} = (N+1-n)^2\,, \quad B_{n,n+1} = n^2, \tag{C.22c}$$

with all other matrix elements vanishing ("tridiagonal matrices"). Note the simple symmetry properties of these formulas under the transformation $n \mapsto N+1-n$. Then the $2\,N$ roots w_j of (C.1) are *all integers.* $\boxdot$

Conjecture C-11a. With the two $N \times N$ matrices A and B defined in *Proposition C-11*

$$\det[w^2 + w\,A + B] = w\,(w+N) \prod_{k=1}^{N-1} (w+k)^2. \ \boxdot \tag{C.23}$$

Conjecture C-11b. The *tridiagonal* $N \times N$ matrix M defined componentwise as follows,

$$M_{n,n} = N\,(N-1) - (n-1)^2 - (N-n)^2\,, \tag{C.24a}$$

$$M_{n,n-1} = (n-1)^2\,, \quad M_{n,n+1} = (N-n)^2\,, \tag{C.24b}$$

has the N eigenvalues $n\,(n-1)$, $n = 1, \ldots, N$, i.e.

$$\det[w - M] = \prod_{n=1}^{N} [w - n\,(n-1)]\,. \;\boxdot \tag{C.25}$$

This *Conjecture C-11b* has now been proven [22].

In [59] the following findings are obtained.

Proposition C-12.. Let the *tridiagonal* $N \times N$ matrix M be defined componentwise as follows:

$$M_{n,n} = 2\,n\,(N+1-n), n = 1, \ldots N, \tag{C.26a}$$

$$M_{n,n+1} = -n\,(N+1-n), n = 1, \ldots N-1, \tag{C.26b}$$

$$M_{n,n-1} = -n\,(N+1-n), n = 2, \ldots N, \tag{C.26c}$$

with all other elements vanishing ("tridiagonal property"). Note that this definition entails the symmetry properties

$$M_{n,n} = M_{N-n,N-n}, n = 1, \ldots N; \quad M_{n,n+1} = M_{N+1-n,N-n}, n = 1, \ldots N-1. \tag{C.26d}$$

Then the N eigenvalues μ_n of this matrix have the following *Diophantine* property,

$$\mu_n = k_n\,(k_n + 1), \quad n = 1, \ldots, N, \tag{C.27}$$

where the numbers k_n are *integers.* $\boxdot$

Conjecture C-12a. The *tridiagonal* $M \times M$ matrix M defined in *Proposition C-12* has the N eigenvalues

$$\mu_n = n\,(n+1)\,, \quad n = 1, \ldots, N, \tag{C.28a}$$

i.e. there holds the determinantal formula

$$\det[\mu - M] = \prod_{n=1}^{N} [\mu - n\,(n+1)]\,. \;\boxdot \tag{C.28b}$$

This *Conjecture C-12a* has now been proven [22].

Conjecture C-12b. Let the *tridiagonal* $N \times N$ matrix M be defined componentwise as follows,

$$M_{n,n} = \frac{2\,n\,(N+1-n)\,\left[(N+1-2\,n)^2 - 3\right]}{(N+2-2\,n)\,(N-2\,n)}, \quad n = 1, \ldots, N, \tag{C.29a}$$

$$M_{n,n+1} = -\frac{n\,(N+1-n)\,(N+2-2\,n)}{(N-2\,n)}, \quad n = 1, \ldots, N-1, \tag{C.29b}$$

$$M_{n,n-1} = -\frac{n\,(N+1-n)\,(N-2\,n)}{(N+2-2\,n)}, \quad n = 2, \ldots, N, \tag{C.29c}$$

of course with all other matrix elements vanishing ("tridiagonal property"). Note that the elements of this matrix have the symmetry properties (C.26d). Then the N eigenvalues of this matrix are given by the formula

$$\mu_n = n\,(n+1), n = 1, \ldots, N-1; \quad \mu_N = (N+1)\,(N+2), \tag{C.30a}$$

i.e. there holds the determinantal formula

$$\det\left[\mu - M\right] = \left[\mu - (N+1)\,(N+2)\right] \prod_{n=1}^{N-1} \left[\mu - n\,(n+1)\right]. \; \boxdot \tag{C.30b}$$

Let us end this Appendix C by mentioning that further developments of the results reported in [22], entailing additional *Diophantine* findings, are reported in [23] (and see also [24] [25]).

REFERENCES

[1] Ablowitz, M. J. and Clarkson, P. A., *Solitons, nonlinear evolution equations and inverse scattering*, Cambridge University Press, Cambridge, 1991. [7.N]

[2] Ablowitz, M. J., Kaup, D. J., Newell, A. C. and Segur, H., *The inverse scattering transform—Fourier analysis for nonlinear problems*, Stud. Appl. Math. **53** (1974) 249–315. [7.N]

[3] Adler, M., *Some finite-dimensional integrable systems*, in the Proceedings of the Conference on the Theory and Application of Solitons (Tucson, Arizona, January 1976), edited by Flaschka, H. and MaLaughlin, D. W., Rocky Mountain J. Math. **8** (1, 2) (1978), pp. 237–244. [1.N]

[4] Adler, M., *Some finite-dimensional integrable systems and their scattering behavior*, Commun. Math. Phys. **55** (1977) 195–230. [1.N]

[5] Arnlind, J. and Hoppe, J., *Eigenvalue-dynamics off Calogero-Moser*, Lett. Math. Phys. **68** (2004) 121–129. [4.N]

[6] Babelon, O. and Talon, M., *The symplectic structure of the spin Calogero model*, Phys. Lett. **A 236** (1997) 462–468. [4.N]

[7] Barucchi, G. and Regge, T., *Conformal properties of a class of exactly solvable N-body problems in space dimension one*, J. Math. Phys. **18** (1977) 1149–1153. [4.N]

[8] Billey, E., Avan, J. and Babelon, O., *The r-matrix structure of the Euler-Calogero-Moser model*, Phys. Lett. **A 186** (1994) 114–118. [4.N]

[9] Billey, E., Avan, J. and Babelon, O., *Exact Yangian symmetry in the classical Euler-Calogero-Moser model*, Phys. Lett. **A 188** (1994) 263–271. [4.N]

[10] Boiti, M., Leon, J.-P., Manna, M. and Pempinelli, F., *On the spectral transform of a Korteweg-de Vries equation in two spatial dimensions*, Inverse Probl. **2** (1987) 271–279. [7.N]

[11] Boiti, M., Leon, J.-P. and Pempinelli, F., *Spectral transform for a two spatial dimension extension of the dispersive long wave equation*, Inverse Probl. **3** (1987), 371–387. [7.N]

[12] Bracken, P., *Bäcklund transformations for several cases of a type of generalized KdV equation*, Int. J. Math. & Math. Sci. **63** (2004) 3369–3377. [7.N]

[13] Bracken, P., *Specific solutions of the generalized Korteweg-de Vries equation with possible physical applications*, Cent. Eur. J. Phys. **3** (2005) 127–138. [7.N]

[14] Bruschi, M. and Calogero, F., *The Lax representation for an integrable class of relativistic dynamical systems*, Commun. Math. Phys. **109** (1987) 481–492. [4.N]

[15] Bruschi, M. and Calogero, F., *Solvable and/or integrable and/or linearizable N-body problems in ordinary (three-dimensional) space. I*, J. Nonlinear Math. Phys. **7** (2000) 303–386. [2.N, 4.N]

[16] Bruschi, M. and Calogero, F., *On the integrability of certain matrix evolution equations*, Phys. Lett. **A 273** (2000) 167–172. [4.N]

[17] Bruschi, M. and Calogero, F., *Integrable systems of quartic oscillators*, Phys. Lett. **A 273** (2000) 173–182. [2.N, 4.N]

[18] Bruschi, M. and Calogero, F., *Convenient parametrizations of matrices in terms of vectors*, Phys. Lett. **A 327** (2004) 312–319. [2.N, 4.N]

[19] Bruschi, M. and Calogero, F., *Integrable systems of quartic oscillators. II*, Phys. Lett. **A 327** (2004) 320–326. [2.N, 4.N]

[20] Bruschi, M. and Calogero, F., *Novel solvable variants of the goldfish many-body model*, J. Math. Phys. **47** (2006) 022703:1–25. [4.2.2, 4.N, C]

[21] Bruschi, M. and Calogero, F., *Goldfishing: a new solvable many-body problem*, J. Math. Phys. **47** (2006) 042901:1–35. [4.2.2, 4.N]

[22] Bruschi, M., Calogero, F. and Droghei, R., *Proof of certain Diophantine conjectures and identification of remarkable classes of orthogonal polynomials*, J. Phys. A: Math. Theor. **40** (2007) 3815–3829. [8.N, C]

[23] Bruschi, M., Calogero, F. and Droghei, R., *Tridiagonal matrices, orthogonal polynomials and Diophantine relations: I*, J. Phys. A: Math. Theor. **40** (2007) 9793–9817. [8.N, C]

[24] Bruschi, M., Calogero, F. and Droghei, R., *Tridiagonal matrices, orthogonal polynomials and Diophantine relations: II*, J. Phys. A: Math. Theor. **40** (2007) 14759–14772. [8.N, C]

[25] Bruschi, M., Calogero, F. and Droghei, R. *Tridiagonal matrices, orthogonal polynomials and Diophantine relations: III. Properties of the polynomials of the Askey scheme,* J. Phys. A: Math. Theory (submitted to, December 2007). [8.N, C]

[26] Bruschi, M. and Ragnisco, O., *On a new integrable Hamiltonian system with nearest-neighbour interaction*, Inverse Probl. **5** (1989) 983–998. [4.N]

[27] Bruzón, M. S., Gandarias, M. L., Muriel, C., Ramirez, J., Saez, S. and Romero, F.R., *The Calogero–Bogoyavlenskii–Schiff equation in 2+1 dimensions*, Theor. Math. Phys. **137** (2003) 1367–1377. [7.N]

[28] Calogero, F., *Solution of the one-dimensional N-body problem with quadratic and/or inversely quadratic pair potentials*, J. Math. Phys. **12** (1971) 419–436; Erratum: J. Math. Phys. **37** (1996) 3646. [1.N]

[29] Calogero, F., *Motion of poles and zeros of special solutions of nonlinear and linear partial differential equations, and related "solvable" many-body problems*, Nuovo Cimento **43B** (1978) 177–241. [1.N, 4.2.2, 4.N, 7.1, 7.N]

[30] Calogero, F., *Matrices, differential operators and polynomials*, J. Math. Phys. **22** (1981) 919–932. [C]

[31] Calogero, F., *The evolution of the partial differential equation* $u_t = u_{xxx} + 3\left(u_{xx}\, u^2 + 3\, u_x^2\right) + 3\, u_x\, u^4$, J. Math. Phys. **28** (1987) 538–555. [7.N]

[32] Calogero, F., *A class of C-integrable PDEs in multidimensions*, Inverse Probl. **10** (1994) 1231–1234. [7.N]

[33] Calogero, F., *A class of integrable Hamiltonian systems whose solutions are (perhaps) all completely periodic*, J. Math. Phys. **38** (1997) 5711–5719. [2.N]

[34] Calogero, F., *Integrable and solvable many-body problems in the plane via complexification*, J. Math. Phys. **39** (1998) 5268–5291. [1.N, 4.N]

[35] Calogero, F., *Tricks of the trade: relating and deriving solvable and integrable dynamical systems*, in: *Calogero-Moser-Sutherland Models*, edited by van Diejen, J. F. and Vinet, L., Proceedings of the Workshop on Calogero-Moser-Sutherland Models held in Montreal, 10–15 March 1997, CRM Series in Mathematical Physics, Springer, Berlin, 2000, pp. 93–116. [1.N]

[36] Calogero, F., *The "neatest" many-body problem amenable to exact treatments (a "goldfish"?)*, Physica **D 152–153** (2001) 78–84. [1.N, 4.2.2, 4.N]

[37] Calogero, F., *Classical many-body problems amenable to exact treatments*, Lecture Notes in Physics Monograph **m 66**, Springer, Heidelberg, 2001. [1.N, 2.N, 4.2.2, 4.N, 5.1, 7.1, C]

[38] Calogero, F., *Periodic solutions of a system of complex ODEs*, Phys. Lett. **A 293** (2002) 146–150. [1.N, 2.N, 4.N]

[39] Calogero, F., *On a modified version of a solvable ODE due to Painlevé*, J. Phys. A: Math. Gen. **35** (2002) 985–992. [3.N]

[40] Calogero, F., *A complex deformation of the classical gravitational many-body problem that features a lot of completely periodic motions*, J. Phys. A: Math. Gen. **35** (2002) 3619–3627. [4.4, 4.N]

[41] Calogero, F., *On modified versions of some solvable ordinary differential equations due to Chazy*, J. Phys. A: Math. Gen. **35** (2002) 4249–4256. [3.N]

[42] Calogero, F., *Partially superintegrable (indeed isochronous) systems are not rare*, in: *New Trends in Integrability and Partial Solvability*, edited by Shabat, A. B., Gonzalez-Lopez, A., Manas, M., Martinez Alonso, L. and Rodriguez, M. A., NATO Science Series, II. Mathematics, Physics and Chemistry, vol. **132**, Proceedings of the NATO Advanced Research Workshop held in Cadiz, Spain, 2–16 June 2002, Kluwer, 2004, pp. 49–77. [4.N]

[43] Calogero, F., *Differential equations featuring many periodic solutions*, in: *Geometry and integrability*, edited by Mason L. and Nutku Y., London Mathematical Society Lecture Notes, vol. **295**, Cambridge University Press, 2003, pp. 9–21. [1.N]

[44] Calogero, F., *On the quantization of two other nonlinear harmonic oscillators*, Phys. Lett. **A 319** (2003) 240–245. [5.N, 7.N]

[45] Calogero, F., *Solution of the goldfish N-body problem in the plane with (only) nearest-neighbor coupling constants all equal to minus one half*, J. Nonlinear Math. Phys. **11** (2004) 1–11. [4.2.2, 4.N]

[46] Calogero, F., *Two new classes of isochronous Hamiltonian systems*, J. Nonlinear Math. Phys. **11** (2004) 208–222. [4.2.4, 4.N]

[47] Calogero, F., *A technique to identify solvable dynamical systems, and a solvable generalization of the goldfish many-body problem*, J. Math. Phys. **45** (2004) 2266–2279. [4.2.2, 4.N, B]

[48] Calogero, F., *A technique to identify solvable dynamical systems, and another solvable extension of the goldfish many-body problem*, J. Math. Phys. **45** (2004) 4661–4678. [4.2.2, 4.N, C]

[49] Calogero, F., *On the quantization of yet another two nonlinear harmonic oscillators*, J. Nonlinear Math. Phys. **11** (2004) 1–6. [5.N, 7.N]

[50] Calogero, F., *Isochronous dynamical systems*, Appl. Anal. **85** (2006) 5–22. [4.N]

[51] Calogero, F., *Isochronous systems*, in: *Proceedings* of the Sixth International Conference on Geometry, Integrability and Quantization, Varna, 3–10, June, 2004, edited by Mladenov, I. M. and Hirshfeld, A. C., Sofia, 2005 (ISBN954-84952-9-5), pp. 11–61. [1.N, 7.N]

[52] Calogero, F., *Integrable systems: overview*, in: *Encyclopedia of Mathematical Physics*, edited by Françoise, J.-P., Naber, G. and Tsou Sheung Tsun, Elsevier, Oxford, 2006 (ISBN 978-0-1251-2666-3), vol. **3**, pp. 106–122. [1.N]

[53] Calogero, F., *Isochronous systems*, in: *Encyclopedia of Mathematical Physics* edited by Françoise, J.-P., Naber, G. and Tsou Sheung Tsun, Elsevier, Oxford, 2006 (ISBN 978-0-1251-2666-3), vol. **3**, pp. 166–172. [1.N]

[54] Calogero, F., *Isochronous systems and their quantization*, Theor. Math. Phys. **152** (1) (2007) 882–893. [1.N]

[55] Calogero, F. and Degasperis, A., *Spectral transform and solitons: tools to solve and investigate nonlinear evolution equations*, Amsterdam, North Holland, 1982. [7.N]

[56] Calogero, F. and Degasperis, A., *On the quantization of Newton-equivalent Hamiltonians*, Amer. J. Phys. **72** (2004) 1202–1203. [7.N]

[57] Calogero, F. and Degasperis, A., *New integrable PDEs of boomeronic type*, J. Phys. A: Math. Gen. **39** (2006) 8349–8376. [2.N, 4.N]

[58] Calogero, F., Di Cerbo, L. and Droghei, R., *On isochronous Bruschi-Ragnisco-Ruijsenaars-Toda lattices: equilibrium configurations, behavior in their neighborhood, diophantine relations and conjectures*, J. Phys. A: Math. Gen. **39** (2006) 313–325. [4.2.2, 4.N, C]

[59] Calogero, F., Di Cerbo, L. and Droghei, R., *On isochronous Shabat-Yamilov lattices: equilibrium configurations, behavior in their neighborhood, diophantine relations and conjectures*, Phys. Lett. **A 355** (2006) 262–270. [4.2.2, 4.N, C]

[60] Calogero, F., Euler, M. and Euler, N., *New evolution PDEs with lots of isochronous solutions*, to be published [7.2, 7.N]

[61] Calogero, F. and Françoise J.-P., *Periodic solutions of a many-rotator problem in the plane*, Inverse Probl. **17** (2001) 1–8. [1.N, 3.N, 4.N]

[62] Calogero F. and Françoise, J.-P., *Periodic motions galore: how to modify nonlinear evolution equations so that they feature a lot of periodic solutions*, J. Nonlinear Math. Phys. **9** (2002) 99–125. [2.N, 3.N]

[63] Calogero, F. and Françoise, J.-P., *Nonlinear evolution ODEs featuring many periodic solutions*, Theor. Mat. Fis. **137** (2003) 1663–1675. [3.N]

[64] Calogero, F. and Françoise, J.-P., *Isochronous motions galore: nonlinearly coupled oscillators with lots of isochronous solutions.* In: *Superintegrability in Classical and Quantum Systems*, Proceedings of the Workshop on Superintegrability in Classical and Quantum Systems, Centre de Recherches Mathématiques (CRM), Université de Montréal, September 16–21 (2003), CRM Proceedings & Lecture Notes, vol. **37**, American Mathematical Society, 2004, pp. 15–27. [4.2.3, 4.N]

[65] Calogero, F. and Françoise J.-P., *New solvable many-body problems in the plane*, J. Nonlinear Math. Phys. **13** (2006) 231–254. [4.2.2, 4.N]

[66] Calogero F., Françoise, J.-P. and Sommacal, M., *Periodic solutions of a many-rotator problem in the plane. II. Analysis of various motions*, J. Nonlinear Math. Phys. **10** (2003) 157–214. [1.N, 2.N, 4.N]

[67] Calogero F., Françoise, J.-P. and Sommacal, M., *Solvable nonlinear evolution PDEs in multidimensional space involving trigonometric functions*, J. Phys. A: Math. Theor. **40** (2007) F363-F368. [7.N]

[68] Calogero F., Françoise, J.-P. and Sommacal, M., *Solvable nonlinear evolution PDEs in multidimensional space involving elliptic functions*, J. Phys. A: Math. Theor. **40** (2007) F705-F711. [7.N]

[69] Calogero, F. and Gómez-Ullate, D., *Two novel classes of solvable many-body problems of goldfish type with constraints*, J. Phys. A: Math. Theor. **40** (2007) 5335–5353. http://arxiv.org/abs/nlin.SI/0701046. [4.2.2, 4.N]

[70] Calogero, F. and Gómez-Ullate, D., *A new class of solvable many-body problems with constraints, associated with an exceptional polynomial subspace of codimension two*, J. Phys. A: Math. Theor. **40** (2007) F573–F580. [4.2.2, 4.N]

[71] Calogero, F. and Gómez-Ullate, D., *Asymptotically isochronous systems*, Physica **D** (submitted to). [6]

[72] Calogero, F., Gómez-Ullate, D., Santini, P. M. and Sommacal, M., *On the transition from regular to irregular motions, explained as travel on Riemann surfaces*, J. Phys. A: Math. Gen. **38** (2005) 8873–8896; arXiv:nlin.SI/0507024 v1 13 Jul 2005. [2.N, 8.N]

[73] Calogero, F., Gómez-Ullate, D., Santini, P. and Sommacal, M., *Deterministic chaos as travel on a Riemann surface*, (in preparation). [2.N, 8.N]

[74] Calogero, F. and Graffi, S., On the *quantization of a nonlinear Hamiltonian oscillator*, Phys. Lett. **A 313** (2003) 356–362. [5.N, 8.N]

[75] Calogero, F. and Inozemtsev, V. I., *Nonlinear harmonic oscillators*, J. Phys. A: Math. Gen. **35** (2002) 10365–10375. [1.N, 2.N, 4.N, 5.5.3]

[76] Calogero, F. and Iona, S., *Novel solvable extension of the goldfish many-body model*, J. Math. Phys. **46** (2005) 103515. [4.2.2, 4.N, B]

[77] Calogero, F. and Ji Xiaoda, *Solvable* (*nonrelativistic, classical*) *n-body problems in multidimensions. I*, J. Math. Phys. **35** (1994) 710–733. [4.N]

[78] Calogero, F. and Ji Xiaoda, *Solvable* (*nonrelativistic, classical*) *n-body problems in multidimensions. II*, in: *National Workshop on Nonlinear Dynamics*, edited by Costato, M., Degasperis, A. and Milani, M., Conference Proceedings vol. **48**, Pavullo nel Frignano (Modena), Italy, 19–22 May 1994; Società Italiana di Fisica, Bologna, 1995, pp. 21–32. [4.N]

[79] Calogero, F. and Langmann, E., *Goldfishing by gauge theory*, J. Math. Phys. **47** (2006) 082702:1–23. [4.2.2, 4.N, C]

[80] Calogero, F. and Leyvraz, F., *Isochronous and partially isochronous Hamiltonian systems are not rare*, J. Math. Phys. **47** (2006) 042901:1–23. [5.1, 5.N]

[81] Calogero, F. and Leyvraz, F., *On a class of Hamiltonians with (classical) isochronous motions and (quantal) equispaced spectra*, J. Phys. A: Math. Gen. **39** (2006) 11803–11824. [5.N, 8.N]

[82] Calogero, F. and Leyvraz, F., *Isochronous extension of the Hamiltonian describing free motion in the Poincaré half-plane: classical and quantum treatments*, J. Math. Phys. **48** (2007) 092903:1–15. [5.4, 5.N, 8.N]

[83] Calogero, F. and Leyvraz, F., *On a new technique to manufacture isochronous Hamiltonian systems: classical and quantal treatments*, J. Nonlinear Math. Phys. **14** (2007) 505–529. [5.N, 8.N]

[84] Calogero, F. and Leyvraz, F., *General technique to produce isochronous Hamiltonians*, J. Phys. A: Math. Theor. **40** (2007) 12931–12944. [5.5, 5.N, 8, 8.N]

[85] Calogero, F. and Leyvraz, F., *Spontaneous reversal of irreversible processes in a many-body Hamiltonian evolution*, New J. Phys (submitted to). [5.N, 8, 8.N]

[86] Calogero, F. and Sommacal, M., *Periodic solutions of a system of complex ODEs. II. Higher periods*, J. Nonlinear Math. Phys. **9** (2002) 483–516. [1.N, 2.N, 4.N]

[87] Calogero, F. and Sommacal, M., *Solvable nonlinear evolution PDEs in multidimensional space*, SIGMA **2** (2006) 088 (17 pages); nlin.SI/0612019. [7.N]

[88] Cavalcante, J. A. and Tenenblat, K., *Conservation law for nonlinear evolution equations*, J. Math. Phys. **29** (1988) 1044–1049. [7.N]

[89] Dorfman, I., *Dirac structures and integrability of nonlinear evolution equations*, Wiley, New York, 1993. [7.N]

[90] Dulac, H., *Solution d'un système d'équations différentièlles dans le voisinage des valeurs singulières*, Bulletin Societé Math. France **40** (1912) 324–383. [4.N]

[91] Erdélyi, A. (editor), *Higher transcendental functions*, McGraw Hill, New York, 1953. [3.N, C]

[92] Euler, M., Euler, N., Lindblon, O. and Perrson, L.-E., *The higher dimensional Bateman equations and Painlevè analysis of nonintegrable wave equations*, Symm. Nonlinear Math. Phys. **1** (1997)185–192. [7.N]

[93] Fairlie, D. B. and Leznov, A. N., *General solution of the Monge-Ampère equation in n-dimensional space*, J. Geom. Phys. **16** (1995) 385–390. [7.N]

[94] Françoise, J.-P., *Géométrie Analytique et Systémes Dynamiques*, PUF, Paris, 1995. [4.N]

[95] Fedorov, Yu. and Gómez-Ullate, D., *A class of dynamical systems whose solutions travel on the Riemann surface of an hyperelliptic function*, Physica **D 227** (2007) 120–134. [2.N, 3.1, 8.N]

[96] Fordy, A. P. and Gibbons, J., *Factorization of operators. II*, J. Math Phys. **22** (1981) 1170–1175. [7.N]

[97] Gaeta, G., *Resonant normal forms as constrained linear systems*, Mod. Phys. Lett. **A17** (2002) 583–597. [4.N]

[98] Gibbons, J. and Hermsen, T., *A generalization of the Calogero-Moser system*, Physica **11D** (1984) 337–348. [4.2.2, 4.N]

[99] Gómez-Ullate, D., Hone, A. N. H. and Sommacal, M., *New many-body problems in the plane with periodic solutions*, New J. Phys. **6** (24) (2004) 1–23. [4.2.2, 4.N]

[100] Gómez-Ullate, D. and Sommacal, M., *Periods of the goldfish many-body problem*, J. Nonlinear Math. Phys. **12** Suppl. **1** (2005) 351–362. [4.2.2, 7.1]

[101] Grinevich, P. and Santini, P. M., *Newtonian dynamics in the plane corresponding to straight and cyclic motions on the hyperelliptic curve* $\mu^2 = v^n - 1,\ n \in \mathbb{Z}$*: ergodicity, isochrony, periodicity and fractals*, Physica **D 232** (2007) 22–33. [2.N, 3.1, 8.N]

[102] Habibullin, I. T. and Kazakova, T. G., *Boundary conditions for integrable discrete chains*, J. Phys. A: Math. Gen. **34** (2001) 10369–10376. [4.2.2, 4.N]

[103] Habibullin, I. T. and Vildanov, A. N., *Integrable boundary conditions for nonlinear lattices*, in: *CRM Proceedings and Lecture Notes* **25** (2000) 173–180. [4.2.2, 4.N]

[104] Hopf, E., *The partial differential equation* $u_t + u\,u_x = \mu\,u_{x\,x}$, Commun. Pure Appl. Math. **3** (1950) 201–230. [7.N].

[105] Ince, E. L., *Ordinary differential equations*, Dover, New York, 1956. [2.2, 2.N, 4.1.1, 4.N]

[106] Inozemtsev, V. I., *Matrix analogues of elliptic functions*, Funct. Anal. Appl. **23** (1990) 323–325 [Russian original: Funct. Anal. Pril. **23** (1989) 81–82]. [2.N, 4.N]

[107] Ibragimov, N. H., *CRC Handbook of Lie group analysis of differential equations*, CRC Press, Boca Raton-Ann Arbor-London-Tokyo, 1994–1996. [7.N]

[108] Iona, S. and Calogero, F., *Integrable systems of quartic oscillators in ordinary (three-dimensional) space*, J. Phys. A: Math. Gen. **35** (2002) 3091–3098. [2.N, 4.N]

[109] Krichever, I., Babelon, O., Billey, E. and Talon, M., *Spin generalization of the Calogero-Moser system and the matrix KP equation*, Amer. Math. Soc. Transl. **170** (1995) 83–119. [4.N]

[110] Kadomtsev, B. B. and Petviashvili, V. I., *On the stability of solitary waves in weakly dispersive media*, Sov. Phys. Dokl. **15** (1970) 539–541. [7.N]

[111] Lax, P. D., *Integrals of nonlinear equations of evolution and solitary waves*, Commun. Pure Appl. Math. **21** (1968) 467–490. [1.N]

[112] Mariani, M., *Identificazione e studio di sistemi dinamici ed equazioni di evoluzione nonlineari che posseggono molte soluzioni completamente periodiche (isocrone)*, Dissertation for the "Laurea in Fisica", University of Rome "La Sapienza", 2003 (unpublished). [4.N, 7.2, 7.N]

[113] Mariani, M. and Calogero, F., *Isochronous PDEs*, Yadernaya Fizika (Physics of Atomic Nuclei) **68** (2005) 899–908. [7.N]

[114] Mariani, M. and Calogero, F., *A modified Schwarzian Korteweg de Vries equation in 2+1 dimensions with lots of periodic solutions*, Yadernaya Fizika (Physics of Atomic Nuclei) **68** (2005) 958–968. [7.N]

[115] Moser, J., *Three integrable Hamiltonian systems connected with isospectral deformations*, Adv. Math. **16** (1975) 197–220. [1.N]

[116] Nakamura, K. and Lakshmanan, M., *Complete integrability in a quantum description of chaotic systems*, Phys. Rev. Lett. **57** (1986) 1661–1664. [4.N]

[117] Nekrasov, N., *Holomorphic bundle and many-body systems*, Commun. Math. Phys. **180** (1996) 587–603. [4.N]

[118] Olshanetsky, M. A. and Perelomov, A. M., *Explicit solution of the Calogero model in the classical case and geodesic flows on symmetric spaces of zero curvature*, Lett. Nuovo Cimento **16** (1976) 333–339. [4.N]

[119] Olshanetsky, M. A. and Perelomov, A. M., *Completely integrable Hamiltonian systems connected with semi-simple Lie algebras*, Invent. Math. **37** (1976) 93–108. [4.N]

[120] Olshanetsky, M. A. and Perelomov, A. M., *Explicit solutions of some completely integrable systems*, Lett. Nuovo Cimento **17** (1976) 97–101. [1.N, 4.N]

[121] Olshanetsky, M. A. and Perelomov, A. M., *Classical integrable finite-dimensional systems related to Lie algebras*, Phys. Rep. **71** (1981) 313–400. [4.N]

[122] Pechukas, P., *Distribution of energy eigenvalues in the irregular spectrum*, Phys. Rev. Lett. **51** (1983) 943–946. [4.N]

[123] Perelomov, A. M., *Completely integrable classical systems connected with semi-simple Lie algebras*, ITEF preprint **27** (1976). [1.N, 4.N]

[124] Perelomov, A. M., *The simple relation between certain dynamical systems*, Commun. Math. Phys. **63** (1978) 9–11. [1.N]

[125] Perelomov, A. M., *Integrable systems of classical mechanics and Lie algebras*, Birkhauser, Basel, 1990. [1.N, 4.N]

[126] Ramirez, J., Bruzòn, M. S., Muriel, C, and Gandarias, M. L., *The Schwarzian Korteweg-de Vries equation in (2+1) dimensions*, J. Phys. A: Math. Gen. **36** (2003) 1467–1484. [7.N]

[127] Ruijsenaars, S. N. M., *Relativistic toda systems*, Commun. Math. Phys. **133** (1990) 217–247. [4.N]

[128] Shabat, A. B. and Yamilov, R. I., *Symmetries of nonlinear chains*, Leningrad Math. J. **2** (1991) 377–400 [Russian original version: Algebra i Analiz **2** (1990) 183–208]. [4.2.2, 4.N]

[129] Shabat, A. B. and Yamilov, R. I., *To a transformation theory of two- dimensional integrable systems*, Phys. Lett. **A 227** (1997) 15–23. [4.2.2, 4.N]

[130] Strogatz, S., *SYNC: how order emerges from chaos in the universe, nature and daily life*, Hyperion, New York, 2003, ISBN 0-7868-8721-4. [8.N]

[131] Toda, K. and Yu, S., *The investigation into the Schwarz-Korteweg-de-Vries equation and the Schwarz derivative in (2+1) dimensions*, J. Math. Phys. **41** (2000) 4747–4751. [7.N]

[132] Toda, M., *Theory of nonlinear lattices*, Springer Series in Solid-State Sciences **20**, Springer, Berlin, 1981. [4.N]

[133] van Bemmelen, T. and Kersten, P., *Nonlocal symmetries and recursion operators of the Landau-Lifshitz equation*, J. Math. Phys. **32** (1991) 1709–1716. [7.N]

[134] Wadati, M., Konno, K. and Ichikawa, Y. H., *New integrable nonlinear evolution equations*, J. Phys. Soc. Japan **47** (1979) 1698–1700. [7.N]

[135] Walcher, S., *On differential equations in normal forms*, Math. Ann. **291** (1991) 293–314. [4.N]

[136] Wang, J. P., *A list of 1+1 dimensional integrable equations and their properties*, J. Nonlinear Math. Phys. **9**, Suppl. **1** (2002) 213–233. [7.N]

[137] Wang, M., Zhou, Y. and Li, Z., *A nonlinear transformation of the dispersive long wave equations in 2+1 dimensions and its applications*, J. Nonlinear Math. Phys. **5** (1998) 120–125. [7.N]

[138] Wojciechowski, S., *An integrable marriage of the Euler equations with the Calogero-Moser system*, Phys. Lett. **A 111** (1984) 101–103. [4.2.2, 4.N]

[139] Wojciechowski, S., *Generalized integrable many-body systems in one dimension*, Phys. Scripta **34** (1986) 304–308. [4.2.2, 4.N]

[140] Yamilov, R. I., *Classification of Toda type scalar lattices*, in: *Nonlinear evolution equations & dynamical systems—NEEDS '92*, edited by Makhankov, V., Puzin, I. and Pashaev, O., World Scientific, Singapore, 1993, pp. 423–431. [4.2.2, 4.N]

[141] Yukawa, T., *New approach to the statistical properties of energy levels*, Phys. Rev. Lett. **54** (1985) 1883–1886. [4.N]

[142] Yukawa, T., *Lax form of the quantum mechanical eigenvalue problem*, Phys. Lett. **A 116** (1986) 227–230. [4.N]

[143] Zakharov V. E., *On the dressing method*, in: *Inverse Methods in Action*, edited by Sabatier P. C., Springer, Heidelberg, 1990, pp. 602–623. [1.N, 4.2.2, 4.N]

[144] Zakharov, V. E. and Shabat, A., B., *Exact solution of two-dimensional and one-dimensional self-modulation of waves in nonlinear media*, Soviet Phys. JETP **34** (1972) 62–69 [Russian version: Zh. Eksp. Teor. Fiz. **61** (1971) 118–143]. [7.N]

INDEX

[Words appearing in the *Contents* (see page ix), or occurring *too* frequently – such as *periodic, isochronous, integrable, dynamical system, Hamiltonian, Newtonian, equation of motion, time, ODE, PDE* and the like – are, as a rule, *not* listed]